John **Widdowson** (SERIES EDITOR)
Rebecca **Blackshaw**
Meryl **King**
Simon **Oakes**
Sarah **Wheeler**
Michael **Witherick**

AQA GCSE (9–1)

SECOND EDITION

GEOGRAPHY

Approval message from AQA

This textbook has been approved by AQA for use with our qualification. This means that we have checked that it broadly covers the specification and we are satisfied with the overall quality. Full details of our approval process can be found on our website.

We approve textbooks because we know how important it is for teachers and students to have the right resources to support their teaching and learning. However, the publisher is ultimately responsible for the editorial control and quality of this book.

Please note that when teaching the *AQA GCSE Geography* course, you must refer to AQA's specification as your definitive source of information. While this book has been written to match the specification, it cannot provide complete coverage of every aspect of the course.

A wide range of other useful resources can be found on the relevant subject pages of our website: www.aqa.org.uk.

HODDER
EDUCATION
AN HACHETTE UK COMPANY

Orders: please contact Hachette UK Distribution, Hely Hutchinson Centre, Milton Road, Didcot, Oxfordshire, OX11 7HH. Telephone: +44 (0)1235 827827. Email education@hachette.co.uk Lines are open from 9 a.m. to 5 p.m., Monday to Friday. You can also order through our website: www.hoddereducation.co.uk

ISBN: 978 1 5104 7751 3

© John Widdowson, Rebecca Blackshaw, Meryl King, Sarah Wheeler, Simon Oakes, Michael Witherick 2020

First published in 2016

This second edition published in 2020 by

Hodder Education,

An Hachette UK Company

Carmelite House

50 Victoria Embankment

London EC4Y 0DZ

www.hoddereducation.co.uk

Impression number   10 9 8 7 6 5
Year   2023 2022

Cover photo © Imago Photo – stock.adobe.com

Illustrations by Barking Dog and Aptara

Typeset in India by Aptara Inc

Printed in Dubai

A catalogue record for this title is available from the British Library.

Contents

Text and photo credits   v

Introduction   vi

Paper 1: Living with the Physical Environment

Section A: The challenge of natural hazards

Chapter 1 Natural hazards   2

Chapter 2 Tectonic hazards   4

Chapter 3 Weather hazards   22

Chapter 4 Climate change   44

Question practice   56

Section B: The living world

Chapter 5 Ecosystems   60

Chapter 6 Tropical rainforests   68

Chapter 7 Hot deserts   84

Chapter 8 Cold environments   100

Question practice   116

Section C: Physical landscapes in the UK

Chapter 9 The physical diversity of the UK   120

Chapter 10 Coastal landscapes   122

Chapter 11 River landscapes   152

Chapter 12 Glacial landscapes   182

Question practice   200

Contents

Paper 2: Challenges in the Human Environment

Section A: Urban challenges

Chapter 13 The global pattern of urban change   204

Chapter 14 Urban growth in Nigeria   208

Chapter 15 Urban challenges in the UK   224

Chapter 16 Sustainable development of urban areas   248

Question practice   254

Section B: The changing economic world

Chapter 17 Economic development and quality of life   258

Chapter 18 Reducing the global development gap   268

Chapter 19 Economic development in Nigeria   276

Chapter 20 Economic change in the UK   290

Question practice   312

Section C: The challenge of resource management

Chapter 21 Global resource management   316

Chapter 22 Resources in the UK   320

Chapter 23 Food   334

Chapter 24 Water   352

Chapter 25 Energy   366

Question practice   380

Paper 3: Geographical Applications

Chapter 26 Issue evaluation   384

Question practice   391

Chapter 27 Fieldwork and geographical enquiry   392

Glossary   402

Index   414

Photo credits   418

Text credits

Figure 4.6 Changes in solar energy falling on the Earth's surface. Science Museum Group; **Figure 6.8** From 'Tropical Rainforest Conservation' by Butler, Rhett A. Mongabay.com. San Francisco. Retrieved from https://www.mongabay.com/images/rainforests/total_forest_cover.gif; **Figure 6.9** From 'The biggest rainforest news stories in 2018' by Rhett A. Butler. Mongabay. Retrieved from https://news.mongabay.com/2018/12/the-biggest-rainforest-news-stories-in-2018/; **Figure 7.31** The Great Wall Initiative, Supporting Resilient Livelihoods and Landscapes in the Sahel, © Global Environment Facility, retrieved from: https://www.thegef.org/sites/default/files/publications/gef_great_green_wall_initiative_august_2019_EN_0.pdf Used with permission; **Figure 8.14** Physical Geography, Figure PS.9 – Permafrost, © Alaska Humanities Forum, Retrieved from: http://www.akhistorycourse.org/geography/physical-geography/; **Figure 10.54** Extract from 'Norfolk coastal erosion (2008)' published on 17 April 2008. Retrieved from https://www.theguardian.com/environment/gallery/2008/apr/17/1. Reproduced with the permissions from Guardian News & Media Limited; **Figure 13.3** A graph to show world urbanisation for different world areas from 1950 to 2050. Cool Geography. Retrieved from http://www.coolgeography.co.uk/A-level/AQA/Year%2013/World%20Cities/Urbanisation/Urbanisation.htm. Used with permission; **Figure 15.22** London's Sectors: More Detailed Jobs Data Working Paper 65. Greater London Authority (GLA), Retrieved from https://www.london.gov.uk/sites/default/files/londons_sectors_-_more_detailed_jobs_data.pdf. Used with permission; **Figure 15.34** From Nationwide's House Price Index, ©2020 Nationwide Building Society Retrieved from: https://www.nationwide.co.uk/about/house-price-index/headlines; **Figure 15.38** From London Atmospheric Emissions Inventory 2016. Greater London Authority. Retrieved from http://content.tfl.gov.uk/borough-pollution-map.pdf. Used with permission; **Figure 15.42** Extract from *AA Street by Street: London* (ISBN: 9780749551872), published by AA Publishing, 2007; **Figure 16.9** Bristol Cycle Strategy (2015), Bristol City Council. Retrieved from https://betterbybike.info/wp/wp-content/uploads/2015/02/Bristol-Cycle-Strategy-FINAL.pdf. Used with permission; **Figure 19.19** 'Chinese Investments Offers in Africa since 2010' is republished with the permission of Stratfor, a leading global geopolitical intelligence and advisory firm. © 2020 Stratfor Enterprises, LLC. Retrieved from https://worldview.stratfor.com/article/chinese-investments-africa. Used with permission; **Figure 20.8** From 'Where growth happens' published on 02 Sep 2014. Grant Thornton, https://www.grantthornton.co.uk/insights/where-growth-happens/. Used with permission; **Figure 21.7** World Water Development Report 4, © United Nations, Retrieved from https://www.un.org/waterforlifedecade/scarcity.shtml; **Figure 21.8** Sound Vision Productions. Map of Energy consumption per person by country. An Energy Journal from http://burnanenergyjournal.com/wp-content/uploads/2013/03/WorldMap_EnergyConsumptionPerCapita2010_v4_BargraphKey.jpg; **Figure 23.1** Daily per capita caloric supply, 2013. Global Change Data Lab. Retrieved from https://ourworldindata.org/grapher/daily-per-capita-caloric-supply; **Figure 24.16** Lily Kuo (2014). 'China is moving more than a River Thames of water across the country to deal with water scarcity', Quartz Media, Inc. Retrieved from https://qz.com/158815/chinas-so-bad-at-water-conservation-that-it-had-to-launch-the-most-impressive-water-pipeline-project-ever-built/. Used with permission; **Figure 25.2 (b)** Map showing distribution of energy usage based on oil consumption. BP Global. Retrieved from http://beodom.com/en/education/entries/peak-oil-the-energy-crisis-is-here-and-it-will-last. Used with permission; **Figure 25.3** Map showing countries at risk of energy insecurity – http://riskmanagementmonitor.com/the-countries-most-at-risk-for-energy-security/ © Maplecroft 2011; **Figure 25.4** Euan Mearns. Global Energy Trends – BP Statistical Review 2014. BP Global. Retrived from http://euanmearns.com/global-energy-trends-bp-statistical-review-2014/; **Figure 26.5** Extract from 'Heathrow third runway protesters vow to step up campaign' by Matthew Taylor (2018), published on 08 Jun 2018. The Guardian. Retrieved from https://www.theguardian.com/environment/2018/jun/08/heathrow-third-runway-protesters-vow-to-go-on-hunger-strike. Reproduced with the permissions from Guardian News & Media Limited.

Maps: **Figures 9.1, 18.2** and **18.5** Copyright Philip's.

Maps: **Figures 11.11, 11.14, 12.16, 12.17, 20.12, 26.7** and **Figures 1** and **2** on pages 200–201, © Crown copyright and database rights 2020 Hodder Education under licence to OS. Licence number 100036470.

Photo credits

p. 2 *l* © AP/Shutterstock.com, *r* © PHILMACD|PHOTOGRAPHY/Alamy Stock Photo; **p. 7** © Ragnar Th. Sigurdsson/age Fotostock/Superstock; **p. 9** © ASP GeoImaging/NASA/Alamy Stock Photo; **p. 11** © age fotostock/Superstock; **p. 13** © Insidefoto srl/Alamy Stock Photo; **p. 15** *l* © AWL Images/Getty Images, *c* © NurPhoto/Getty Images *r* © Wolfgang Rattay/Reuters; **p. 18** © Ragnar Th Sigurdsson/ARCTIC IMAGES/Alamy Stock Photo; **p. 19** *t* © Park, W. B./Cartoon Stock, *b* © Nicolas Marino/Mauritius images GmbH/Alamy Stock Photo; **p. 20** © ODI/Alamy Stock Photo; **p. 21** © Sigit Pamungkas/Reuters; **p. 24** © YASUYOSHI CHIBA/AFP/Getty Images; **p. 26** © NOAA (National Oceanic and Atmospheric Administration); **p. 29** © Lance Cpl. Niles Lee/AB Forces News Collection/Alamy Stock Photo; **p. 33** © ODD ANDERSEN/AFP/Getty Images; **p. 34** © Tom Miller/NASA (National Aeronautics and Space Administration); **p. 35** © ArendTrent/Thinkstock/iStock/Getty Images; **p. 36** © Keith morris news/Alamy Stock Photo; **p. 37** © Network Rail; **p. 39** © Michael Scott/Alamy Stock Photo; **p. 40** © Ashley Cooper/Alamy Stock Photo; **p. 45** *t* © Petty Officer 2nd Class Cory J. Mendenhall/AB Forces News Collection/Alamy Stock Photo, *b* © Museum of London/Heritage Image Partnership Ltd/Alamy Stock Photo; **p. 46** © Stocktrek Images/Stocktrek/Superstock; **p. 47** © InterNetwork Media/Photodisc/Getty Images; **p. 53** © Lynch, Mark/Cartoon Stock; **p. 54** © Charles O. Cecil/Alamy Stock Photo; **p. 63** © rumandawi/123RF; **p. 65** *l* © Michael Turner/123RF.com, *r* © yair leibovich/123RF.com; **p. 69** © Rumandawi/123RF.com; **p. 71** © MikeLane45/iStock/Getty Images; **p. 75** © Andre Penner/AP/Shutterstock.com; **p. 76** © Uwe Bergwitz/stock.adobe.com; **p. 77** © Ueslei Marcelino/Reuters; **p. 78** © Per-Anders Pettersson/Corbis News/Getty Images; **p. 79** © Medicshots/Alamy Stock Photo; **p. 80** © Photofusion/Shutterstock.com; **p. 81** © Christian Ender/Getty Images **p. 82** © Christopher Scott/Alamy Stock Photo; **p. 83** © MShieldsPhotos/Alamy Stock Photo; **p. 86** © Nicolas Marino/Mauritius images GmbH/Alamy Stock Photo; **p. 87** © JAY//stock.adobe.com; **p. 88** © Inga spence/Alamy Stock Photo; **p. 89** © John Moore/Getty Images News/Getty Images; **p. 90** © zrfphoto/Thinkstock/iStock/Getty Images; **p. 91** *t* © Simon Oakes, *b* © Zeality, https://commons.wikimedia.org/wiki/File:Taos_plaza_la_fonda.jpg, https://creativecommons.org/licenses/by/2.5/; **p. 92** © Radek Hofman / Alamy Stock Photo **p. 93** *t* © Tim Roberts Photography/Shutterstock.com; *b* © Karl Weatherly/Age fotostock/Alamy Stock Photo; **p. 94** © Weber Raphael/Prisma by Dukas Presseagentur GmbH/Alamy Stock Photo; **p. 97** © THOMAS COEX/AFP/Getty Images; **p. 98** © Universal Images Group /Getty Images; **p. 99** © Kristin Mosher/Danita Delimont/Alamy Stock Photo; **p. 100** © Erectus/Fotolia.com; **p. 102** *l* © Hunta/stock.adobe.com, *c* © Kimberly Walker/Robertharding/Alamy Stock Photo, *r* © Peter Prokosch, http://www.grida.no/photolib/detail/larch-forest-and-swamps-close-to-the-tree-line-taigatundra-boundary-in-southern-taimyr-russia-july-1993_bc94; **p. 105** © Luis Sinco/Los Angeles Times/Getty Images; **p. 107** *t* © Steven Kazlowski/RGB Ventures/SuperStock/Alamy Stock Photo, *b* © Egmont Strigl/imageBROKER/Alamy Stock Photo;

Continued on page 418.

Introduction to this book

The AQA GCSE Geography course has three main units, with two or three sections within each part. Most of your study is compulsory, although there are some optional topics. Your teacher will tell you which options you will be studying.

Unit 3.1 Living with the physical environment

- Section A Chapters 1–4
- Section B Chapters 5–8
  You will study one of either:
  - 7 Hot deserts **or**
  - 8 Cold environments
- Section C Chapters 9–12
  You will study either:
  - Chapter 10 Coastal landscapes and Chapter 11 River landscapes **or**
  - Chapter 10 Coastal landscapes and Chapter 12 Glacial landscapes **or**
  - Chapter 11 River landscapes and Chapter 12 Glacial landscapes

Unit 3.2 Challenges in the human environment

- Section A Chapters 13–16
- Section B Chapters 17–20
- Section C Chapters 21–25
  You will study one of:
  - Chapter 23 Food **or**
  - Chapter 24 Water **or**
  - Chapter 25 Energy

Unit 3.3 Geographical applications

- Section A Chapter 26 Issue evaluation
- Section B Chapter 27 Fieldwork and geographical enquiry

Unit 3.4 Geographical skills

- Throughout

Features of this book

The following features of this book have been designed to help you make the most of your course.

- **Key learning** boxes at the top of each spread provide a useful overview of the learning objectives.
- **Activities**, linked to the text and resources on each spread, to develop your understanding.
- **Geographical Skills** boxes to give you opportunities to practise and develop your skills.
- **Get out there!** Fieldwork suggestions to show you how the knowledge you learn can be applied to your fieldwork.
- **AQA-specific key terms** in purple and other key terms in black throughout the chapters, all defined in the glossary to help you learn and demonstrate your geographical vocabulary.
- **Case studies and examples** with up-to-date information to illustrate geographical concepts.
- **Practice questions** in the Question practice pages, which include a range of topical questions and practical advice from skilled teachers to help you prepare for your exam.

Case studies and examples in the UK used in this book

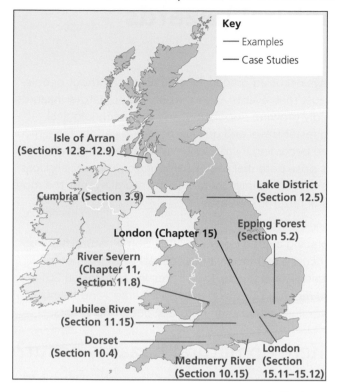

Key
— Examples
— Case Studies

Isle of Arran (Sections 12.8–12.9)

Cumbria (Section 3.9)

Lake District (Section 12.5)

London (Chapter 15)

Epping Forest (Section 5.2)

River Severn (Chapter 11, Section 11.8)

Jubilee River (Section 11.15)

Dorset (Section 10.4)

Medmerry River (Section 10.15)

London (Section 15.11–15.12)

Case studies and examples across the world used in this book

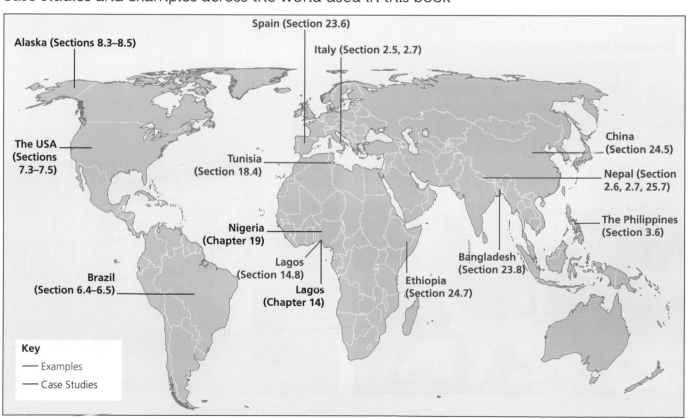

Spain (Section 23.6)

Italy (Section 2.5, 2.7)

Alaska (Sections 8.3–8.5)

The USA (Sections 7.3–7.5)

China (Section 24.5)

Tunisia (Section 18.4)

Nepal (Section 2.6, 2.7, 25.7)

Nigeria (Chapter 19)

The Philippines (Section 3.6)

Lagos (Section 14.8)

Brazil (Section 6.4–6.5)

Lagos (Chapter 14)

Ethiopia (Section 24.7)

Bangladesh (Section 23.8)

Key
— Examples
— Case Studies

1 Natural hazards

Defining natural hazards

What is a natural hazard?

Natural events have always occurred on our dynamic Earth. Without people, natural events would be just that, events – there would be no natural 'hazards'. Yet in a world with a rapidly growing population, and with technological developments leading to faster travel and quicker communication, it is difficult to ignore that humans are becoming increasingly vulnerable to **natural hazards**. Natural hazards pose potential risk of damage to property, and loss of life. The more humans that come into contact with natural events, the more the potential risk of natural hazards increases.

How are different types of natural hazard classified?

Natural hazards are most commonly classified by their physical processes, that is, what caused the hazard to occur. These processes include:

- **tectonic hazards**, such as **earthquakes** or tsunamis, which involve movement of tectonic plates in the Earth's crust
- **atmospheric hazards**, such as tropical storms
- **geomorphological hazards**, such as flooding, which occur on the Earth's surface
- **biological hazards**, such as forest fires, which involve living organisms.

However, these categories are closely linked. For example, tsunamis are a tectonic hazard, but can also be caused by a **landslide** (a geomorphological hazard)

displacing a large body of water. Some natural hazards are caused by human influence rather than a natural process. For example, forest fires in California in 2014 were recorded as being caused by arson and falling power lines rather than naturally occurring events.

Where do natural hazards occur?

Some regions around the world are more vulnerable to natural hazards than others. The different colours of the countries in Figure 1.4 show the likelihood of a natural disaster occurring, based on historical data.

The year 2018 was a bad one for natural hazards. There were volcanic eruptions in Indonesia and Guatemala, earthquakes and tsunamis in Indonesia, flooding across East Africa and wildfires in Greece. Some of the biggest natural hazards that occurred around the world in 2018 are located on the map in Figure 1.4.

▲ **Figure 1.1** Avalanche, Italy, 2017

▲ **Figure 1.2** Flooding, York, 2015

What factors affect hazard risk?

The frequency and magnitude (strength) of natural hazards are increasing due to human influences such as the enhanced greenhouse effect (which increases the risk of more **extreme weather** such as drought) and **deforestation** (which increases the risk of hazards such as flooding). The main factors that affect **hazard risk** are:

- Wealth – a risk tends to decrease with increased wealth, as people can afford to prepare for and respond to natural hazards.
- Population growth – by 2024, 8 billion people are expected to populate the planet, which inevitably increases hazard risk as there are more people to interact with natural events.

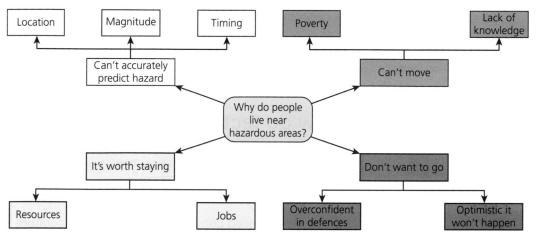

▲ Figure 1.3 Why people live near hazardous areas

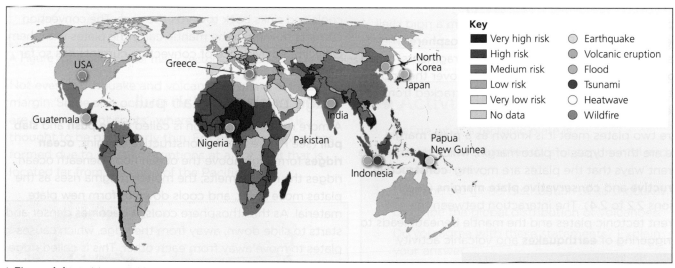

▲ Figure 1.4 World natural hazard risk and biggest natural disasters in 2018

→ Activities

1 Distinguish between a natural hazard and a natural event.

2 Categorise the following natural hazards into tectonic, atmospheric, geomorphological or biological: earthquakes, volcanic eruptions, flooding, landslides, tornados, avalanches, forest fires; hurricanes, typhoons or cyclones.

3 Explain why some hazards are more difficult to categorise than others.

4 Describe the pattern of natural hazard risk in Figure 1.4.

5 Use Figure 1.3 to explain why people can make themselves more vulnerable to natural hazards.

6 Suggest how humans could have caused the hazards in Figure 1.1 and Figure 1.2.

⊙ KEY LEARNING

➤ How plates at constructive margins move

➤ Why earthquakes and volcanoes are found at constructive plate margins

Constructive plate margins

How do plates move at constructive margins?

Constructive plate margins occur when tectonic plates move apart from each other. Most tectonic plates move a few centimetres a year. This may not sound much, but over time this has meant that whole continents have moved position. As shown in Figure 2.2 in Section 2.1, the Eurasian Plate and North American Plate form a constructive plate margin: they are moving away from each other at a rate of six centimetres per year.

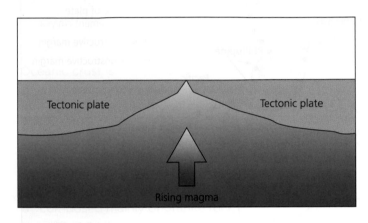

Why are earthquakes and volcanoes found at constructive plate margins?

1   At constructive margins, the upper part of the mantle melts and the hot molten magma rises. The East Pacific Rise, on the other hand, is a faster moving constructive plate margin. It separates the Pacific Plate from the North American Plate and Nazca Plate at a rate of about ten centimetres per year.

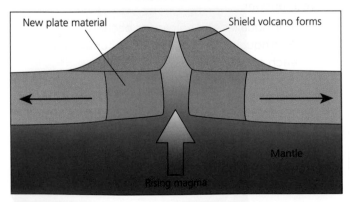

2   As the tectonic plates are moved away from each other, this releases pressure and therefore molten magma rises between. It cools down to form solid rock. As the rock cools, it becomes denser and sinks, causing the tectonic plates to move further (see ridge push on page 4).

3   Much of the magma never reaches the surface but it is buoyant enough to push up the crust at constructive margins to form ridge and rift features. In a few places the magma erupts on to the surface, producing a basic lava that is runny and spreads out before solidifying. Over many eruptions, a volcano that typically is low, has a wide base and gentle slopes, known as a **shield volcano**, is formed.

▲ Figure 2.4 Constructive plate margins

The Mid-Atlantic Ridge is located along a constructive plate margin. The North American plate is moving away from the Eurasian plate (Figure 2.5). The magma has risen and caused uplifting of the Earth's crust, forming the ridge as a range of underwater mountains.

Iceland is the world's largest volcanic island situated on the Mid-Atlantic Ridge. Magma has risen to form underwater volcanoes, which have grown above sea level to form this volcanic island. The Westman Islands are 15 islands all created by volcanic activity originating on the seabed (Figure 2.5).

Rift valleys are steep-sided valleys that form at constructive plate margins. The strain of the tectonic plates moving away from each other is splitting Iceland in two, causing cracks or faults to form on either side. As the sides of the rift move and stretch apart, sections drop down to form rift valleys.

Iceland's Thingvellir National Park has a visitor centre with a path leading to the Almannagjá fault where a stretch of the plate margin can be viewed (Figure 2.6). Hundreds of small earthquakes occur in Iceland on a weekly basis.

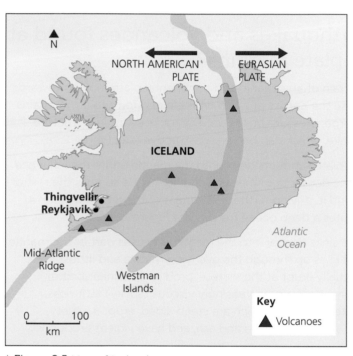

▲ Figure 2.5 Map of Iceland

▲ Figure 2.6 Mid-Atlantic Ridge, Thingvellir, Iceland

→ Activities

1 Describe the direction in which plates move at constructive plate margins.

2 Use Figure 2.2 in Section 2.1 (page 5) to identify the plate names at two different constructive plate margins.

3 Are earthquakes experienced at constructive plate margins? Explain your answer.

4 What evidence can you find on the map (Figure 2.5) and the photograph (Figure 2.6) that Iceland lies on a constructive plate margin?

5 Explain why there are so many volcanoes located in Iceland.

Geographical skills

Draw a sketch of Figure 2.6 and label human and physical features.

➤ How plates at destructive margins move

➤ Why earthquakes and volcanoes are found at destructive plate margins

Destructive plate margins

How do plates move at destructive margins?

Destructive plate margins occur when tectonic plates move towards each other and collide. The effect this has depends on the type of plate:

■ If two continental plates collide, they are both buoyant and so cannot sink into the mantle. As a result, compression forces the plates to collide and form mountains.

■ If an oceanic and a continental plate collide, the denser oceanic plate is **subducted** and sinks under the continental plate, into the Earth's mantle, where it is recycled, causing earthquakes, fold mountains and volcanoes.

Why are earthquakes and volcanoes found at destructive plate margins?

The pressure and strain of an oceanic and continental plate moving towards each other can cause the Earth's crust to crumple and form fold mountains. As the plates converge, pressure builds up. The rocks eventually fracture, causing an earthquake, which can be very destructive (see Figure 2.7a).

The denser oceanic plate subducts down into the mantle under the influence of gravity. The plate is denser than the surrounding mantle so pulls the rest of the plate along behind it, driving further movement of the tectonic plate. At the surface, this creates a deep ocean trench (see Figure 2.7b).

As the oceanic plate sinks deeper into the mantle, it causes part of the mantle to melt. Hot magma rises up through the overlying mantle and lithosphere, and some can eventually erupt at the surface, producing a linear belt of volcanoes. The magma becomes increasingly viscous (sticky) as it rises, producing **composite volcanoes** which are steep-sided, erupt a variety of materials such as sticky acidic lava and ash, and have violent eruptions (see Figure 2.7c).

Japan's volcanoes

Japan is prone to earthquakes and volcanic eruptions. It has 118 **active volcanoes** currently erupting or showing signs of eruption (ten per cent of the global total), more than almost anywhere else in the world. Japan's band of volcanoes form part of the Ring of Fire (see page 5) surrounding the Pacific Ocean. There are so many because Japan lies on the margin of four plates: the Eurasian, North American, Pacific and Philippine (see Figure 2.2 in Section 2.1). The Pacific plate subducts beneath the North American plate and the Philippine plate, which then subducts beneath the Eurasian plate. Many parts of Japan have experienced earthquakes due to the pressure built up in the plates as they move at this destructive plate margin.

Also formed at the destructive plate margin of the Pacific and Philippine plates is an ocean trench, known as the Mariana Trench. It is 10,994 metres deep – deeper than the tallest mountain, Mount Everest (8,848 metres). The Mariana Trench is the deepest known part of the Earth's oceans.

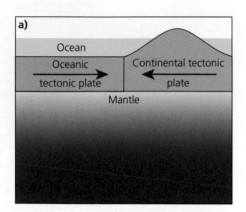

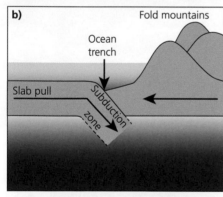

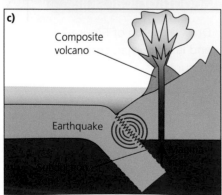

▲ Figure 2.7 Destructive plate margins

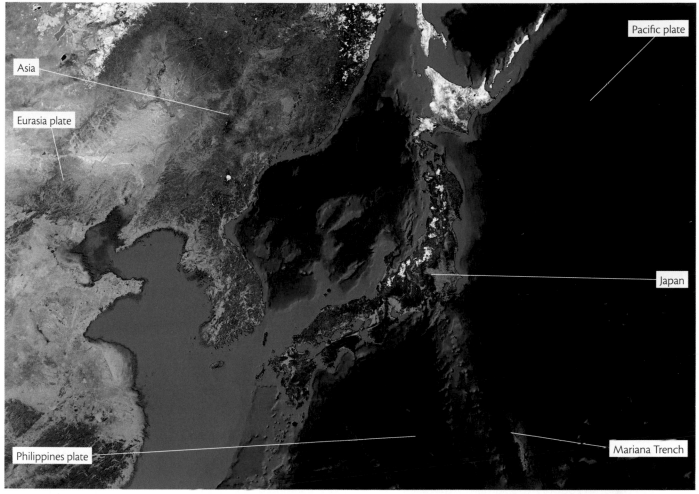

Asia

Eurasia plate

Pacific plate

Japan

Philippines plate

Mariana Trench

▲ Figure 2.8 Satellite image of Japan. The darker the blue, the greater the ocean's depth

→ Activities

1 Describe the direction in which plates move at destructive plate margins.

2 Use Figure 2.2 in Section 2.1 to determine which of the following are destructive plate margins:
   - Eurasian and Philippine plates
   - Nazca and Pacific plates
   - Nazca and South American plates
   - North American and Eurasian plates.

3 What landforms are found at destructive plate margins?

4 Explain how earthquakes and volcanoes are formed at destructive plate margins.

Geographical skills

Draw an annotated sketch map of Figure 2.8. Annotate the image to show where you would expect earthquakes, volcanoes and mountains to be formed – explain why.

➤ How plates at conservative margins move

➤ Why earthquakes are found at conservative plate margins

Conservative plate margins

How do plates move at conservative margins?

A conservative plate margin occurs when tectonic plates move parallel to each other. The two plates can move side by side, either in the same direction but at different speeds, or simply in the opposite direction to one another (see Figure 2.9).

Why are earthquakes found at conservative plate margins?

One theory is that pressure might build up at the margin of the tectonic plates as they are pulled along behind a plate being subducted elsewhere. As the plates move past each other, friction causes them to become stuck. Pressure builds up and up until eventually the rock fractures and pressure is released, sending out huge amounts of energy, causing an earthquake. However, volcanoes are not formed at conservative plate margins. Magma cannot rise to fill a gap as there is no gap created between the tectonic plates, and therefore there is no new land formed. Neither is there any land destroyed, because there is no tectonic plate subducted into the mantle.

The San Andreas Fault stretches 800 kilometres through the state of California in the USA. It is found along the margin between the North American plate and the Pacific plate. These tectonic plates are sliding past each other in roughly the same northwest direction, but at different speeds (see Figure 2.10). The North American plate moves at approximately 6 centimetres per year, whereas the Pacific plate moves at approximately 10 centimetres per year. Fifteen to twenty million years ago, Los Angeles would have been south of where San Diego is now. If the plates continue to move at the same speed, in 20 million years' time, Hollywood in Los Angeles will be adjacent to the Golden Gate Bridge in San Francisco.

California experiences thousands of small earthquakes every year. One of the biggest earthquakes to hit California was in San Francisco in 1906. Scientists have estimated that the 1906 San Francisco earthquake measured a magnitude of 8.3 on the **Richter scale**. (The largest earthquake ever recorded was in Chile, in 1960, which measured 9.5.) Approximately 700 people died and the damage caused was estimated to cost over US$500 million.

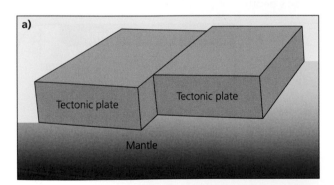

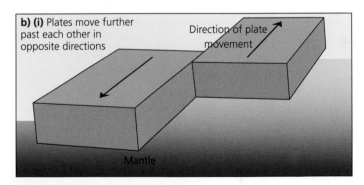

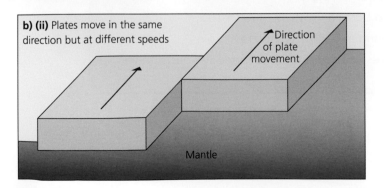

▲ Figure 2.9 Movement of tectonic plates at a conservative plate margin

The US Geological Survey have run computer models to predict the next major earthquake on the San Andreas Fault. They report that there is most likely to be an earthquake similar to the 1906 San Francisco earthquake at intervals of about 200 years. There is only a small chance (about 2 per cent) that an earthquake like this would happen within the next 30 years.

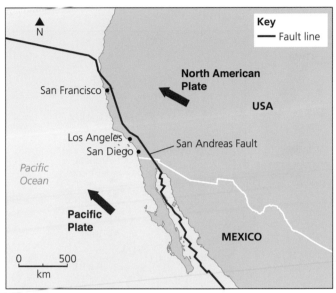

▲ Figure 2.10 Map showing the San Andreas Fault, California, USA

▲ Figure 2.11 The San Andreas Fault

→ Activities

1   Use Figure 2.2 in section 2.1 to identify a different conservative plate margin from the San Andreas Fault.

2   a)   Are the following statements about conservative margins true or false?
   i)   Two tectonic plates move away from each other.
   ii)   Two tectonic plates move parallel past each other.
   iii)   Two tectonic plates move towards each other.
   iv)   Earthquakes are found at conservative margins.
   v)   Volcanoes are found at conservative margins
   b)   Explain your answer for each statement.

3   Describe and explain the similarities between the photographs in Figure 2.5 in Section 2.2 and Figure 2.11 on this page.

4   Complete the following summary table about the three types of plate margins.

Margin type	Type of plates (oceanic or continental or either)	Direction of plate movement	Earthquakes (tick or cross)	Volcanoes (tick or cross)
Constructive				
Destructive				
Conservative				

Geographical skills

The Richter scale is logarithmic. This means each increase of one on the scale means the power is increased by 10, not 1. So a magnitude 6 earthquake is 10 times more powerful than a magnitude 5 earthquake.

Calculate how much more powerful a magnitude 8 earthquake is than a magnitude 4 earthquake.

Example

⊗ KEY LEARNING

➤ Primary and secondary effects of an earthquake

➤ How people responded to the Amatrice earthquake

Earthquake in Amatrice, Italy (2016)

What were the effects of the earthquake?

On 24 August 2016, an earthquake measuring a magnitude of 6.2 on the Richter scale struck central Italy. The earthquake's epicentre was halfway between the towns of Amatrice and Norcia, at a shallow depth of 5.1 km.

Amatrice is a town in the Apennine Mountains in central Italy, approximately 30 miles north of L'Aquila, which also experienced a major earthquake in 2009 (see Figure 2.12). The Amatrice earthquake was felt as far as 100 miles away, in Rome. Amatrice experienced a range of impacts which affected the wealth of the area and of the community (**economic impacts**), the lives of members of the community (**social impacts**) and the **landscape** (**environmental impacts**).

Primary effects

As a direct result of the earthquake, 299 people were killed, 400 were injured and 4454 were made homeless. It struck at 3.36 a.m., so most people were asleep in buildings which collapsed, with no time to evacuate:

- Towns and villages in the regions of Umbria, Lazio and Marche suffered most damage.
- Around 293 historic buildings were damaged or destroyed, including the fourteenth-century Basilica of St. Benedict in Norcia, and the Church of Sant'Agostino and the Basilica of San Francesco in Amatrice.
- Over half the buildings were destroyed in Amatrice, 80 per cent of the historic old town, despite their reinforcements.
- Amatrice's Romolo Capranica school completely collapsed (considered a substandard construction).
- Even though the government had allocated €1 billion for building improvements to be made following the 2009 L'Aquila earthquake, many properties did not meet seismic building standards (across Italy, this is true of around 70 per cent of buildings).
- The earthquake struck during the summer holiday season, so the population was much higher than normal. The death toll included tourists celebrating an annual food festival.
- Overall, the EU estimated the total damage at €21.9 billion.

▲ **Figure 2.12** Amatrice in the region of Lazio, Italy

Secondary effects

Some effects of the earthquake occurred later and indirectly as a result of the initial earthquake itself. These are some of the **secondary effects**:

- The centre of Amatrice town was cordoned off due to the unsafe buildings. Parts of the centre were made 'red zones', reducing business, tourism and income.
- Farmers struggled to earn a living as 90 per cent of sheep, goats and cattle barns (and their milking systems) were destroyed.
- Landslides blocked roads and reduced access to the area.
- Residents suffered psychological damage, especially as they know they live in the most seismically active area in Europe.
- The press reported individuals were arrested for looting properties in Amatrice.

What were the responses to the earthquake?

Immediate responses

There were a range of **immediate responses**. For those made homeless, 10,000 were accommodated in 58 tent camps. Converted sports halls and hotels on the Adriatic coast provided shelter for 4000. Additionally:

- Rescue workers, including the Italian Red Cross, 5000 soldiers and Alpine guides, arrived within an hour, searching for survivors and providing food and tents. Twelve helicopters and 70 dog teams were involved in the rescue effort.
- Patients at Amatrice hospital were transferred to a nearby hospital in Rieti as Amatrice hospital was severely damaged. A temporary hospital was set up. The national blood donation service appealed for new donors to ensure demand was met.
- Facebook set up their safety check feature for people in the area to let friends and family know they were safe. The Italian Red Cross requested locals remove wi-fi passwords so rescue teams could communicate more easily.
- Italian Prime Minister Matteo Renzi announced €50 million for the emergency response, for reconstruction work to begin immediately and taxes for residents to be cancelled.
- British chef Jamie Oliver pledged donations for every plate of Amatrice's famous Amatriciana pasta served in his restaurants. All proceeds from visits to museums and archaeologist sites on 28 August throughout Italy went to a fund to help rebuild the historic sites damaged in the earthquake.

However, rescue efforts were hampered due to blocked roads, a damaged bridge, the mountainous terrain and over 2000 aftershocks which caused even more damage.

▲ **Figure 2.13** Aftermath of the earthquake in Amatrice, Italy, 2016

Long-term responses

There were several **long-term responses**:

- Students attended classes in neighbouring schools while 12 classrooms in prefabricated buildings were constructed in Amatrice so children could return to school there.
- A €42 million government initiative called 'Italian Homes' sought to rebuild villages with buildings of the same character, but now earthquake-proof. Tax incentives allow 65 per cent of total renovation costs to be used as tax breaks to help reduce the cost of rebuilding. However, this was not enough to ensure a significant increase in buildings that are safe. Prime Minister Matteo Renzi stated it was 'absurd' to think Italy could build completely earthquake-proof buildings. The Italian press criticised the government over building regulations, especially as some recently renovated buildings had also collapsed.
- Six months after the earthquake, the Italian government promised to move people from temporary camps into lightweight wooden houses.
- A year later, 2.4 million tons of rubble and debris remained in affected areas.

→ Activities

1. Use Figure 2.12 to describe the location of Amatrice.
2. Define the primary and secondary effects of an earthquake.
3. a) Give three examples of primary effects likely in any earthquake around the world.
   b) Give three examples of secondary effects likely in any earthquake around the world.

4. Complete the table of effects for this earthquake.

Economic	Environmental	Social

5. Explain how four of the responses described would help to manage the effects.
6. Justify why both immediate and long-term responses were needed.

Example

⭐ KEY LEARNING

➤ Primary and secondary effects of the earthquake

➤ How people responded to the Gorkha earthquake

Earthquake in Gorkha, Nepal (2015)

What were the effects of the earthquake?

On Saturday 25 April 2015, at 11.56 a.m. (local time), a 7.8 magnitude earthquake struck the Gorkha district in Nepal (Figure 2.14). The earthquake's epicentre was in Barpak, 80 kilometres northwest of the capital, Kathmandu.

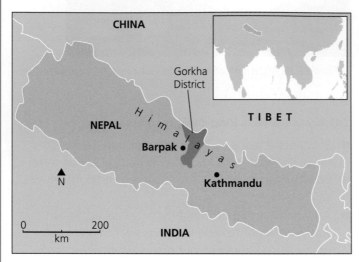

▲ **Figure 2.14** Nepal, Asia

Primary effects

The immediate **primary effects** included:

- A total of 8,841 killed, over 16,800 injured and 1 million made homeless.
- Historic buildings and temples in Kathmandu, including the iconic Dharahara Tower (Figure 2.16), a UNESCO World Heritage Site, in which 200 people were estimated to be trapped, were destroyed; there were no compulsory building standards in Nepal, so many modern buildings also collapsed.
- The destruction of 26 hospitals and 50 per cent of schools. (Save the Children estimated 29,000 more people would have been killed if the earthquake had struck during school hours, rather than on a Saturday.)
- A reduced supply of water, food and electricity.
- 352 aftershocks, including a second earthquake on 12 May 2015 measuring 7.3 magnitude.

What were the responses to the earthquake?

Immediate responses

Nepal requested international help. The UK's Distasters Emergency Committee (DEC) raised US$126 million by September 2015 to provide emergency aid and start rebuilding the worst-hit areas.

Temporary shelters were set up. The Red Cross provided tents for 225,000 people. The United Nations (UN) health agency and the World Health Organization (WHO) distributed medical supplies to the worst-affected districts. This was important as the monsoon season had arrived early, increasing the risk of **waterborne diseases**.

Nepal's mountainous terrain and inadequate roads made it difficult for aid to reach remote villages. 315,000 people were cut off by road and 75,000 were additionally unreachable by air. Sherpas were used to hike relief supplies to remote areas. Facebook launched a safety feature so people could indicate

Secondary effects

The earthquake triggered an avalanche on Mount Everest. It swept through Everest Base Camp, which is used by international climbing expeditions. Out of the 19 who died, several were tourists and the rest were native people from Nepal called Sherpas. Sherpas work as porters, guides and cooks. They are aware of the dangers of Mount Everest, but tourism can provide them with an income to help lift them out of poverty.

In 2014, the World Travel and Tourism Council reported that tourism was 8.9 per cent of Nepal's GDP and provided 1.1 million jobs. It was expected to increase by 5.8 per cent in 2015, but until Nepal has recovered from the earthquake, tourism, employment and income will shrink.

The earthquake happened just before the monsoon season, when rice is planted. Rice is Nepal's staple diet, and two-thirds of the population depend on farming. Rice seed stored in homes was ruined in the rubble, causing food shortages and income loss.

they were 'safe'. Several companies did not charge for telephone calls.

Long-term responses

Nepal's government (along with the UN, EU, World Bank, Japan International Cooperation Agency and Asian Development Bank) carried out a Post-Disaster Needs Assessment. It reported that 23 areas required rebuilding, such as housing, schools, roads, monuments and agriculture. Eight months after the earthquake, the Office for the Coordination of Humanitarian Affairs (OCHA) reported that US$274 million of aid had been committed to the recovery efforts.

The Durbar Square heritage sites were reopened in June 2015 in time to encourage tourists back for the tourism season. Mount Everest was reopened for tourists by August 2015 after some stretches of trail were re-routed. By February 2016, the Tourism Ministry extended the climbing permits that had been purchased in 2015 to be valid until 2017, so that climbers would return and attempt Everest again.

A recovery phase started six months later by the Food and Agriculture Organization of the United Nations (FAO). To expand crop production and growing seasons, individuals were trained how to maintain and repair **irrigation** channels damaged by landslides in the earthquake.

However, by November 2018, a community perception report (funded by UK Aid) stated that 34 per cent of people affected by the earthquake were still living in temporary shelters or in homes that were unrepaired.

Nepal's recovery needs are US$6.7 billion, roughly a third of the economy. Early estimates suggest that an additional three per cent of the population has been pushed into poverty as a direct result of the earthquakes. This translates into as many as one million more poor people.

▲ Figure 2.15 Post-Disaster Needs Assessment (*Source: Worldbank.org press-release 23/06/2015*)

▲ Figure 2.16 Dharahara (Bhimsen) Tower, Kathmandu, before and after the earthquake

▲ Figure 2.17 Tents set up near Kathmandu airport, Nepal

→ Activities

1 Use Figure 2.14 to describe the location of Nepal.

2 Draw a sketch of the photos in Figure 2.16 or Figure 2.17 that shows the effects of the earthquake. Annotate the effects.

3 Identify the economic, environmental and social effects for the Gorkha earthquake.

4 Why do you think the tents in Figure 2.16 were set up near Kathmandu airport?

5 Explain three immediate and three long-term responses to the Gorkha earthquake.

6 Research other 'before and after' images of the earthquake. Describe the differences.

➤ Where people live in relation to earthquakes and volcanoes

➤ Why people live in areas at risk of tectonic hazards

Risking it

Do people live in areas prone to volcanic eruptions and earthquakes?

Despite all the dangers, millions of people still live in hazard-prone areas. As the world's population rises, more people will live in volcanic and earthquake-prone areas. Figure 2.20 shows that large cities, with over 1 million inhabitants, can be found in the same locations (and within close proximity) to earthquakes. Approximately eight per cent of the 7.7 billion people who live in the world live near volcanoes, and 50 per cent of the 330 million people in the USA are living at risk of earthquakes.

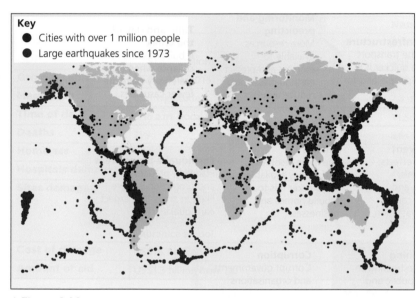

Key
● Cities with over 1 million people
● Large earthquakes since 1973

▲ Figure 2.20 Cities with over 1 million inhabitants and earthquakes

Why do people continue to live in hazardous areas?

Geothermal energy

In volcanically active areas, **geothermal energy** is a major source of electrical power: steam is heated by hot magma in permeable rock, then boreholes are drilled into the rock to harness the super-heated steam to turn turbines at power stations. It is renewable energy – it will not run out, and it will reduce greenhouse gases and the likely effects of **climate change**.

Hellisheidarvirkjun (or Hellisheidi) power plant is the largest geothermal power station in Iceland and the second largest in the world. It provides electricity and hot water for the capital, Reykjavik. Geothermal energy produces approximately 30 per cent of Iceland's total electricity.

Farming

Lava and ash erupting from volcanoes kill livestock and destroy crops and vegetation. After thousands of years, weathering of this lava releases minerals and leaves behind extremely fertile soil, rich in nutrients. Land can be farmed productively in these areas to provide a source of food and income. Volcanic soils are found on less than one per cent of the Earth's surface, but support 10 per cent of the world's population.

▲ Figure 2.21 Hellisheidarvirkjun geothermal power plant, Iceland

Mining

Settlements develop where valuable minerals are found, as jobs are created in the mining industry. It is not just dormant and extinct volcanoes that are mined, but also active volcanoes.

Kawah Ijen is an active volcano in East Java, Indonesia. Its crater is one of the biggest sulphuric lakes in the world. Sulphur is sold, for example, to bleach sugar, make matches, medicines and fertiliser. However, mining in active volcanoes is dangerous:

- Miners can afford little protective clothing.
- Hydrogen sulphide and sulphur dioxide gases burn their eyes and throats and cause respiratory diseases.
- In the last 40 years, 74 miners have died from the fumes.
- Loads of sulphur weighing 100 kilograms are carried up and down the rocky and slippery mountain paths.

Nevertheless, miners can earn an average of $6 per day (more than on a coffee plantation), so miners continue to live and work in dangerous areas.

Tourism

Tourists visit volcanoes for the spectacular and unique views, relaxing hot springs, adventure and, for thrill seekers, the sense of danger. More than 100 million people visit volcanic sites every year. The revenue they generate benefits the locals and the countries they are in.

Family, friends, feelings and freedom

People do not wish to leave because their friends and family are there. It is often cheaper and easier to stay, especially when the risks may not be perceived as dangerous enough or residents are in denial that a disaster may occur. Some people may have a lack of choice due to poverty, which means they may not have the funds to enable them to move to a new location to live.

▲ Figure 2.22 A cartoon showing how people may be persuaded to move away from hazardous areas. The speech bubble says, 'Hey – great news! I've finally decided to sell you my house on the island!'

▲ Figure 2.23 Mining sulphur at Kawah Ijen crater, Indonesia

→ Activities

1 How does the map in Figure 2.20 demonstrate that people are at risk from earthquakes?

2 Why would a farmer consider the benefits of living near a volcano different in the short term rather than in the long term?

3 Draw a sketch of the photograph in Figure 2.23. Annotate it with the following labels:

crater, lake, sulphur, hydrogen sulphide and sulphur dioxide gases, little protective clothing, hand-carried loads of sulphur, rocky paths.

4 Describe three benefits of living near a volcano.

5 Explain why it may have taken time for the character in Figure 2.22 to make that decision.

✪ KEY LEARNING

➤ How the risks of earthquakes can be reduced

➤ How the risks of volcanic eruptions can be reduced

Risk management

Can the risks of earthquakes be reduced?

Prediction, **monitoring**, **protection**, and **planning** all aim to reduce the damage that earthquakes and volcanic eruptions cause to people and property.

Monitoring and prediction

It is possible to predict the general locations where earthquakes are most likely to happen, as they occur along plate margins. However, it is extremely difficult to predict their time, date and exact location. The following show some ways that technology is used to try to monitor and predict tectonic hazards:

- **Seismologists** use radon detection devices to measure radon gas in the soil and **groundwater**, which escapes from cracks in the Earth's surface.
- Sensitive seismometers are used to measure tremors or **foreshocks** before the main earthquakes.
- Earthquake locations and their times are mapped to spot patterns and predict when the next earthquake will occur.
- Smart phones have GPS (Global Positioning System) receivers and accelerators built in. They can detect movements in the ground, which are analysed to potentially warn others further away.
- Animals are believed to act strangely when an earthquake is impending.

Protection

Buildings made of brick or buildings with no reinforcement collapse easily during an earthquake. Designing buildings and strengthening roads and bridges to withstand earthquakes provides protection. This is also called **mitigation**. Figure 2.24 shows features of earthquake-resistant designs.

Unfortunately, earthquake-resistant buildings and **infrastructure** are extremely expensive, so it is usually not possible to adapt existing buildings. The aim of earthquake-resistant buildings is to ensure that people are not injured or killed – so although this might be achieved, the building may still need to be repaired or even rebuilt after an earthquake.

Planning

Planning and preparing what to do during and after an earthquake helps the authorities, emergency services and individuals to act quickly and calmly, so there is less chaos and fewer injuries and deaths:

- Furniture and objects can be fastened down so they are prevented from toppling over.
- Residents can learn how to turn off the main gas, electricity and water supplies to their properties.
- Preparing emergency aid supplies, how they would be distributed and where evacuation centres will be saves lives, as food, water, medicine and shelter are accessed faster.
- On 1 September each year, the Japanese practise earthquake drills on a national training day. This marks the anniversary of the Tokyo earthquake in 1923, which killed 156,000 people.
- The American Red Cross provides an earthquake safety checklist to help people plan and prepare for earthquakes in their homes, at work and in schools.

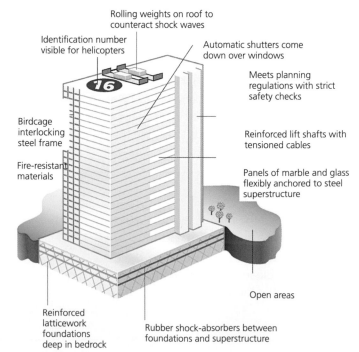

Rolling weights on roof to counteract shock waves

Identification number visible for helicopters

Automatic shutters come down over windows

Meets planning regulations with strict safety checks

Birdcage interlocking steel frame

Reinforced lift shafts with tensioned cables

Fire-resistant materials

Panels of marble and glass flexibly anchored to steel superstructure

Open areas

Reinforced latticework foundations deep in bedrock

Rubber shock-absorbers between foundations and superstructure

➤ Figure 2.24 Earthquake-resistant building design

Can the risks of volcanic eruptions be reduced?

Monitoring and prediction

It is easier to predict volcanic eruptions than earthquakes. Volcanoes usually give advance warning signals that they are going to erupt. However, the exact time and day of the eruption is still difficult to predict.

- Satellites (GPS) and tiltmeters monitor ground deformation (changes in the volcano's surface).
- Seismometers measure small earthquakes and tremors.
- Thermal heat sensors detect changes in the temperature of the volcano's surface.
- Gas-trapping bottles and satellites measure radon and sulphur gases released.
- Scientists measure the temperature of water in streams and rivers to see if it has increased.

▲ **Figure 2.25** Monitoring volcanic activity

Planning

An evacuation plan is one of the most effective methods of protection against an eruption. Authorities and emergency services need to prepare emergency shelter, food supplies and form evacuation strategies. Exclusion zones can be designated so that no one is allowed to enter where people are considered vulnerable and in danger. Additionally, residents can be educated about preventing unnecessary injury and loss of life. They can practise advice to cover their eyes, nose and mouth to prevent being irritated by gas fumes. If residents are not evacuated, they are taught to seek shelter or go indoors to avoid the dangers of falling ash and rock.

Protection

Protecting against a volcanic eruption is extremely difficult. Buildings cannot be designed to withstand the lava flows, **lahars** or weight of debris and ash falling on roofs, especially if this mixes with water. Therefore, people need to **evacuate** their homes to a safe location under the instruction of the authorities.

→ Activities

1. Define the terms prediction, protection, planning and monitoring.

2. True or false? (a) Earthquakes can be predicted. (b) Volcanoes can be predicted. Explain your answers.

3. Use Figure 2.24 to describe how each feature in the earthquake-resistant building reduces injury and loss of life.

4. Describe how the risks of living in an earthquake-prone area may be lessened by (a) individuals and (b) governments.

5. Devise your own earthquake safety checklist and explain your reason for each item.

6. Copy and complete the information in the following table.

Monitoring and predicting a volcanic eruption		
Changes a volcanologist would observe	Equipment used to monitor volcano	Explain why the changes mean a volcanic eruption is imminent

7. Suggest what the man in Figure 2.25 is doing.

8. 'Predicting tectonic hazards is a waste of time.' To what extent do you agree with this statement?

9. Why are prediction, protection and planning each important factors in reducing the risks of a tectonic hazard?

10. Suggest why monitoring, prediction, protection and planning may differ in different parts of the world.

3 Weather hazards

Global atmospheric circulation

What are the features of global atmospheric circulation?

Global atmospheric circulation helps to explain the location of world climate zones (see Figure 3.1) and the distribution of weather hazards. In Chapter 5, you will also learn how the Earth's climate zones govern the pattern of **global ecosystems**.

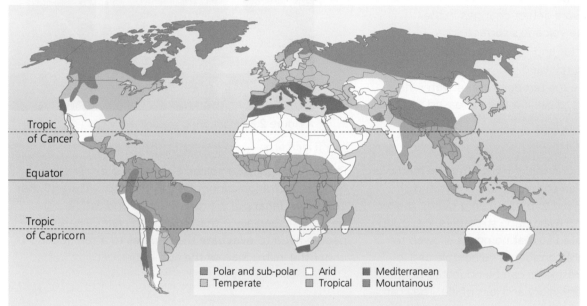

◄ Figure 3.1 World climate zones

■ Polar and sub-polar ☐ Arid ■ Mediterranean
☐ Temperate ■ Tropical ■ Mountainous

The most important influence on worldwide variations in climate is **latitude**. Because of the curved surface of the Earth, the Equator receives much higher **insolation** than the **polar** latitudes. The parallel rays of the Sun are spread thinly when they strike the Earth's surface at high latitudes, whereas at low latitudes sunlight is more highly concentrated (Figure 3.2).

As a result, air at the Equator is heated strongly. It becomes less dense and rises to a high altitude. This creates a global climate zone of low pressure: the equatorial zone. After rising, the air spreads out and begins to flow towards the North and South Poles.

Meanwhile, the low insolation received at polar latitudes results in colder, dense air and high pressure. As the air sinks towards ground level, it spreads out and flows towards the Equator.

Taken together, the low pressure belt at the Equator and the high pressure belt at the Poles provide the basis for a simple **convection cell** to operate. Global atmospheric circulation, however, is a little more complicated.

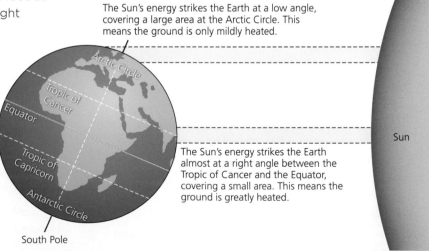

The Sun's energy strikes the Earth at a low angle, covering a large area at the Arctic Circle. This means the ground is only mildly heated.

The Sun's energy strikes the Earth almost at a right angle between the Tropic of Cancer and the Equator, covering a small area. This means the ground is greatly heated.

▲ Figure 3.2 Solar heating of the Earth varies with latitude (showing the position of the Earth on 21 June, the Summer Solstice)

There are three convection cells, not just one, as Figure 3.3 illustrates. As well as pressure belts at the Poles, there are areas of high pressure at the Tropics of Cancer and Capricorn. As the air sinks towards the ground, it warms up. The result is high pressure and hot, dry desert conditions. This circulation of air between the Tropics and the Equator is called the Hadley cell.

Global circulation involves three cells because the Earth rotates on its axis, generating strong, high-altitude winds which wrap around the planet like belts. These winds flow towards the east, as the Earth spins, and interact with the convection cells. Figure 3.3 shows two particularly strong high-altitude currents of air called **jet streams**.

The exact position of the jet streams and convection cells changes with the seasons, which take place because the Earth is tilted on its axis. Each year, as the planet journeys around the Sun, insolation rises and falls at each latitude. In high polar latitudes, the Sun does not even rise during the winter. In southern Europe, temperatures rise steeply in summer before falling in winter.

→ Activities

1 Study Figure 3.1.
   a) Identify the climate zone which the UK belongs to.
   b) Identify all the different climate zones found in (i) the USA and (ii) Russia. (Use an atlas or online map to show you the countrys' borders.)

2 a) State what is meant by insolation.
   b) Using Figure 3.2, explain one reason why the Equator receives higher insolation than polar latitudes.

3 a) State what is meant by an arid climate.
   b) Explain why arid conditions exist in (i) parts of the tropics and (ii) some polar regions.

How do global pressure and surface winds influence precipitation?

Global pressure and surface wind patterns influence **precipitation** in several important ways:

■ Rainfall is high and constant throughout the year near the Equator. As hot air rises, it cools slightly. Water vapour is converted into droplets of rain.

■ The low-pressure zone around the Equator is called the intertropical convergence zone (ITCZ). Air rises and triggers bursts of torrential rain. Sometimes, the ITCZ grows a 'wave' of low pressure which extends further than usual. Tropical storms develop along these waves. Once they gain energy, they can travel even further away from the Equator (pages 24–25).

■ Rainfall is often higher in coastal areas in Western Europe due to the movement of the Polar jet stream over the Atlantic. Rain-bearing weather systems called depressions (also known as **cyclones**) follow the Polar jet stream, often bringing stormy conditions to the UK's west coast (see page 38).

■ Rainfall is often low around the Tropics of Capricorn and Cancer. Dry air descends there as part of the Hadley cell, resulting in **arid** conditions.

■ Precipitation is also very low in polar regions and falls mostly as snow, as cold air cannot hold much water vapour.

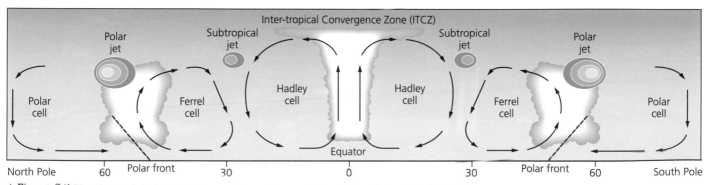

▲ **Figure 3.3** The three global convection cells and the position of the high-altitude jet streams

➤ How tropical storms form
➤ The structure and features of tropical storms

In a spin

How does a tropical storm form?

Tropical storm formation follows a particular sequence:

1 Air is heated above the surface of warm tropical oceans. The warm air rises rapidly under the low-pressure conditions.
2 The rising air draws up more air and large volumes of moisture from the ocean, causing strong winds.
3 The **Coriolis effect** causes the air to spin upwards around a calm central eye of the storm.
4 As the air rises, it cools and condenses to form large, towering cumulonimbus clouds, which generate torrential rainfall. The heat given off when the air cools powers the tropical storm.
5 Cold air sinks in the eye, therefore there is no cloud, so it is drier and much calmer.
6 The tropical storm travels across the ocean in the **prevailing wind**.
7 When the tropical storm meets land it is no longer fuelled by the source of moisture and heat from the ocean, so it loses power and weakens.

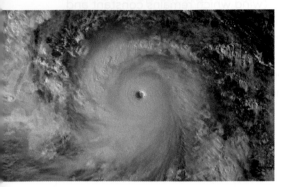

▲ Figure 3.6 Satellite image of a tropical storm (Typhoon Haiyan, 2013)

A	B	C	D	E
At the start of a tropical storm, the temperature and air pressure fall. Air rises and clouds begin to form. It becomes windy.	As the tropical storm continues, the air pressure falls more rapidly, wind increases, cumulonimbus cloud forms and there is heavy rainfall.	There is a period of calm with no wind or rain at the eye of the storm. The Sun appears, so it gets warmer. Air pressure is very low.	Wind and heavy rainfall increase dramatically again, the temperature drops and air pressure begins to rise.	As the tropical storm ends, the air pressure and temperature rise. Wind and rainfall subside.

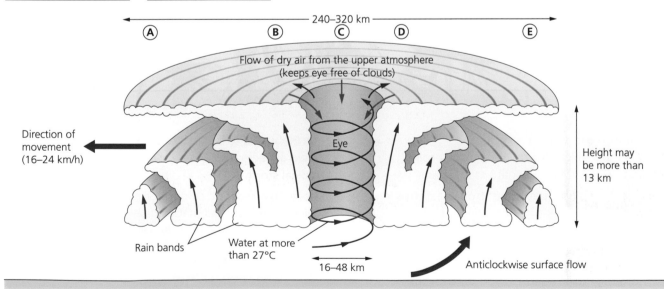

▲ Figure 3.7 Cross-section through a tropical storm

What are the structure and features of a tropical storm?

Figure 3.7 shows a cross-section of the structure of a tropical storm. The satellite image in Figure 3.6 shows the swirling wind and cloud around the central circular eye of the storm where there is no cloud.

Why does a tropical storm spin?

The Coriolis effect bends and spins the warm rising air. The spinning can be seen in satellite images (such as in Figure 3.6 on page 26). Hurricanes in the northern hemisphere bend to the right, which causes the clouds to swirl anticlockwise, whereas cyclones in the southern hemisphere swirl in a clockwise direction.

What direction do tropical storms travel?

Tropical storms travel from east to west due to the direction in which the Earth spins. When they hit land, they lose the energy source from the sea that powered them. As they pass over land, friction also slows them down. As they lose energy they change direction. This exact direction and speed is unknown. However, tropical storms in the northern hemisphere track north and tropical storms in the southern hemisphere track south (Figure 3.5 on page 25). An average tropical storm has a lifespan of approximately one to two weeks.

→ Activities

1 Draw a sequence of at least three diagrams, with captions, to show the formation of a tropical storm.
2 What is the eye of the storm?
3 Using the satellite image on page 26, state which hemisphere the tropical storm is in and how you reached your answer.
4 Sketch a larger version of the cross-section of a tropical storm in Figure 3.7. Annotate it with the sequence of weather conditions that would be experienced. (Include information about wind, rain, clouds, temperature and air pressure.)

5 Write a short paragraph to explain what causes tropical storms to spin.
6 What happens to a tropical storm when it reaches land? Explain why.

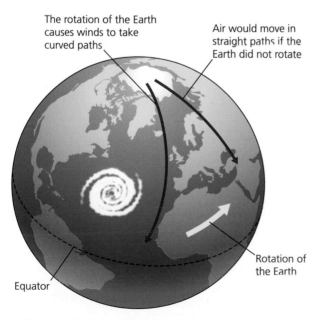

▲ Figure 3.8 The Coriolis effect

What is the Coriolis effect?

Winds blow from areas of high pressure to areas of low pressure. They do not blow in straight lines across the Earth but are affected by the Coriolis effect. As the Earth rotates it causes the wind to bend. This is because the Earth has a curvature, with the Equator far wider than the poles. Therefore the Earth has to spin faster at the Equator. This difference in speed means that wind bends as it blows across the Earth. This is known as the Coriolis effect.

Climate change and tropical storms

How might climate change affect tropical storms?

Climate change (see Chapter 4) will alter the conditions that cause tropical storms to form.

Higher storm surges

As the temperature increases, sea levels will rise due to **thermal expansion**. The impact of rising sea levels will mean **storm surges** are expected to become higher.

Increased heavy rainfall

A warmer atmosphere will mean the air can hold more moisture so heavy rainfall is expected to increase. Historic amounts of rainfall of over 1524 millimetres were recorded in south-eastern Texas during Hurricane Harvey in 2017 (see Figure 3.10).

More destructive flooding

As heavy rainfall is expected to increase, this will mean there is likely to be increased flooding during tropical storms, which will be more destructive.

The number and intensity of tropical storms varies greatly from year to year. This makes it more challenging for scientists to detect trends in the future frequency and intensity of tropical storms due to climate change. Since the 1970s, satellite technology has made it possible to track tropical storm frequency and intensity more consistently. Longer-term trends are more complicated, although scientists use hurricane models to project future trends.

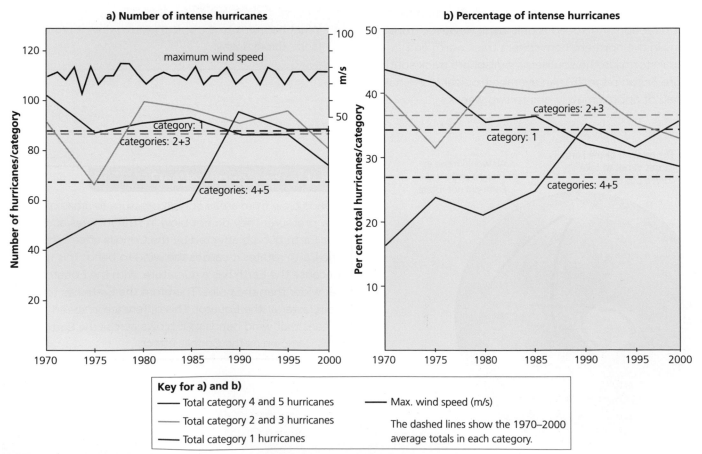

Key for a) and b)

—— Total category 4 and 5 hurricanes   —— Max. wind speed (m/s)

—— Total category 2 and 3 hurricanes   The dashed lines show the 1970–2000

—— Total category 1 hurricanes   average totals in each category.

▲ Figure 3.9 The number, and percentage, of intense hurricanes according to the Saffir-Simpson scale (categories 1 to 5)

Intensity

The impact of climate change on tropical storms is unknown. However, there is evidence of a link between warmer oceans and the intensity (destructive power) of tropical storms. The strength of tropical storms is classified using the Saffir-Simpson hurricane wind scale. Category 5 is the most intense and 1 the least intense (see Figure 3.11 on page 31).

Tropical storms are expected to become more intense, by 2–11 per cent, by 2100. The number of the most severe category 4 or 5 tropical storms (see Section 3.5) has increased since the 1970s (Figure 3.9). Predictions suggest that every 1°C increase in tropical sea surface temperatures will mean a 3–5 per cent increase in wind speed.

Frequency and distribution

Scientists are uncertain whether climate change will lead to an increase in tropical storms. They mostly agree that the overall frequency of tropical storms is expected to either remain the same, or decrease, as a result of climate change – although the number of more severe tropical storms (categories 4 and 5) will probably increase, while category 1–3 storms will decrease (see Figure 3.9). The regions where tropical storms are experienced are not expected to change significantly as a result of climate change.

Uncertainty

The reasons for all this uncertainty is due to:

- Inaccurate data: wind speed monitoring has only become more accurate in recent decades, so the use of previous data – which is less accurate – to decide how tropical storms are affected by climate change is questionable.
- Unreliable predictions: predicting the impact of climate change is unreliable, as the rate of and impact of climate change in the future is uncertain.
- Other impacts: potential risk to life and property has already increased due to population growth and building in coastal locations, even without factoring in climate change.

→ **Activities**

1 What conditions will climate change affect that cause tropical storms to form?
2 Is climate change expected to affect:
   a) the intensity of tropical storms?
   b) the frequency of tropical storms?
   c) the distribution of tropical storms?
   In each case, explain why.
3 Explain how climate change may make the impact of tropical storms worse.
4 What makes the link between climate change and tropical storms uncertain?
5 Using Figure 3.10, suggest how increased rainfall would make tropical storms more destructive.

▲ **Figure 3.10** Marines transporting supplies in the Hurricane Harvey relief effort

Example

⭐ KEY LEARNING

➤ Primary and secondary effects of a typhoon

➤ The immediate and long-term responses

In numbers

- 6,190 people died
- 14.1 million people affected, of which 4.8 million already lived in poverty
- US$12 billion overall damage
- Over 1 million farmers and 600,000 hectares of agricultural land affected
- 1.1 million tonnes of crops destroyed
- 1.1 million houses damaged
- 4.1 million people homeless

The trouble with Typhoon Haiyan

What were the primary effects?

On 8 November 2013 at 4.40 a.m. local time, a category 5 typhoon struck the Philippines. Typhoon Haiyan, also known as Typhoon Yolanda, originated in the northwest Pacific Ocean. It was incredibly powerful, with recorded wind speeds of up to 314 kilometres per hour.

The strong winds battered people's homes and even the evacuation centre. Those made homeless were mainly in the Western and Eastern Visayas. Power was interrupted, the airport was badly damaged and roads were blocked by trees and debris. Leyte and Tacloban had a five-metre storm surge, and 400 millimetres of heavy rainfall flooded one kilometre inland. Ninety per cent of the city of Tacloban was destroyed (Figure 3.14).

Coconut, rice and sugarcane production made up 12.7 per cent of the Philippines' GDP before the typhoon. The harvest season had just ended before it struck, but rice and seed stocks were lost in the storm surges. The damage to rice cost US$53 million and 75 per cent of farmers and fishers lost their income. The UN totalled the recovery costs for agriculture and fishing at US$724 million.

What were the secondary effects?

An oil barge ran aground at Estancia, causing an 800,000 litre oil leak. Most of this washed ashore, contaminating 10 hectares of mangroves (see Figure 3.17, page 35). Survivors struggling to get food took whatever supplies they could get, including taking rice from a government warehouse where eight were killed when a wall collapsed. By 2014, rice prices had risen by 11.9 per cent.

The flooding caused surface and groundwater to be contaminated with seawater, chemicals from industry and agriculture, and sewage systems.

▼ Figure 3.12 Tracking co-ordinates of Typhoon Haiyan

Date	Time	Lat.	Long.	Wind (mph)
05/11	00:00	6°N	146°E	75
05/11	12:00	7°N	143°E	105
06/11	00:00	7°N	140°E	150
06/11	12:00	8°N	136°E	160
07/11	00:00	9°N	133°E	175
07/11	12:00	10°N	129°E	190
08/11	00:00	11°N	125°E	185
08/11	12:00	12°N	121°E	155
09/11	00:00	12°N	116°E	135
09/11	12:00	15°N	113°E	115
10/11	00:00	16°N	110°E	100
10/11	12:00	19°N	108°E	85
11/11	00:00	22°N	107°E	70

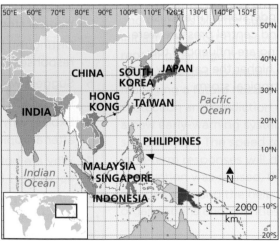

▲ Figure 3.13 Location map of the Philippines

This meant fishing had to stop in Estancia. The likelihood of infectious diseases spreading increased.

What were the immediate responses?

The president televised a warning. The authorities evacuated 800,000 people. Many sought refuge in an indoor stadium in Tacloban. Although this had a reinforced roof, they died when it was flooded. The government sent out essential equipment and medical supplies. In one region these supplies were washed away.

Emergency aid supplies arrived three days later by plane once the main airport reopened. It was a week before power was fully or partially restored. Within two weeks, over 1 million food packs and 250,000 litres of water were distributed. A curfew was imposed two days after the typhoon to reduce looting.

Thirty-three countries and international organisations pledged help and sent rescue operations. Celebrities, and large multinational organisations, such as Coca-Cola, Walmart, Apple and FIFA, donated and used their influence to raise awareness and encourage public donations. Over US$1.5 billion was pledged in aid from other countries, including those in Southeast Asia.

What were the long-term responses?

In July 2014, the Philippine government declared it was working towards long-term recovery, 'building back better' so that buildings would not just be rebuilt, but upgraded and therefore future-proof. The government also has:

- a 'no build zone' along the coast in Eastern Visayas, which was later changed to a 'no dwelling zone' so commercial buildings can be built
- a new storm-surge warning system
- replanted mangroves
- plans to build the Tacloban-Palo-Tanauan Road Dike.

Restoring previous livelihoods has been hard, especially for farmers and fishers. It is expected to take 5–10 years for new coconut trees to grow and bear fruit. Rebuilding permanent homes was also slow, with 100,000 families still in temporary accommodation in 2015.

▲ Figure 3.14 Tacloban city after Typhoon Haiyan, 2013

→ Activities

1. Study Figure 3.14. What evidence is there of (a) strong winds, (b) torrential rainfall and (c) storm surges?
2. Describe the following effects of the typhoon: (a) social (b) environmental (c) economic.
3. Suggest reasons why Typhoon Haiyan was so destructive.
4. Do you think people in the Philippines were prepared for the typhoon? Explain your answer.

Geographical skills

1. Use Figure 3.12 to:
   a) Plot the path of Typhoon Haiyan on a map using the latitude and longitude co-ordinates.
   b) Label the countries and islands Typhoon Haiyan passed through on your map. Give your map a suitable title.
   c) Draw a line graph to show the speed of wind over the duration of the typhoon.
   d) State a reason why a line graph is more suitable than a bar graph.

Example

⭐ KEY LEARNING

➤ The causes of record rainfall and flooding in Cumbria in December 2015

➤ The environmental, economic and social impacts for people and places

Record rainfall and flooding in Cumbria, 2015

What caused the record rainfall and flooding in Cumbria in December 2015?

Scientists say the floods that hit northern England during Storm Desmond in December 2015 were the most extreme in 600 years, based on the strength of the floodwaters. At Honister Pass in Borrowdale, 341 millimetres of rain fell in just 24 hours, setting a new record for the UK.

More than 4,000 homes were affected by river flooding and surface water flooding in the settlements of Kendal, Carlisle, Appleby, Keswick and Cockermouth (Figure 3.21). The cause was a very deep Atlantic depression called Storm Desmond. Many places received over one month's worth of precipitation in just two days, on 5 and 6 December (Figure 3.23). Wind speeds reached 220 km/hr in upland areas.

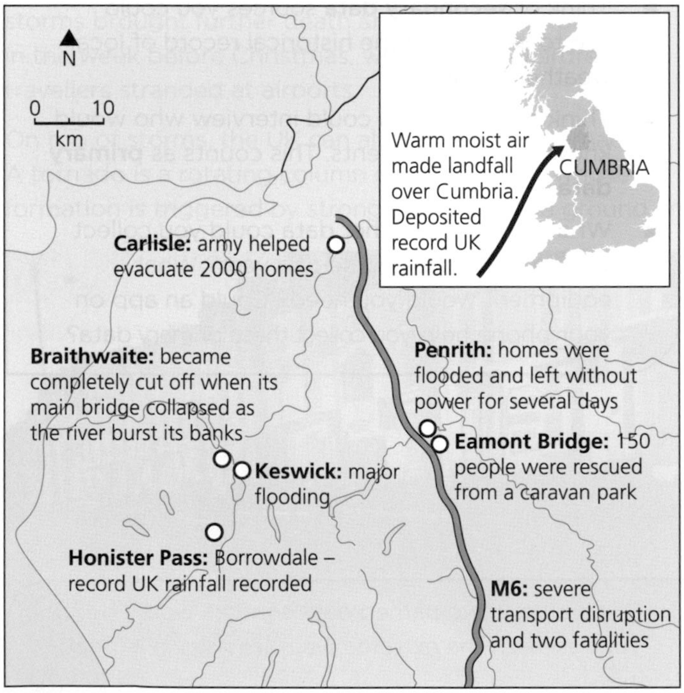

▼ Figure 3.22 Rainfall totals at selected Cumbria gauging stations, 4–6 December 2015

Site name	Highest 24-hour rainfall recorded during Storm Desmond (mm)	Total rainfall, 4–6 December (mm)
Brothers Waters	293	419
Dale Head Hall	262	377
Thirlmere	323	450
Honister Pass	341	428

▼ Figure 3.23 Storm Desmond formed part of an even 'bigger picture' of extreme weather in 2015–16. The UK was hit by a series of storms – Desmond was the fourth to arrive. (Can you spot a pattern in the way the storms were named?)

Date	Storm name
12–13 November 2015	Storm Abigail
17–18 November 2015	Storm Barney
29 November 2015	Storm Clodagh
5–6 December 2015	Storm Desmond

▲ Figure 3.21 Selected physical and human impacts of Storm Desmond on Cumbria

Extreme rainfall causes flooding

Storm Desmond was just part of a longer period of unusually warm, wet and stormy weather during the winter of 2015–16. Warm conditions in the North Atlantic Ocean helped feed moisture-laden air and storms towards the UK, following a stronger-than-normal jet stream. As Storm Desmond travelled north-eastwards over the mountains of Cumbria, large quantities of rainfall were released. Cumbria had already received prolonged heavy rainfall before the arrival of Storm Desmond, due to previous storms (Figure 3.23). The soil was therefore already very wet and new rainfall could not soak in. It flowed straight down the steep slopes of the Lake District into its rivers.

What were the environmental, economic and social impacts for places and people?

A range of harmful short-term and long-term impacts were experienced throughout Cumbria immediately during and after the storm and floods (Figure 3.24). Occasionally, there were positive effects too.

■ *Environmental*: Over 100 bridges were damaged, including 600-year-old Eamont Bridge. The slopes of Helvellyn collapsed, damaging the A591 road. In Carlisle, home to 110,000 people, the River Eden broke its banks after reaching a record flow level. Widespread landslides and riverbank **erosion** led to the deaths of many cattle in rural areas.

■ *Economic*: Disruption to the rail and road network meant people could not work. (The village of Braithwaite was cut off completely.) There were longer-term economic costs too. Hundreds of businesses in Allerdale and Carlisle remained closed for over a year. According to Cumbria County Council, the recovery cost £500 million and the insurance bill reached £1.3 billion.

■ *Social*: More than 700 families were unable to return to their homes for around two years. Many people experienced great stress and reduced mental well-being due to their life-changing losses. Positively, however, some communities were strengthened by the shared experience of struggle and recovery. This is called **community cohesion**.

▲ Figure 3.24 Flooded streets in Cumbria, December 2015

→ Activities

1　Study Figure 3.21.
   a) Identify the direction the storm had travelled from.
   b) State one physical and two human impacts of the storm.

2　Study Figure 3.22.
   a) Identify which site has the highest 24-hour rainfall.
   b) Calculate the range of values for total rainfall.

3　Using information on these pages and your own understanding, assess the consequences of Storm Desmond for communities in Cumbria. In your answer, you should:
   ■ offer a view about which is the most damaging impact, and why
   ■ distinguish between short-term and long-term impacts
   ■ highlight any positive impacts.

   If you mention physical impacts (such as record rainfall) you also need to mention how communities may have been affected.

4　Carry out additional research by clicking on '2015' and then 'Flooding in Cumbria December 2015' at www.metoffice.gov.uk/weather/learn-about/past-uk-weather-events.

Responding to the risk of extreme weather

How have management strategies reduced the impacts of extreme weather in Cumbria?

Storm Desmond formed part of the wettest and stormiest three-month period in the UK since 1910. Lasting from November 2015 to January 2016, conditions were unusually warm too. Rivers across much of the country flooded for extended periods. This all followed severe flooding of southern England in the winter of 2013–14 and of Cumbria previously in 2009. The evidence suggests that **flood risk** may be increasing due to climate change.

Since 2016, the UK government has provided over £150 million to support recovery in Cumbria. In the short-term, government money was used as emergency funding for households and businesses, and for repairing roads and bridges, including Eamont Bridge and the A591. The rest of the money has been spent on long-term strategies to help prepare Cumbria for flooding in the future.

■ Most importantly, flood defences have been strengthened throughout the region (Figure 3.25). In Carlisle, work has been carried out to raise flood **embankments**, and to strengthen and repair flood walls. The aim is to protect the city from flood levels equivalent to those that were observed in Storm Desmond. In Kendal, where 2,000 homes were flooded, £24 million has been found for new flood defences.

■ Affordable flood insurance has been available for households across Cumbria.

■ To make individual homes more resilient, a £5,000 grant was given to every flooded household, to help protect that home better in the future.

■ In Keswick, some badly damaged houses were knocked down and rebuilt altogether.

■ A decision was made to raise the height of the replacement buildings by 1 metre to offer greater future flood protection, helping make people feel safer.

■ Despite this, some floodplain residents in Keswick want to move away but cannot because property values have dropped – not many people living in other places want to move to an area that has repeatedly suffered major flooding.

The Environment Agency (EA) continues to play an important role in Cumbria, providing residents with improved **flood warning** information. This increases safety by giving people more time to evacuate and to protect their own properties (by closing windows or placing sandbags against doors, for instance). The EA website gives clear, visual information about the severity of expected storms and floods. Many people in the region have also asked the EA to send future flood warning messages directly to their phones.

What more can be done to manage the risk of extreme weather in the UK?

In response to the UK's major **hydro-meteorological hazards**, a range of 'top-down' actions are sometimes taken to protect communities (Figure 3.27). There are 'bottom-up' actions to consider too – personal actions people take to increase their **human resilience** to hazards.

Some bottom-up steps are focused on reducing individual exposure to harm. In the last few years, some risks have

▲ Figure 3.25 Flood defence work in Cumbria

been lessened by communication via social media like Facebook and WhatsApp (Figure 3.26). During winter storms, railway companies tweet information to customers who, in turn, retweet them to others. That way, people have time to change their route to work or can choose to return home. Some passengers even tweet pictures of fallen trees to the railway companies in order to alert them to damage.

Additionally, some homeowners in flood-prone areas 'future-proof' their homes by very sensibly having stone tiles rather than carpets in their ground floor and basement rooms. They know that although their insurance policy will cover any damage to carpets, it could take a very long (and stressful) time to get the problem sorted – especially if the insurance companies are dealing with thousands of similar claims.

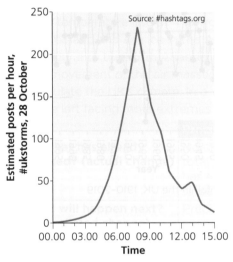

▲ Figure 3.26 How social media users helped to spread information about the St Jude storm of 28 October 2013

→ **Activities**

1   Outline three different flood management strategies that are used in Cumbria.
2   a) Identify the maximum number of posts in Figure 3.26.
    b) Using Figure 3.26 and your own understanding, suggest two ways in which social media could also be used by citizens to help their local community cope with a major flood.
3   a) Using Figure 3.27, identify three organisations or types of business who are working to reduce extreme weather risks for people.
    b) Using your own understanding, suggest one other top-down action which could be added to each box in Figure 3.27.

Drought

A hosepipe ban can be put in place in affected regions. During the 1976 drought, the Minister for Drought sent 'hosepipe patrols' in search of breakers of the hosepipe ban. Offenders can be fined.	Water companies can apply to the government for an official Drought Order in an extreme drought. Water supplies to houses are turned off and members of the public take their turn queuing in the street at standpipes.	Water companies can encourage people to have a water meter fitted (this tends to stop people from leaving taps running) and do more to repair old water pipes (London has 6,000 km of 150-year-old pipes).

Storms

The UK Met Office is constantly improving its weather predictions. Extreme storm events can be predicted days in advance. The UK has little need of tornado warnings due to their low frequency (30 a year in the UK, typically small and short-lived).	Severe weather warnings can now be issued using a whole range of media (in the past, only television and radio were available). Smartphones allow people to receive weather warnings while away from home.	Airlines and rail companies cancel their services when very strong winds are forecast in order to minimise the safety risk to their customers.

Flooding

The Thames Barrier was completed at great cost in 1982 to protect London from any future storm surges of similar magnitude to the 1953 event.	The EA constantly monitors ground moisture levels in river basins using its own communications system. This helps the EA to make accurate flood predictions and give timely evacuation orders.	A new agreement called Flood Re was reached in 2013 between UK insurance companies and the government. Any new housing on floodplains faces higher insurance bills. This should deter construction in risky places.

Cold weather

The responsibility for clearing roads of snow and ice lies with local councils. Councils in the UK are legally responsible for safety along nearly 500,000 km of road.	National organisations like Public Health England make announcements in the media warning people to take great care while out and about in blizzard conditions.	Charities raise awareness about the heightened health risks for older people during cold conditions. They work to raise public awareness of this issue.

▲ Figure 3.27 Top-down actions to manage extreme weather risks

4 Climate change

➤ What the Quaternary period is
➤ Changes in climate through time
➤ Evidence of climate change

Climate change? Prove it!

What is the Quaternary period?

The Earth is believed to be 4.55 billion years old. Studying the Earth has led us to devise a geological timeline that divides its history into a series of eras, periods and epochs. The period of time that stretches from 2.6 million years ago to the present day is called the **Quaternary period**, which is in the Cenozoic era. This period marks a time when there was a global drop in temperature and the most recent ice age began. (It is thought that the Earth has experienced five ice ages in its history.) The Quaternary period is split into two epochs, the **Pleistocene epoch** and the **Holocene epoch**.

How has climate changed?

The Quaternary period is often called an ice age due to a permanent ice sheet on Antarctica.

During the Pleistocene epoch:

■ There were cold **glacial episodes** lasting approximately 100,000 years.
■ Thick ice expanded, covering vast areas of continents, but then retreated, as each glacial episode was followed by a warmer **interglacial episode**.
■ The warmer intervals were much shorter, lasting for approximately 10,000 years.

The Holocene epoch began when the last glacial expansion ended and the current interglacial episode started. This is what we live in today. There are still sheets of ice covering Greenland and Antarctica, but our climate has remained relatively stable.

Figure 4.2 shows average global temperatures during the last 400,000 years. The spikes represent interglacials (warmer times with less ice) and the troughs show glacials (colder times with more ice).

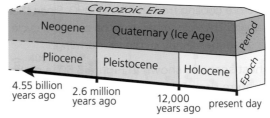

▲ Figure 4.1 Geological timeline of the Quaternary period

What is the evidence for climate change?

Climate change is the long-term change in the weather. Global climate change occurs very slowly over thousands of years. Since 1914, the Met Office has recorded reliable climate change data using weather stations, satellites, weather balloons, radar and ocean buoys. The Earth's average surface air temperature has increased by approximately 1°C over the last 100 years. In addition:

■ sea levels have risen by 19 centimetres since 1900 and are expected to continue to rise – this is due to thermal expansion and ice sheets melting
■ ocean temperatures are the warmest they have been since 1850, and the world's glaciers and ice sheets are decreasing in size
  ■ NASA data show that since 2002, the volume of ice lost in Antarctica is 134 billion tonnes per year, and 287 billion tonnes per year in Greenland.

For the era before there were reliable data records, we need to take clues from **proxy data** (natural recorders), such as tree rings, fossil pollen, ice cores and ocean **sediments** to estimate what the climate was like. However, these records are not as reliable, because these only indicate climate change rather than providing direct evidence of accurate temperatures.

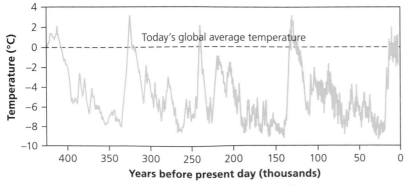

▲ Figure 4.2 Trends in average global temperatures (400,000 years ago to the present day)

Ice cores

Antarctic **ice cores** are crucial in understanding long-term climate change. Antarctica is a wilderness with no permanent residents, so the layers of ice remain unaltered. They act like time capsules, holding information as different layers build up over thousands of years. The deeper the ice that is drilled, the older the snow (see Figure 4.3). Records go back to about 800,000 years ago.

Oxygen isotopes in ice cores are commonly used to estimate the temperatures at the time. The isotopes are atoms with different numbers of neutrons. There are three different oxygen isotopes. The ratio of two types of oxygen isotopes are measured to work out what the climate was like. Additionally, when ice cores are melted, trapped carbon dioxide and methane are released, which can be compared to present levels to see the differences between climate then and now.

▲ Figure 4.3 Measuring the data held in ice cores

Ocean sediments

As with ice, the deeper the sediment, the older it is. The billions of tonnes of sediment deposited at the bottom of the sea also act as a timeline for providing evidence of climate change. Organisms and remains of plankton reveal information such as past surface water temperatures, and levels of oxygen and nutrients.

Paintings and diaries

During 1300–1870, parts of Europe and North America experienced much colder winters than today. Written observations suggest evidence of climate change such as crops failing, sea ice preventing ships from landing in Iceland, and winter 'frost fairs' held on the frozen River Thames (see Figure 4.4).

▲ Figure 4.4 'A Frost Fair on the Thames at Temple Stairs', c1684. Artist: Abraham Hondius

→ Activities

1 Sketch the graph in Figure 4.2. Label the glacial episodes and interglacial episodes.

2 Describe the trend in average global temperatures shown in Figure 4.2.

3 Contrast the Holocene epoch with the Pleistocene epoch.

4 If there was no human effect on climate, how would you expect the climate to change over the next several thousand years?

5 List the different ways that evidence is found for climate change.

6 Why do natural recorders (or proxy data) have to be used to show evidence of climate change?

7 Using Figure 4.4, describe how paintings can show evidence of climate change.

8 Why might people argue that some evidence for climate change is better than others?

9 Research how tree rings and fossil pollen provide evidence of climate change.

✪ KEY LEARNING

➤ The possible natural causes of climate change

Climate change as a natural phenomenon

Is climate change a natural phenomenon?

Geological evidence suggests that climate change has been happening throughout the Quaternary period, before humans were present. This suggests that long-term climate change is a result of natural causes.

Solar energy output

Solar energy output is measured by observing **sunspots** on the Sun's surface. Sunspots are caused by magnetic activity inside the Sun, resulting in dark patches on its surface (see Figure 4.5). The output increased slightly from 1900 to 1940. Satellites have recorded the intensity of solar energy output using radiometers since 1978. These data show that solar output has barely changed in the last 50 years; in fact, it has decreased slightly (see Figure 4.6). Therefore, solar output cannot be responsible for causing the climate change seen from the 1970s.

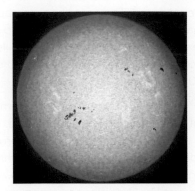

▲ Figure 4.5 Sunspots on the Sun's surface

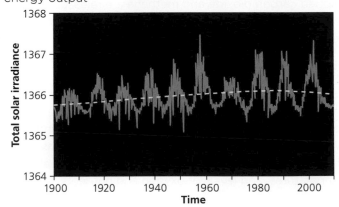

▲ Figure 4.6 Changes in solar energy falling on the Earth's surface

Orbital changes

The distribution of the Sun's energy on the Earth changes due to the Earth's orbit:

■ The Earth's orbit is an ellipse. The Sun is not perfectly in the centre and the ellipse changes shape every 100,000 years (see 'Eccentricity' on Figure 4.7). This means the distance between the Earth and the Sun changes as the Earth orbits. As the Earth orbits closer to the Sun, the climate becomes warmer, and the opposite happens as it orbits away.

■ The Earth's axis is tilted on an angle. The angle of the tilt changes due to the gravitational pull of the Moon. When the tilt angle increases, this can exaggerate the climate, so summers get warmer and winters get colder. The tilt angle moves back and forth every 41,000 years (see 'Axial tilt' on Figure 4.7).

■ The Earth is not a perfect sphere, so as the Earth spins, it wobbles on its axis in a 26,000-year cycle (see 'Precession' on Figure 4.7).

Together, these three **orbital changes** vary the distribution of the Sun's energy on the Earth. This can mean a significant impact on climate change.

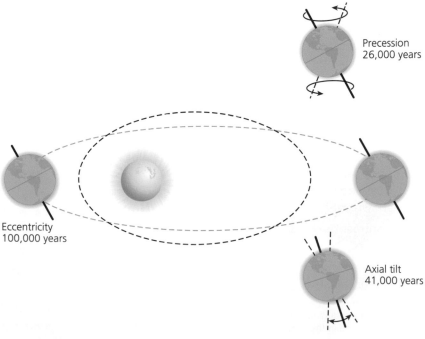

▲ Figure 4.7 Orbital changes of the Earth (the Milankovitch cycles)

Volcanic activity

Volcanic eruptions can temporarily cause climate change. In June 1991, Mount Pinatubo in the Philippines erupted. An ash cloud was thrown vertically 40 kilometres into the stratosphere and carried around the world for about three weeks. This is extremely important in understanding the impact the volcanic eruption had on climate change. Approximately 20 million tonnes of sulphur dioxide (SO_2) was released by Mount Pinatubo (see Figure 4.8). When SO_2 mixes with water vapour, it becomes a volcanic (sulphate) aerosol. Volcanic aerosols reflect the sunlight away and reduce the Sun's heat energy entering the Earth's atmosphere. Following Mount Pinatubo's eruption, global temperatures dropped by approximately 0.5°C.

Carbon dioxide (a greenhouse gas) also erupted from Mount Pinatubo. Carbon dioxide should help to trap the Sun's heat in the Earth's atmosphere. Instead, the temperature dropped, as the cooling effect of the sulphate aerosols was greater.

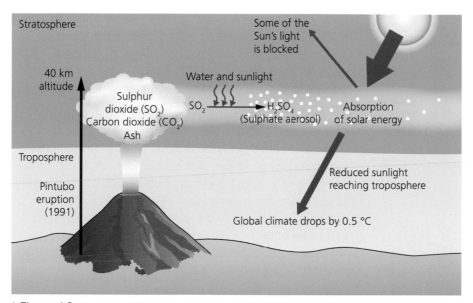

▲ Figure 4.8 The Mount Pinatubo eruption in the Philippines, 1991

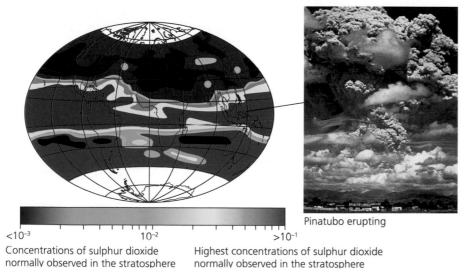

Pinatubo erupting

<10⁻³   10⁻²   >10⁻¹

Concentrations of sulphur dioxide normally observed in the stratosphere

Highest concentrations of sulphur dioxide normally observed in the stratosphere

▲ Figure 4.9 Sulphur dioxide levels during the 40 days after Mount Pinatubo's eruption

→ **Activities**

1 Why are orbital changes, volcanic eruptions and solar output categorised as natural causes of climate change?

2 'Solar output is responsible for climate change.' True or false? Explain your answer.

3 Explain how shifts in the Earth's orbit can cause climate change.

4 Use Figure 4.8 to describe how volcanic eruptions cause climate change.

5 Use Figure 4.9 to describe the location of sulphur dioxide across the world following the Mount Pinatubo eruption.

6 How important do you think natural causes are in explaining climate change? Explain your opinion.

➤ How the greenhouse effect works
➤ How humans have contributed to climate change

Climate change: our fault?

What is the greenhouse effect?

The greenhouse effect is a naturally occurring phenomenon that keeps the Earth warm enough for life to exist. It is thought that without the greenhouse effect, the Earth would be approximately 33°C colder and, therefore, life as we know it would not exist. Solar radiation enters the Earth's atmosphere.

The heat is reflected from the Earth's surface. The natural layer of greenhouse gases allows some heat to be reflected out of the Earth's atmosphere, but some of the Sun's infrared heat is trapped, which keeps the Earth warm enough for life.

However, there is an **enhanced greenhouse effect** whereby human activity has increased the layer of existing greenhouse gases. Activities which generate more greenhouse gases include burning fossil fuels in industry, agriculture, transport, heating and deforestation. Less heat escapes from the Earth and more is trapped by the thicker layer of greenhouse gases, which means the Earth warms up even more.

Are humans causing climate change?

Scientists have measured and proved that natural causes are responsible for climate change, yet natural causes cannot account for the increases in temperature since the 1970s (see Figure 3.29 in Section 3.11). The link between increasing carbon dioxide levels and increasing global temperatures can be seen in Figure 4.11. Carbon dioxide emissions have increased since the Industrial Revolution, and especially since the 1970s. The Intergovernmental Panel on Climate Change (IPCC) reports that it is very likely that this increase in carbon dioxide is the main cause of climate change.

Despite volcanoes naturally releasing carbon dioxide, it is thought that humans generate more than 130 times the volume of carbon dioxide than volcanoes. Greenhouse gases consist of 77 per cent carbon dioxide (CO_2), 14 per cent methane (CH_4), 8 per cent nitrous oxide (N_2O) and 1 per cent chlorofluorocarbons (CFCs). Each methane molecule has 25 times the global warming potential over 100 years of carbon dioxide. Nitrous oxide is worse, at 125 times. This is despite carbon dioxide making up the majority of greenhouse gases.

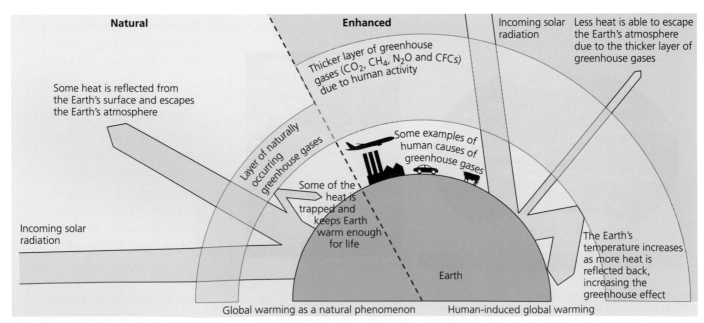

▲ Figure 4.10 The greenhouse effect: natural and enhanced

How do humans cause climate change?

Fossil fuels

Fossil fuels account for the majority of global greenhouse gas emissions – over 50 per cent. Burning these releases carbon dioxide into the atmosphere. Fossil fuels are used in transportation, building, heating homes, and the manufacturing industry. Additionally, they are burnt in power stations to generate electricity. As wealth increases and the world's population grows, people are demanding more and more energy, which increases the level of fossil fuels used and carbon dioxide produced.

Agriculture

Agriculture contributes to approximately 20 per cent of global greenhouse gas emissions. It produces large volumes of methane: cattle produce it during digestion, and microbes produce it as they decay organic matter under the water of flooded rice paddy fields.

As the world's population increases, more food is required. When countries increase their standard of life, there is almost always an increasing demand for meat (Section 23.1). If current population rates continue, it is inevitable that large-scale agriculture's contribution to climate change will continue to grow.

Deforestation

Deforestation is the clearing of forests on a huge scale. If this continues at the current rate, the world's forests could disappear completely within a hundred years.

There are several reasons why forests are cut down:
- clearing land for agriculture so that farmers have space to plant crops and graze livestock
- **logging** for wood and paper products
- building roads to access remote areas
- making room for the expansion of urban areas.

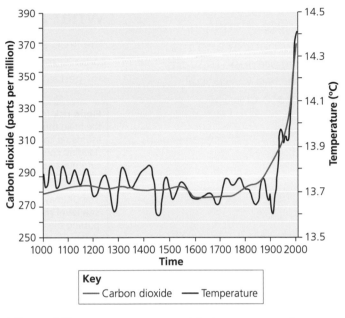

▲ Figure 4.11 Carbon dioxide and global temperature change

Continent	% of world emissions from agriculture
Asia	42%
Americas	25%
Africa	15%
Europe	13%
Oceania	4%

▲ Figure 4.12 World agriculture emissions (average 1990–2017)

During the process of **photosynthesis**, trees absorb carbon dioxide, which reduces the amount of carbon dioxide in the atmosphere. The process of deforestation leaves fewer trees to absorb carbon dioxide. Therefore, the enhanced greenhouse gases contribute to rapid climate change. When trees are burnt to clear an area, such as with **slash and burn**, the carbon dioxide that has been stored is also released, which again contributes to climate change.

→ Activities

1 Describe the relationship between temperature and atmospheric carbon dioxide in Figure 4.11.

2 Write a sequence of numbered statements to explain the greenhouse effect.

3 How do (a) fossil fuels, (b) agriculture and (c) deforestation each contribute to climate change?

4 'Humans are to blame for climate change.' To what extent do you think this statement is true?

➤ The likely effects of climate change

➤ How people and the environment may be affected by global climate change

The effects of climate change

What are the likely effects of climate change?

The IPCC states that the cost of damage caused by climate change is likely to be 'significant and to increase over time'. The effects of climate change are not certain. The likely effects will vary and be uneven globally and regionally.

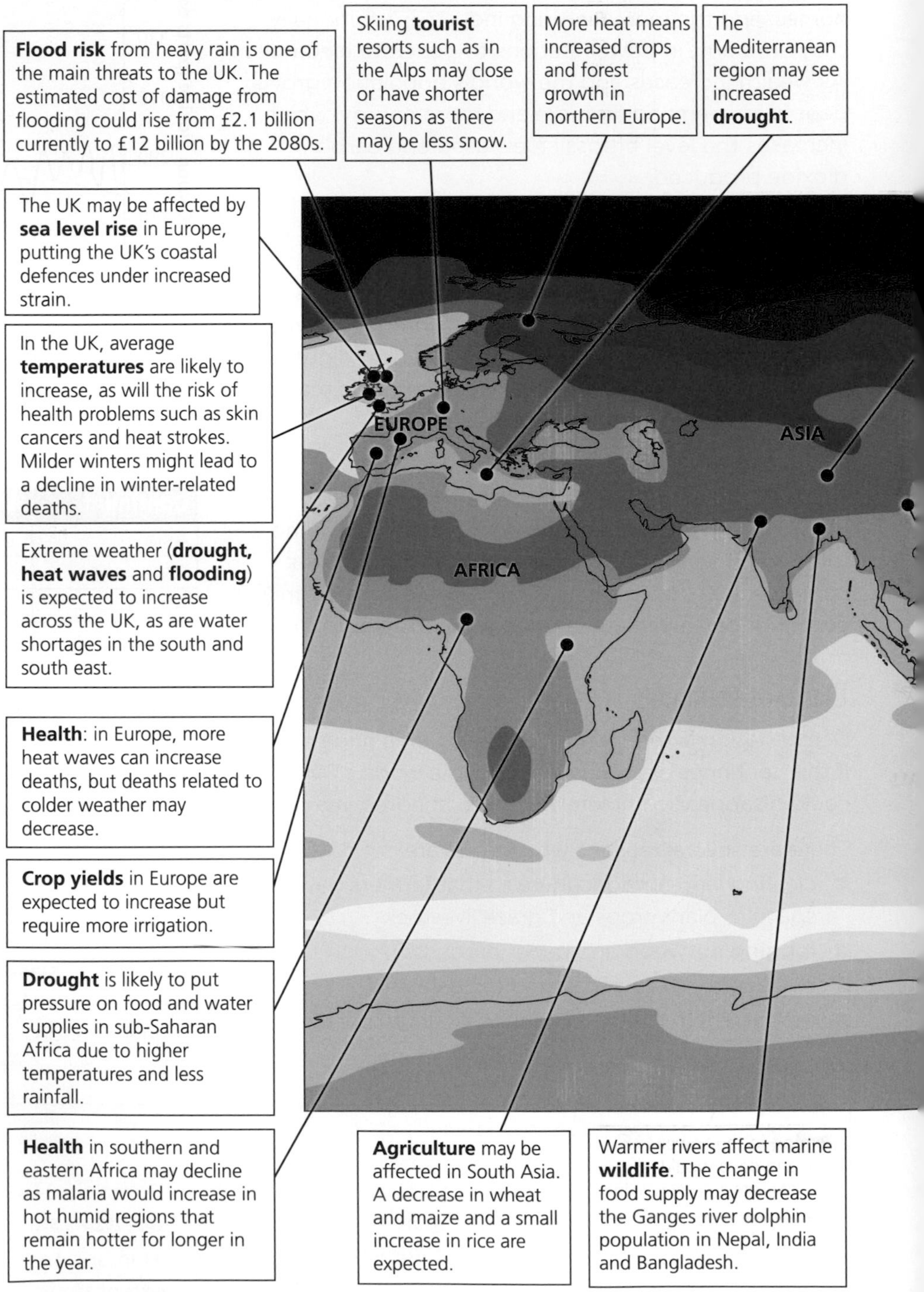

Flood risk from heavy rain is one of the main threats to the UK. The estimated cost of damage from flooding could rise from £2.1 billion currently to £12 billion by the 2080s.

Skiing **tourist** resorts such as in the Alps may close or have shorter seasons as there may be less snow.

More heat means increased crops and forest growth in northern Europe.

The Mediterranean region may see increased **drought**.

The UK may be affected by **sea level rise** in Europe, putting the UK's coastal defences under increased strain.

In the UK, average **temperatures** are likely to increase, as will the risk of health problems such as skin cancers and heat strokes. Milder winters might lead to a decline in winter-related deaths.

Extreme weather (**drought, heat waves** and **flooding**) is expected to increase across the UK, as are water shortages in the south and south east.

Health: in Europe, more heat waves can increase deaths, but deaths related to colder weather may decrease.

Crop yields in Europe are expected to increase but require more irrigation.

Drought is likely to put pressure on food and water supplies in sub-Saharan Africa due to higher temperatures and less rainfall.

Health in southern and eastern Africa may decline as malaria would increase in hot humid regions that remain hotter for longer in the year.

Agriculture may be affected in South Asia. A decrease in wheat and maize and a small increase in rice are expected.

Warmer rivers affect marine **wildlife**. The change in food supply may decrease the Ganges river dolphin population in Nepal, India and Bangladesh.

EUROPE

ASIA

AFRICA

➤ Activities

1   State two positive and two negative effects of climate change, using Figure 4.13.

2   Complete the following table (notice the impacts are plural):

Social impacts	
Environmental impacts	

3   Describe the global pattern of temperature change shown in Figure 4.13.

4   How might the effects in one part of the world impact on another? Give at least one example.

▲ Figure 4.13 Global effects of climate change

How might people and the environment be affected by climate change around the world?

Expected global effects of climate change can be seen on the map in Figure 4.13, which is centred on the Pacific Ocean.

Around 70 per cent of Asia may be at increased risk of **flooding**.

The **fishing** industry in East Asia is expected to decline due to higher temperature and more acidic sea.

Wildlife declines as polar bears and seals disappear with the loss of habitat as ice melts.

Less ice in the Arctic Ocean would allow more shipping and extraction of **gas and oil reserves**.

The tree line of the Sub-Arctic boreal **forests** is expected to retreat north as temperatures rise.

It is likely that **agriculture** may yield more wheat, soybean and rice but see a decrease in maize yields in North America.

Forests in North America may be affected more by pests, disease and forest fires.

In Central America, maize **crop yields** may fall by up to 12 per cent.

In the Amazon **rainforests**, a modest level of climate change can cause high levels of extinction. Eastern Amazonia may become a savannah with warmer temperatures and less soil moisture.

South America is expected to decrease in maize, soybeans and wheat **crop yields**.

NORTH AMERICA

SOUTH AMERICA

ALIA

ANTARCTICA

Key
Predicted air surface temperature by end of the 21st century

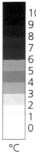

10
9
8
7
6
5
4
3
2
1
0

°C

Fishing in the Lower Mekong delta would decline, affecting 40 million people, due to reduced water flow and sea level rise changing the quality of the water.

Less rainfall may affect **wildlife**, causing food shortages for orangutans in Borneo and Indonesia.

Coral reefs such as the Great Barrier reef could see **biodiversity** lost, and warmer, more acidic (due to CO$_2$ in the atmosphere) water would cause coral bleaching.

Wildlife such as Adélie penguins on the Antarctic Peninsula may continue to decline as ice retreats.

51

1.10 Outline **one** factor affecting hazard risk. [2 marks]

> Identify the factor and then state how it affects hazard risk.

1.11 Study Figure 2.16 from page 15.

Using Figure 2.16 state **two** effects of a tectonic hazard. [2 marks]

> You must refer to effects only seen in the figure.

1.12 Use Figure 3 to complete the table below. [1 mark]

	Magnitude	Deaths
Sichuan, China		

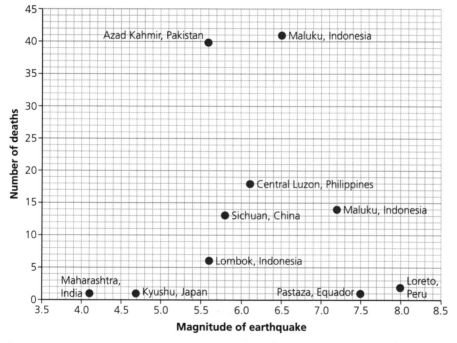

▲ Figure 3 A scattergraph showing the magnitude and the number of deaths for selected earthquakes in 2019

1.13 'As the magnitude of the earthquake increases, so does the number of deaths.'

Do you agree with this statement? Use evidence from Figure 3 to support your answer. [2 marks]

> When asked for evidence from a graph remember to quote figures.

1.14 Explain the formation of a tropical storm. [4 marks]

> Tropical storms follow a particular sequence so show this in the way you structure your answer.

1.15 'Planning for tropical storms can reduce the economic and social impacts.' To what extent do you agree with this statement? [6 marks]

> Think about what else can reduce the impacts of tropical storms and decide which is more effective.

1.16 Using Figure 4.16 on page 54 and your own understanding, explain how changes in agricultural systems can help people to adapt to climate change. [4 marks]

> Make sure you use BOTH the information in the figure AND your own understanding to answer this question. Don't just describe the strategy but link it to how it can help people adapt to climate change.

1.17 Using an example you have studied, assess the measures which could be taken to reduce the risks associated with an extreme weather event in the UK. [9 marks] [+3 SpaG marks]

> Make sure you know what the command word means. Assess means you need to make an informed judgement so in this case consider the advantages and disadvantages of measures taken to reduce the risks of an extreme weather event and come to a conclusion about their effectiveness.

5 Ecosystems

KEY LEARNING

- How the different parts of an ecosystem are linked together
- How food chains and nutrient cycles work

How ecosystems operate

How are the different parts of an ecosystem linked?

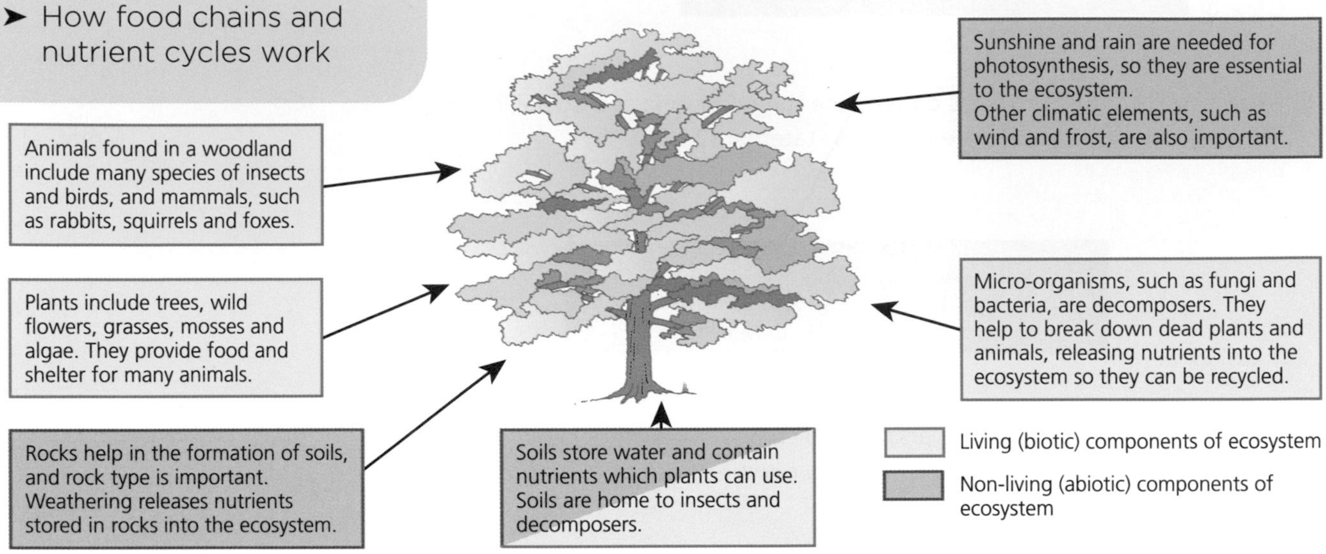

Sunshine and rain are needed for photosynthesis, so they are essential to the ecosystem.
Other climatic elements, such as wind and frost, are also important.
Animals found in a woodland include many species of insects and birds, and mammals, such as rabbits, squirrels and foxes.
Plants include trees, wild flowers, grasses, mosses and algae. They provide food and shelter for many animals.
Micro-organisms, such as fungi and bacteria, are decomposers. They help to break down dead plants and animals, releasing nutrients into the ecosystem so they can be recycled.
Rocks help in the formation of soils, and rock type is important. Weathering releases nutrients stored in rocks into the ecosystem.
Soils store water and contain nutrients which plants can use. Soils are home to insects and decomposers.
Living (biotic) components of ecosystem
Non-living (abiotic) components of ecosystem

Figure 5.1 The biotic and abiotic components of an ecosystem

An **ecosystem** is made up of plants, animals and their surrounding physical environment, including soil, rainwater and sunlight (Figure 5.1). Important interrelationships link together the **biotic** (living) and **abiotic** (non-living) parts of the ecosystem. These interrelationships consist of:

- physical linkages between different parts of the ecosystem (animals eat the plants, for example)
- chemical linkages (mild acids in rainwater speed up the decay of dead leaves, for example).

In ecosystems, plants and animals can migrate from one place to another, bringing change. There are also inputs and outputs from the ecosystem to other places (Figure 5.2). Most importantly of all, ecosystems depend on a constant input of light from the Sun, as well as rain from the atmosphere. In turn, rainwater leaves the ecosystem when it evaporates and returns to the atmosphere or runs into a river.

Ecosystems can be any size:

- local (a small-scale ecosystem is also called a habitat)
- regional (England's Lake District moorland)
- global **biomes** (South America's **tropical rainforest**)
- Earth (some scientists argue that all of the planet's organisms are linked together).

Sunlight
Green plants, e.g. pondweed, flag iris
Primary producers
Algae
Water fleas
Tadpoles
Primary consumers
Dragonfly larvae
Diving beetles
Secondary consumers
Frogs
Goldfish
Tertiary consumers
Grey heron
Quaternary consumers

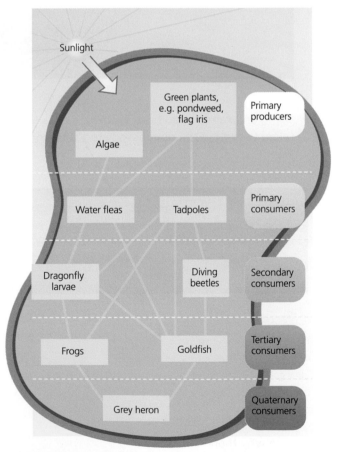

Figure 5.2 The food web of a pond. Sunlight is the main system input. Outputs include insects that hatch in the pond but then migrate elsewhere

How do food chains and nutrient cycles work?

The biotic community of an ecosystem consists of different species of plants and animals in feeding groups:

- plants or primary **producers**: green plants that use photosynthesis and take nutrients from the soil using their roots
- herbivores or primary **consumers**: plant-eating animals (cows or rabbits)
- carnivores (such as foxes or cats) or secondary consumers: these animals feed on herbivores
- top carnivores: these animals will hunt and eat other carnivores as well as the herbivores. They include the largest and fastest hunters, like lions and wolves.

The interrelationships between these feeding groups are shown in the **food chain** in Figure 5.3. The weight of **biomass** gets smaller at each level.

For instance, in Brazil's tropical rainforests (see Chapter 6), there are only 5 kilograms of animal biomass per 40 kilograms of plant biomass. There are two important reasons for this reduction:

- Many parts of plants are simply not eaten by animals, and carnivores do not eat all of their prey (such as the bones). Also, much of what the animals do eat is excreted.
- Energy is lost at each level. Hunters use a lot of kinetic energy: chasing prey can be time-consuming and exhausting; some herbivores search around for plants to eat. Energy is also constantly being used up in respiration. Much of an animal's daily calorie intake is used simply to stay alive, rather than to build new biomass.

The **decomposers** in Figure 5.3 are the organisms that, over time, break down dead organic matter and animal excretions. They include: scavengers (insects that eat dead wood) and detritivores (bacteria). Decomposers help to return nutrients to the soil in the form of an organic substance called humus.

The importance of nutrient cycling

All plants and animals depend on nutrients in food for their health and vitality. Nutrients occur naturally in the environment and are constantly recycled in every ecosystem. Figure 5.4 shows these important pathways.

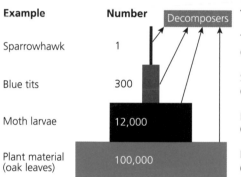

Example	Number		Trophic level
		Decomposers	
Sparrowhawk	1		Tertiary consumers (top carnivores)
Blue tits	300		Secondary consumers (carnivores)
Moth larvae	12,000		Primary consumers (herbivores)
Plant material (oak leaves)	100,000		Primary producers (plants)

▲ Figure 5.3 An ecosystem food chain

→ Activities

1. a) Using Figure 5.2, identify three plant and three animal species that are part of the pond ecosystem.
    b) Explain why sunlight is a vital input for the health of the ecosystem shown in Figure 5.2.
2. Using Figure 5.1, identify three interrelationships between the biotic and abiotic components of the ecosystem.
3. Using Figure 5.3 and your own understanding, explain why there is such a large difference in the numbers of primary producers and top predators.

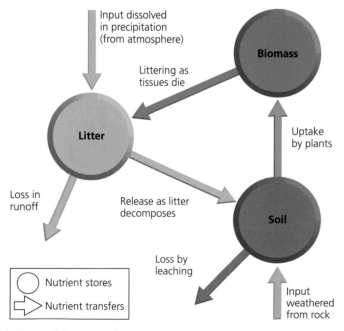

▲ Figure 5.4 The nutrient cycle

⚙ KEY LEARNING

➤ The characteristics of Epping Forest's food web
➤ The interdependence of the ecosystem
➤ The characteristics of Epping Forest's nutrient cycle

Epping Forest ecosystem, UK

What are the characteristics of Epping Forest's food web?

Biodiversity in the forest has remained high, thanks to careful management, so there is a complex **food web** composed of thousands of species (Figure 5.6 shows this in a simplified form).

Key facts

- Located east of London (Figure 5.5), Epping Forest is all that remains of a larger forest that colonised England at the end of the last Ice Age.
- Bogs and ponds in the forest have their own unique species, including 20 kinds of dragonfly.
- For 1000 years, Epping Forest has been managed in a variety of ways: as hunting grounds for royalty, a timber resource and, more recently, for recreation (as Figure 5.5 shows, it is easily accessible).

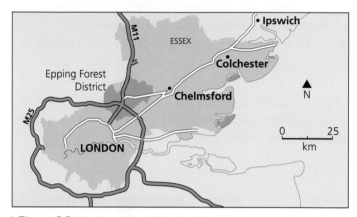

▲ Figure 5.5 The location of Epping Forest

Epping Forest is home to:

- a large number of native tree species, including oak, elm, ash and beech
- a lower shrub layer of holly and hazel at five metres, overlying a field layer of grasses, brambles, bracken, fern and flowering plants; 177 species of moss and lichen grow here. Altogether, there is great diversity of **producer** species (Figure 5.7)

- many insect, mammal and bird consumer species are supported, including 9 amphibian and reptile species and 38 bird species. Studies have also found 700 species of fungi, which are important decomposers.

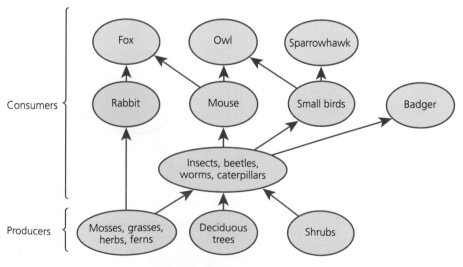

▲ Figure 5.6 Epping Forest's food web

How is the ecosystem interdependent?

The forest's producers, consumers and decomposers are all interdependent. This is most clearly shown by the annual lifecycle of the trees.

Most of the trees are deciduous, meaning that they lose their leaves in winter. This is an **adaptation** to the UK's seasonal climate. Winters are darker and cooler than summers (the mean monthly temperature is 18°C in July but just 5°C in January). As a result, the trees grow broad green leaves in spring. This allows them to maximise photosynthesis during the summer. They shed their leaves in the autumn, and so conserve their energy during winter.

By mid-autumn, the forest floor is covered with a thick layer of leaves. Remarkably, by spring, the leaf litter has all but disappeared: the decomposers and detritivores have done their work (page 61). Nutrients stored in the leaves are converted to humus in the soil, ready to support the new season's plant growth. This will ultimately include the fruits and berries that, in turn, support many primary consumers.

Nutrient cycling demonstrates clearly the interdependence of plants, animals and soil. People and ecosystem components are interdependent too. In the past, coppicing was common (cutting back trees to encourage new growth of wood). Today, visitors pick berries and flowers. In turn, this helps spread the seeds, which stick to their clothing.

▲ Figure 5.7 Leaves on the forest floor and Epping Forest's tree canopy, which in places reaches 30 metres

What explains the characteristics of Epping Forest's nutrient cycle?

In Figure 5.8, a deciduous forest nutrient cycle, the biomass store is large because of the great height of the trees and the dense undergrowth beneath them. The soil store is large too because there is always plenty of humus (organic matter).

The high flow rates between the litter, soil and biomass stores reflect the vigorous cycle of new growth that takes place each year. The forest also loses a lot of nutrients each year, via leaching, during episodes of heavy rainfall.

→ **Activities**

1. a) Identify one producer and one primary consumer in Figure 5.6.
   b) Using Figure 5.6 and your own understanding, suggest how three different food web species might be affected by the removal of rabbits from the ecosystem.
2. Identify three characteristics of the plants shown in Figure 5.7.
3. Using information from these pages, explain two ways in which different parts of an ecosystem are interdependent on one another.
4. Birds, mammals, insects, amphibians and reptiles are all present in Epping Forest. Some are herbivores (primary consumers), whereas others are carnivores (secondary consumers and top carnivores). For revision, draw a table with two columns and add as many named examples as you can of both categories (for instance, the tawny owl is a carnivore). You can research additional facts online.

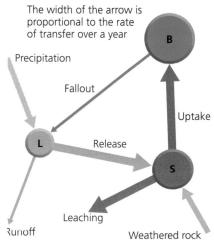

The width of the arrow is proportional to the rate of transfer over a year

Precipitation

Fallout

Uptake

Release

Leaching

Runoff

Weathered rock

▲ Figure 5.8 The nutrient cycle of a deciduous forest (compare with Figure 5.4 on page 61)

Changes affecting ecosystem balance

How do physical and human forces disturb ecosystem balance?

Periods of **extreme weather** or **climate change** can disturb the balance of ecosystems. In the years 1976–77, southern England experienced an 18-month drought that killed many trees. A further 15 million English trees were felled by a great storm in 1987. As a result, population numbers declined for many consumer species in the food chain. Secondary forest growth has since taken place, however, and consumer species have migrated back. The recent recovery of English woodland is an example of ecosystem **resilience**.

Ecosystems are sometimes damaged in permanent ways, especially when human forces are involved, for instance by **deforestation** (page 49). The removal of the forest exposes the soil beneath to rainfall, and so it can be washed away, making it impossible for the ecosystem to recover. This is especially true in tropical rainforest regions, where heavy rain falls most days (Chapter 6). In the longer term, human-induced climate change could threaten the ecosystem balance of many places. For instance, changes in **temperature** and precipitation patterns for southern England might make it harder for ecosystems like Epping Forest to survive in their current form. In some places, grass (rather than trees) may dominate in the future, if climate change predictions are correct.

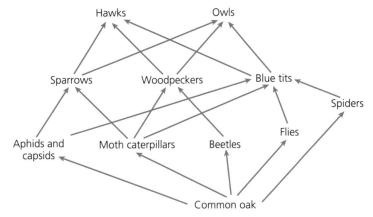

▲ Figure 5.9 A food web supported by an oak tree

How does the loss or gain of one species affect a food web?

Figure 5.9 shows a food web supported by oak woodland. Suppose that the population of beetles is reduced by disease. This would directly impact on the numbers of woodpeckers. With fewer beetles to eat, their numbers may decline. In contrast, we may expect to see more oak tree growth now fewer beetles are feeding on them. In addition to these direct impacts, there are indirect impacts to consider:

■ Owl and hawk numbers may also fall because they eat woodpeckers.

■ Woodpeckers are carnivorous and have multiple food sources. They may just eat more caterpillars instead. However, this could now impact on blue tit numbers. How would this happen? Can you identify more possible food web impacts that could follow?

How can ecosystem balance be restored through management?

Many species have been hunted to extinction, without a full understanding of how this could affect ecosystem balance. In Europe and the USA, killing wolves and bears removed danger to people and their cattle. But fewer carnivores meant that rabbit and deer populations quickly multiplied and began to eat all available vegetation, stripping the land bare, leading to **soil erosion**. The ecosystem lacks balance.

Many scientists believe that 'rewilding' or 'ecosystem restoration' is the best way to restore ecosystem balance. Grey wolves were reintroduced into Yellowstone National Park in the USA in 1995, which has resulted in numerous impacts (Figures 5.11 and 5.12). The wolves have restored balance to the ecosystem and **landscape** (Figure 5.10).

▲ Figure 5.10 Yellowstone National Park

▲ Figure 5.11 A grey wolf

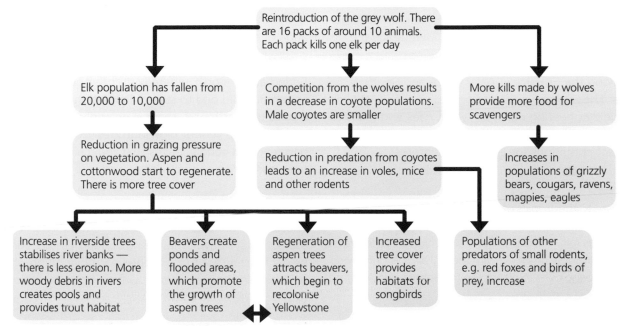

Reintroduction of the grey wolf. There are 16 packs of around 10 animals. Each pack kills one elk per day

Elk population has fallen from 20,000 to 10,000

Competition from the wolves results in a decrease in coyote populations. Male coyotes are smaller

More kills made by wolves provide more food for scavengers

Reduction in grazing pressure on vegetation. Aspen and cottonwood start to regenerate. There is more tree cover

Reduction in predation from coyotes leads to an increase in voles, mice and other rodents

Increases in populations of grizzly bears, cougars, ravens, magpies, eagles

Increase in riverside trees stabilises river banks — there is less erosion. More woody debris in rivers creates pools and provides trout habitat

Beavers create ponds and flooded areas, which promote the growth of aspen trees

Regeneration of aspen trees attracts beavers, which begin to recolonise Yellowstone

Increased tree cover provides habitats for songbirds

Populations of other predators of small rodents, e.g. red foxes and birds of prey, increase

▲ Figure 5.12 Impacts of the reintroduction of the grey wolf to the Yellowstone ecosystem since 1995. Note the interdependence of beavers and aspen as part of the changes.

→ Activities

1 Use Figure 5.9 to identify:
   a) the number of levels in the food web
   b) four named primary consumers and three named secondary consumers.
2 Outline one human and one physical cause of ecosystem disturbance.
3 Use Figure 5.12 to:
   a) identify two species whose numbers rose following the reintroduction of wolves
   b) identify two species whose numbers fell following the reintroduction of wolves
   c) explain two changes to the physical environment which resulted from ecosystem changes.
4 Explain possible costs and benefits of reintroducing dangerous wild animals to the countryside in countries like the USA. You could write about possible conflicts with other land uses such as tourism.

Fieldwork: Get out there!

Look at Figure 5.9. Plan a fieldwork investigation of this small-scale ecosystem. Possible themes to investigate could include the number of different plant and animal species, or evidence of interrelationships and nutrient cycling.

➤ How climate explains the distribution and characteristics of global ecosystems

➤ Altitude, relief and ocean currents

The distribution and characteristics of global ecosystems

How does climate explain the distribution and characteristics of global ecosystems?

Figure 5.13 shows the distribution of the world's large-scale **global ecosystem** or biomes. Figure 5.13 and this page explain how global-scale variations influence the distribution of ecosystems.

Tropical rainforests

Found along the Equator in Asia, Africa and South America. The Sun's rays are concentrated at this latitude (see page 22), heating moist air which rises and leads to heavy rainfall, with little seasonal variation. This creates the perfect conditions for evergreen rainforest.

Deserts

Found close to the Tropics of Cancer and Capricorn. The air that rises over the Equator heads polewards after shedding its moisture as rain. The Sun's rays are still highly concentrated at this low latitude. Combined with the dry air, this brings arid (dry) desert conditions to places like the Sahara and Australia.

Tropical grasslands

Sandwiched between the two extremes of tropical rainforest and desert. Conditions are dry for half of the year, due to the seasonal movement of the Hadley cell (page 23). This seasonal dryness limits the growth of trees.

Temperate grasslands

Short tussock and feather grasses dominate the landscape between 40° and 60° north of the Equator, in the drier centres of continents away from the sea.

Mediterranean

Drought-resistant small trees and evergreen shrubs grow between 30° and 40° north and south of the Equator, but only on the west coasts of continents.

Deciduous forests

These grow in many places at higher latitudes: in western Europe (with its rain-bearing storms) and the east coasts of Asia, North America and New Zealand. The Sun's rays are weaker and so trees shed leaves in the cooler, darker winters.

Coniferous forests

Found at 60° north, where winter temperatures are extremely cold due to lack of **insolation** (page 22). Due to the Earth's tilt, there is no sunlight for some months of the year at high latitudes. Coniferous trees have evolved needle leaves that reduce moisture and heat loss during the cold, dark winter months.

Tundra (or 'cold desert')

These areas are found around the Arctic Circle, where the Sun's rays have little strength. Temperatures are below freezing for most of the year. Only tough, short grasses can survive, often in waterlogged conditions (due to surface ice thawing).

Why are altitude, relief and ocean currents also important?

Although latitude and distance from the sea are the main factors affecting distribution, the following are also important:

■ altitude: temperatures fall by about half a degree for every 100-metre increase in altitude, and tough grasses replace trees on steep mountainsides

■ mountain ranges: in the USA and Asia, inland areas isolated from the sea suffer from low rainfall. This is because winds blowing off the oceans quickly lose their moisture when air is forced to rise upwards over a high mountain range. The drier lands found east of the USA's Rocky Mountains are in a **rain shadow**

■ ocean currents: a cold ocean current flowing along South America's coast helps to create arid conditions in Chile's Atacama Desert because little evaporation takes place over the cold water. In contrast, the warm Gulf Stream ocean current brings mild conditions to western Europe.

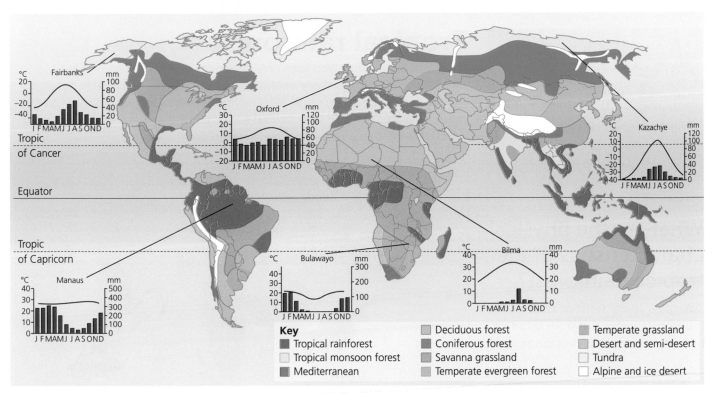

▲ Figure 5.13 World distribution of global ecosystems. In each climate graph, the red line shows monthly temperature averages; the blue bars show precipitation.

Key
- Tropical rainforest
- Tropical monsoon forest
- Mediterranean
- Deciduous forest
- Coniferous forest
- Savanna grassland
- Temperate evergreen forest
- Temperate grassland
- Desert and semi-desert
- Tundra
- Alpine and ice desert

→ Activities

1   Using the climate graphs in Figure 5.13:
   a)   state the minimum and maximum annual temperatures for the tundra ecosystem
   b)   calculate the annual rainfall that deciduous forest needs.
2   Kilimanjaro is a six-kilometre-high mountain located close to the Equator in Africa. At its base, it is surrounded by tropical rainforest.
   a)   Describe how the vegetation on Kilimanjaro might change with increasing altitude.
   b)   Explain reasons for the changes you have described.
3   To help with revision, draw a table showing the three main types of forest. In one column, describe the associated climate. In another column, briefly describe the characteristics of the trees.

Geographical skills

Each global-scale ecosystem in Figure 5.13 has several distributional features. Choose one ecosystem and take the following steps:

1   What is the overall pattern? Does the ecosystem circle the Earth at a particular latitude? Or is it found only in particular regions?
2   Can you name the continents or any particular places where it is found?
3   Are there any unusual features or anomalies? Perhaps the ecosystem is found at an unusually high or low latitude in some places.
4   Can you try to quantify any of your statements? For example: 'Around one-third of South America is tropical rainforest'!

6 Tropical rainforests

✪ KEY LEARNING

➤ The physical characteristics of the tropical rainforest
➤ Interdependence within the ecosystem
➤ Biodiversity and its issues

Tropical rainforest

The tropical rainforest occupies only seven per cent of the world's land surface, but it contains many useful resources. It is also valuable in the fight against global warming (page 76). The main areas of tropical rainforest are in the Amazon basin (Brazil), Central Africa and South East Asia (Figure 6.1), with the largest area of tropical rainforest in the Amazon.

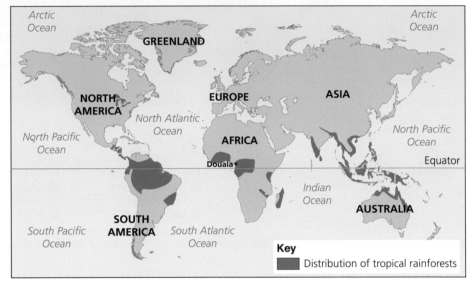

▲ Figure 6.1 The global distribution of tropical rainforests

What are the physical characteristics of the tropical rainforest?

Two main characteristics distinguish the tropical rainforest from other biomes: climate and vegetation.

Climate

As tropical rainforests occur on or close to the Equator, the climate is typically warm and wet. Annual temperatures average around 26 °C and show little variation (Figure 6.2). Annual rainfall usually exceeds 2,500 millimetres. This abundant supply of water feeds huge rivers such as the Amazon in Brazil and the Congo in Central Africa, an impressive physical feature of tropical rainforests.

Vegetation

Tropical rainforests are renowned for their rich vegetation. Particularly spectacular are their very tall trees, typically 30–45 metres high. (For more about vegetation, see Section 6.2.)

Soils

The soils of the tropical rainforest are mainly thin and poor, but there is so much luxuriant vegetation because of the rapid recycling of nutrients (Figure 6.3). Most of the forests' vital nutrients are locked up in:

■ the biomass – living vegetation and animals
■ the litter – dead wood and leaves, and animal remains on the ground.

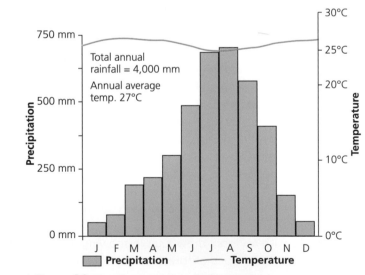

▲ Figure 6.2 The climate at Douala (Cameroon)

The warm, humid conditions cause the litter to decompose very quickly. The little rain that reaches the forest floor often washes away litter nutrients before they become part of the soil. It is not surprising that rainforest soils are rather infertile. Nonetheless, plants can pick up enough nutrients from the soil to survive. Many nutrients are stored in large, thick trees.

How are components of the tropical rainforest interdependent?

Figure 6.4 shows the main components of the tropical rainforest ecosystem: climate (rain and sunlight), soil, vegetation (trees and plants) and animals. The arrows show how they interact to create an interdependence (Section 6.2). The diagram also shows people. indigenous people live as part of, and in harmony with, the ecosystem, protecting 80% of remaining biodiversity. Equally, human activity can badly upset the ecosystem's balance.

Why is biodiversity an issue in tropical rainforests?

Tropical rainforests are renowned for their high level of biodiversity. It is higher than in any other biome. More than two-thirds of the world's plant species are found in these forests. The forests contain around half of the world's known animal species, ranging from mammals (such as monkeys and sloths) and birds, to reptiles (such as snakes and frogs) and insects.

Human exploitation of the rainforests' resources is reducing this rich biodiversity. Many species are becoming endangered; many others have already become extinct. A loss of biodiversity means a decline in ecosystem productivity.

The challenge is this: can the tropical rainforests be used in a sustainable way that does not threaten biodiversity?

▲ **Figure 6.3** The nutrient cycle

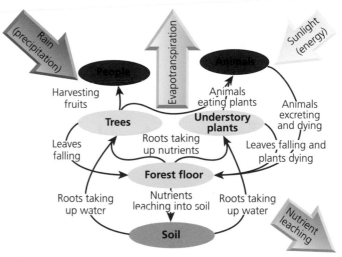

▲ **Figure 6.4** The tropical rainforest ecosystem

Geographical skills

Study Figure 6.2. Describe the features of Douala's climate. Mention:

a) the maximum and minimum monthly temperature and rainfall

b) the mean annual temperature and the total annual rainfall

c) the annual range of temperature and rainfall.

→ Activities

1 What is the difference between an ecosystem and a biome? (Look back at Chapter 5.)

2 Study Figure 6.3.

   a) Describe how nutrients are cycled in tropical rainforests. Mention biomass, litter and soil.

   b) Explain why nutrient cycling happens so rapidly in a tropical climate.

3 Suggest reasons why biodiversity is so high in the tropical rainforests.

Adaptations to the tropical rainforest environment

Let us now look more closely at the adaptations of plants and animals within the tropical rainforest. They provide examples of the interdependence that exists within its ecosystem (Figure 6.4 on page 69).

How have plants adapted to survive in the rainforest?

Factors that help poor soils

In Section 6.1, the soils of the tropical rainforest were described as being very poor. Not good news for plants! So how do they manage to survive and prosper? The answer lies mainly in four factors:

- a rapid cycling of nutrients through the ecosystem (see Figure 6.3) – a sort of fast-food delivery
- the absorption of sunlight, leading to photosynthesis
- the warm, humid climate, which is ideal for plant growth throughout the year
- the ability of plants to adapt as they compete for sunlight and nutrients.

Heat and humidity

The nutrient cycle is one way in which the components of an ecosystem work together. Another is the water cycle (Figure 6.5). This constant recycling of water occurs every day.

The leaves of many trees are waxy and have tips that allow water to run off them. Leaf stems are also flexible to allow leaves to move with the Sun to maximise photosynthesis. Therefore, vegetation copes with both heat and heavy rainfall by:

- using the circulating water as a sort of cooling system
- passing water to the soil or returning it to the atmosphere
- having leaves that can cope with the large amounts of water falling on them.

Competition for sunlight

Although photosynthesis is important for plant growth, plants still need minerals and these come mainly from the soil. The dense vegetation of the tropical rainforest shows four distinct layers (Figure 6.7). In each layer, the plants have adjusted to the physical conditions, particularly to available sunlight. Most sunlight is received by the tops of tall trees and, due to a shading effect, the least sunlight is received close to the forest floor.

In the lowest two layers there is little photosynthesis to convert the small amount of sunlight into plant food. So plants have to rely on other ways of getting their food supply. In most cases, this means from the soil. In some cases, a different strategy is used. For instance, parasitic plants have developed a way of attaching themselves to a host tree or shrub and sharing its supply of food and water.

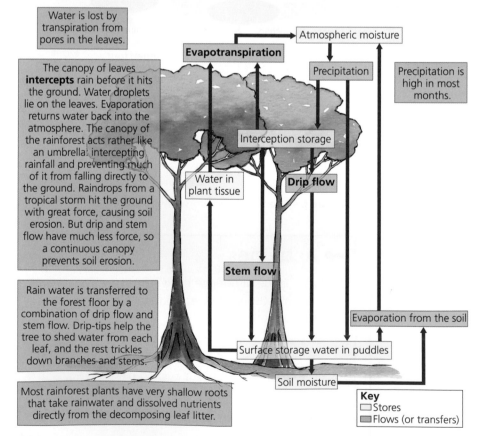

Water is lost by transpiration from pores in the leaves.

The canopy of leaves **intercepts** rain before it hits the ground. Water droplets lie on the leaves. Evaporation returns water back into the atmosphere. The canopy of the rainforest acts rather like an umbrella, intercepting rainfall and preventing much of it from falling directly to the ground. Raindrops from a tropical storm hit the ground with great force, causing soil erosion. But drip and stem flow have much less force, so a continuous canopy prevents soil erosion.

Rain water is transferred to the forest floor by a combination of drip flow and stem flow. Drip-tips help the tree to shed water from each leaf, and the rest trickles down branches and stems.

Most rainforest plants have very shallow roots that take rainwater and dissolved nutrients directly from the decomposing leaf litter.

Evapotranspiration

Atmospheric moisture

Precipitation

Precipitation is high in most months.

Interception storage

Water in plant tissue

Drip flow

Stem flow

Evaporation from the soil

Surface storage water in puddles

Soil moisture

Key
☐ Stores
▨ Flows (or transfers)

▲ Figure 6.5 The water cycle

How have animals adapted to survive?

The climate, vegetation and food chains of the tropical rainforest are ideal for animal life of all sorts. There is plenty of food and water throughout the year.

Competition for food

Because there are so many animals, there is a great deal of competition for food. Some animals are very specialised and live off a specific plant or animal that few others eat. For example, parrots and toucans have developed big strong beaks to crack open hard nuts.

There are relationships between animals and plants that benefit both. Some trees depend on animals to spread the seeds of their fruit. Birds and mammals, such as bats, eat the fruit and travel some distance before the seeds pass through their digestive systems in another part of the forest.

Other survival strategies

Many animals use camouflage to escape becoming prey, and predators use it to help them catch their prey. Some animals are poisonous and use bright colours to warn predators to leave them alone. There are several species of brightly coloured, poisonous arrow frogs. Central and South American communities used to wipe the ends of their arrows on the frogs' skin to transfer the deadly poison.

From all this, it can be seen that the plants and animals of the tropical rainforest are finely tuned to the environment and to each other. Everything works well, but only as long as it is not disturbed by people.

▲ Figure 6.6 Both plants and animals adapt to live in the rainforest

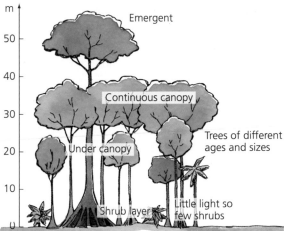

▲ Figure 6.7 The layers of the tropical rainforest

→ Activities

1 Write your own definitions for each of these processes in the water cycle: evapotranspiration, interception, drip flow and stem flow.

2 Make a copy of Figure 6.7. Add the following labels to the diagram to match the layers (emergent, continuous canopy, under canopy, shrub layer):

- This layer is dark and gloomy. There is little vegetation between the tree trunks.
- Sunlight is limited in this layer. Saplings and seedlings wait here for larger trees to die and leave gaps into which they can grow.
- This is the layer in which the upper parts of most trees are found. The leaf environment is home to insects, birds and mammals.
- The tops of the tallest trees are in this layer. Because of their height, these trees are able to get more light than the trees with their tops in the canopy.

3 Suggest how animals contribute to the nutrient and water cycles.

4 a) Research ways in which each of the following have adapted to the environment:
  - three-toed sloths
  - leaf-cutter ants
  - boa constrictors
  - toucans.

b) In which layer of the rainforest would you be most likely to find each of these animals? Give reasons for your answers.

⊗ KEY LEARNING
➤ The scale of deforestation
➤ The recent changes in the rate of deforestation
➤ The situation in Brazil

Deforestation of tropical rainforests

Deforestation (page 49) has a very long history. But it is only over the last 100 years or so that it has begun to have a serious impact on the tropical rainforest.

What is the scale of deforestation?

There are 62 countries with a tropical rainforest within their borders. Figure 6.8 shows the top 30 countries. Few, if any, early records were kept of the original extent of tropical rainforest. The UN Food and Agriculture Organization estimates that about half the world's tropical rainforest has now been cleared. The scale and accelerating rate of the deforestation are truly worrying.

Is the rate of deforestation changing?

Figure 6.9 shows that during the twenty-first century, rates of deforestation in the Amazon have fluctuated. By far the greatest rate of deforestation happened in Brazil, but Bolivia, Peru and Colombia also made significant contributions. Rates declined from 2004 to 2009 with greater government protection of the rainforest, but have increased again since then. The rate of deforestation in Brazil fell to a record low in 2009 and it is estimated that about half of Brazil's remaining rainforest had some form of protected status.

In 2016, there was a particularly big jump in deforestation. Two years later, a new government in Brazil announced a change in policy, encouraging more farmers, miners and loggers to clear the forest. At least 20 per cent of the Amazon rainforest has been cleared since 1970, an area of 761,000 km² (roughly three times the size of the UK).

The rate of reductions elsewhere may also reflect that other countries have already put in place measures to protect their rainforests. In some cases, for example in Mexico, strenuous efforts are being made to save what little is left before it disappears. But it is only the rate

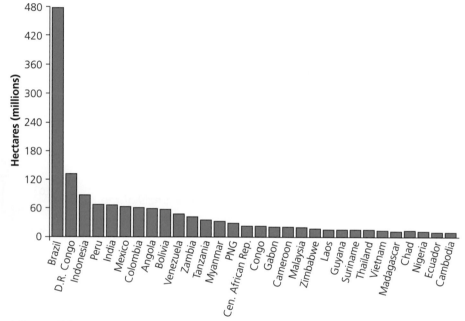

▲ Figure 6.8 Area of tropical rainforest by country, 2015

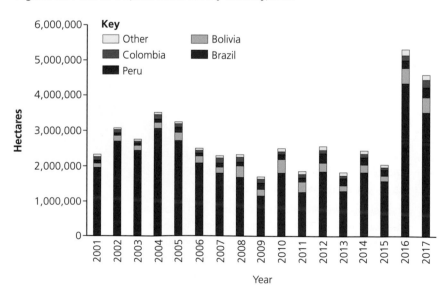

▲ Figure 6.9 Annual forest loss in the Amazon rainforest, 2001–17

that has decreased. Deforestation continues in all the countries shown in Figure 6.8. Indeed, it is still happening in all 62 countries.

Today, the global rate at which the tropical rainforest is being cleared is estimated to be:

- 1 hectare per second
- 60 hectares per minute
- 86,000 hectares per day (an area larger than New York City)
- 31 million hectares per year (an area larger than Poland).

What has happened in Brazil?

The following is an introduction to two important aspects of deforestation: its causes (Section 6.4) and impacts (Section 6.5).

The Brazilian rainforest occupies the huge lowland basin drained by the Amazon and its tributaries. Figure 6.10 clearly shows how much of the tropical rainforest cover was lost up to 2006. It is interesting that the clearance has been to the south of the Amazon. This is the part of the rainforest most accessible from Brazil's main cities, such as Rio de Janeiro, São Paulo and Brasilia.

An important point here is that for centuries the rainforest has been lived in and used by indigenous communities. They have:

- harvested fruits and nuts
- cut wood for fuel
- used timber to build their dwellings
- discovered cures for various illnesses (Section 6.6)
- cleared small areas, by a technique known as slash and burn (page 49), a type of **subsistence farming**.

Slash and burn has done little lasting damage to the forest. When the soil in one small area becomes exhausted, the community moves on and clears another. This is why it is sometimes referred to as 'shifting cultivation'. Once abandoned, the forest is able to regenerate.

Use of the rainforest by people other than those in indigenous communities does cause lasting damage. However, it does not always lead to deforestation. In many cases, it leads to **forest degradation**, where the forest ecosystem is changed and its supply of resources declines (Section 6.6).

→ Activities

1. a) Give your own definition of the word 'deforestation' (use page 49 to help).
   b) What is the difference between deforestation and forest degradation?
2. Explain why we cannot be sure how much of the world's tropical rainforest has been cleared.
3. Study Figure 6.8.
   a) Identify the countries where the Amazon rainforest is found. What is the total area of the Amazon rainforest?
   b) What percentage, roughly, of the world's rainforest is found in the Amazon?
4. Study Figure 6.10.
   a) Describe the distribution of deforested areas of the Amazon shown on the map.
   b) Suggest the reasons for this distribution.

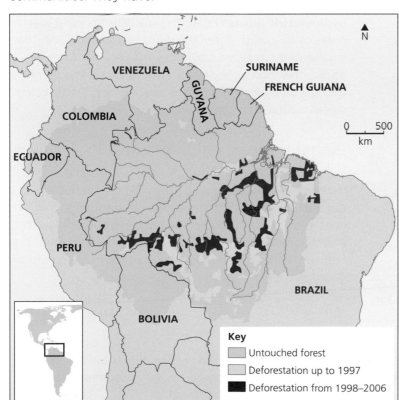

▲ Figure 6.10 Deforestation in the Amazon basin

Geographical skills

Taking data about forest cover from Figure 6.8 and about the annual rate of deforestation from Figure 6.9, calculate the percentage loss of rainforest in 2016 for Brazil, Bolivia, Peru and Colombia. Use this data to create a new graph.

Case study

⭐ KEY LEARNING

➤ How people exploit rainforest resources
➤ The activities that cause deforestation

The tropical rainforest in Brazil: causes of deforestation

Brazil is located in South America. It is the fifth largest country in the world and contains the largest area of tropical rainforest. The tropical rainforest of Brazil, as in other countries, is being exploited in two ways:

- by using its resources, such as timber, water and minerals
- by clearing the forest to make way for other activities, such as growing crops and rearing livestock.

What are the main resource-exploiting activities?

Logging

Figure 6.11 shows the main causes of deforestation in Brazil. You may be surprised that the figure for **logging** is so small. Logging is the first step in the conversion of forest land to other uses. It is the eventual use the cleared land is put to that is recorded in the pie chart. Timber companies are most interested in trees such as mahogany and teak, and sell them to other countries to make furniture (**selective logging**). Smaller trees are often used as wood for fuel or made into pulp or charcoal. Vast areas of rainforest are cleared in one go (clear-felling). It is estimated that 50 per cent of deforestation is illegal. The problem is that it is unregulated and ignores environmental laws. Illegal loggers destroy large areas of forest to get the most valuable trees.

Mineral extraction

It so happens that some of the minerals that high-income countries want are found beneath stretches of tropical rainforest. In the Amazon, **mineral extraction** is mainly about gold. In 1999, there were 10,000 hectares of land being used for gold mining. Today, the area is over 50,000 hectares. The rainforest suffers badly as it is clear-felled. The same applies to the extraction of another mineral, bauxite, from which aluminium is made. Deforestation for mining can also be illegal, bringing few opportunities for local people or any money for the government.

Energy development

An unlimited supply of water and ideal river conditions have encouraged **dams** to be built to generate

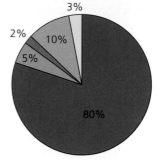

Key
■ Cattle ranching   ■ Small-scale agriculture   ■ Other
□ Logging   ■ Commercial agriculture

▲ Figure 6.11 Causes of deforestation in the Brazilian rainforest

hydroelectric power (HEP). This involves flooding vast areas of rainforest. Often the dams have a short life. The submerged forest gradually rots, making the water very acidic. This then corrodes the HEP turbines. The dams also become blocked with soil washed down deforested slopes by the heavy rain.

Illegal trade in wildlife

Hunting, poaching and trafficking in wildlife and animal parts are still big business in Brazil. Although this is not a direct cause of deforestation, it is endangering species such as the jaguar, the golden-bellied capuchin and the golden lion tamarind. It is also upsetting the natural balance of the rainforest ecosystem and therefore degrading it.

What activities are causing the forest to be cleared?

Subsistence farming (see Section 6.3) is a small contributor to deforestation, however **commercial farming** is much more damaging.

Commercial farming: cattle

Cattle rearing is believed to account for 80 per cent of the tropical rainforest destruction in Brazil (Figure 6.11). The meat is exported to other countries including China and those in the EU. However, the land cannot be used for long. The quality of the pasture quickly declines (Section 6.5). The cattle farmers then have to move on and destroy more rainforest to create new cattle pastures.

Commercial farming: crops

The forest is being cleared to make way for vast plantations, where crops such as bananas, palm oil, soybeans, pineapple, sugar cane, tea and coffee are grown and exported around the world. Soybean production increased sixfold between 1990 and 2019 (Figure 6.13). As with cattle ranching, the soil will not sustain crops for long. After a few years, the farmers have to cut down more rainforest for new plantations. Sugar cane, used for **biofuel**, is beginning to become a major crop.

Road building

Roads are needed to bring in equipment and transport products to markets, but road building means cutting great swathes through the rainforest. Additionally, a road built for one particular commercial activity makes the forest accessible to other exploiters of the tropical rainforest's resources. The Trans-Amazonian Highway began construction in 1972 and is 4,000 kilometres long. Although only a small part of it is paved, it has played an important part in opening up remote areas of the Amazon rainforest.

Settlement and urban growth

All the above activities have a common knock-on effect. They need workers, and workers and their families need homes and services. That, in turn, means clearing the forest to build settlements where these people can live.

The huge challenges presented by these causes of deforestation are their scale, speed and wasteful use of the forest's land and resources.

▲ Figure 6.12 Cattle grazing on deforested land

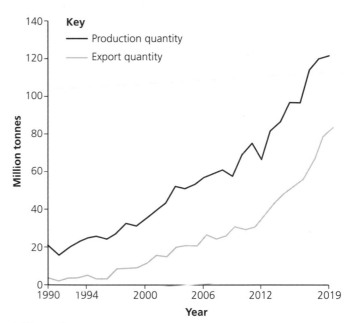
▲ Figure 6.13 Soybean production in Brazil (1990–2010)

Geographical skills

Use Figure 6.13 to calculate the percentage increase between 1990 and 2019 in a) soybean production and b) soybean exports.

→ Activities

1 Study Figure 6.11.
   a) Describe the importance of different causes of deforestation in Brazil.
   b) Explain why logging appears to make such a small contribution to deforestation.
2 Explain why the construction of roads is a serious threat to the tropical rainforest.
3 Outline two ways in which human use of the rainforest is wasteful.

⊙ KEY LEARNING

➤ The global impacts of deforestation
➤ The local impacts of deforestation

The tropical rainforest in Brazil: the impacts of deforestation

What are the global impacts of deforestation?

The main argument for deforestation, and one made by the Brazilian government, is that it leads to economic **development**. Deforestation creates jobs for loggers, farmers, ranchers, miners and others, allowing natural resources to be exploited. Many of these resources are exported, earning money for the Brazilian economy.

However, there is another argument that deforestation only leads to short-term development. Once the trees are chopped down they are gone forever and their loss leads to a decline in biodiversity and soil fertility. These are important for Brazil's long-term development. Figure 6.14 shows the main negative impacts of deforestation, globally and locally.

Global warming

The rainforest is significant at a global level. The tree canopy absorbs carbon dioxide in the atmosphere. This stops as soon as the trees are felled and more carbon dioxide remains in the air. Fire is often used in clearing rainforests, which means that the carbon stored in the wood returns to the atmosphere. In these ways, deforestation is a main contributor to the greenhouse effect, which is a cause of global warming.

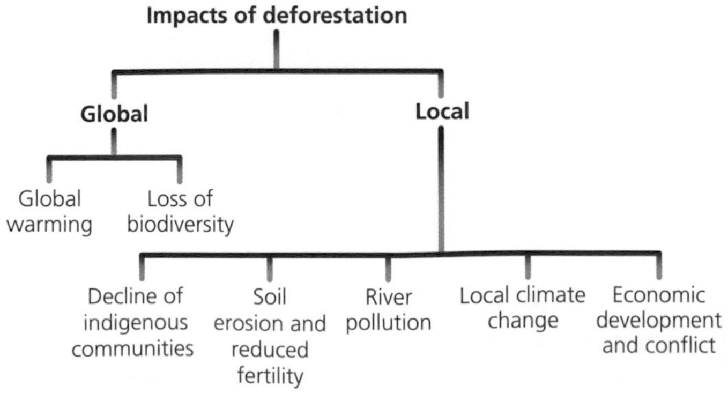

Loss of biodiversity

Clearing tropical rainforests means that:

- the biodiversity will be reduced
- individual species will become endangered and then possibly extinct.

It has been estimated that 137 plant, animal and insect species are being lost every single day due to rainforest deforestation. That amounts to 50,000 species a year. As the rainforest species disappear, so do many possible cures for life-threatening diseases (Section 6.6). New research has shown that parts of the Amazon rainforest could lose between 30 and 45 per cent of their main species by 2030.

What are the local impacts of deforestation?

Local climate change

Deforestation also disrupts the water cycle (Section 6.2). With the felling of trees, evapotranspiration is reduced, and so too the return of moisture to the atmosphere. The local climate becomes drier. The recycling of water is like a cooling system. Once it is reduced, the local climate becomes warmer. The combination of increasing dryness and rising temperatures is not good for humans or human activities, such as agriculture.

Impacts of deforestation

- Global
  - Global warming
  - Loss of biodiversity
- Local
  - Decline of indigenous communities
  - Soil erosion and reduced fertility
  - River pollution
  - Local climate change
  - Economic development and conflict

▲ Figure 6.14 Some impacts of deforestation

▲ Figure 6.15 The frightening scale and devastation of deforestation

Soil erosion and fertility

As soon as any part of the forest cover is cleared, the thin topsoil is quickly removed by heavy rainfall. Bare slopes are particularly prone to soil erosion. Once the topsoil has been removed, there is little hope of anything growing again. Soil erosion also leads to the silting up of river courses (Chapter 11).

Even where the soil is protected, the soil quickly loses what little fertility it had when covered by trees. Grazing and plantations do little or nothing to keep the soil fertile. The decline in soil fertility leads to pastures and plantations being abandoned, so more areas of rainforest are cleared.

River pollution

Gold mining not only causes deforestation, but the mercury used to separate the gold from the ground is allowed to enter the rivers. This poisons fish as well as people living in nearby towns. Rivers are also being polluted by soil erosion.

Decline of indigenous people

Not all Brazilians are benefitting from exploitation of the rainforest resources. Most obviously, indigenous people have a traditional way of life that is closely geared to the rainforest. There are now only around 240 communities left (Figure 6.16), compared with over 330 in 1900. Many

▲ Figure 6.16 Men of the Yawalapiti community fishing

indigenous people have been forced out of the rainforest by the destruction of their environment through the actions of governments and large companies, such as:

- the construction of roads
- logging
- the creation of ranches, plantations and reservoirs
- the opening of mines.

Most displaced people have ended up in towns and cities. Few have adjusted to this very different environment. Addiction to drugs and alcohol has been common, and sadly many have died young. With the loss of these communities have gone centuries of detailed knowledge of the forest, such as the medicinal value of various rainforest species (Section 6.6).

Despite all this, Brazil's tropical forest is still home to an estimated 1 million indigenous people. They still make their living through subsistence farming or hunting and gathering, or through low-impact harvesting of forest products like rubber and nuts.

Economic development and conflict

Disputes between indigenous people and loggers and other developers of the rainforest often end in open conflict. Disputes arise because people have conflicting views about the rainforest, for example between conservationists and developers (Section 6.7).

All these conflicts arise from the different causes of deforestation. The mix of causes varies from country to country, but economic development and population growth are often the main drivers.

→ Activities

1. Make a large copy of Figure 6.14. Add your own notes to explain how each of the impacts of deforestation happen. Use short bullet points.

2. Which of the five local impacts in Figure 6.14 do you think is the most serious? Give reasons.

3. Study Figure 6.15. What evidence can you see of the
   a) causes of deforestation
   b) impact of deforestation?

4. Imagine you are either a rancher, an indigenous person, a government minister or a conservationist. Would you be in favour of deforestation or not? Write a short statement that:
   a) shows your view of the rainforest
   b) identifies how your view might conflict with others.

KEY LEARNING

- The goods the tropical rainforest supplies
- The services the tropical rainforest supplies

The value of the tropical rainforest to people and the environment

Section 6.4 showed that the tropical rainforest is a valuable provider of resources and **economic opportunities**. These fall into two different groups:

- those provided by the rainforest in its natural state
- those provided by the land once it is cleared of its forest cover.

The vast commercial value of the second group, particularly the crops and livestock, is the driving force behind much of the current deforestation. But what is the value of the tropical rainforest itself?

The resources and opportunities offered by the tropical rainforest or any other biome or ecosystem are more widely known as **goods and services**. In this instance, goods are things that can be obtained directly from the rainforest. Services are benefits that the rainforest can offer to both people and the environment (Figure 6.17).

Figure 6.17 Tropical rainforests: goods and services

Goods – of direct value to people	Services – of value to the environment and indirect value to people
• Native food crops (fruit and nuts) • Wild meat and fish • Building materials (timber) • Energy from HEP • Water • Medicines	• Air purification (absorbing CO_2) • Water and nutrient recycling • Protection against soil erosion • Wildlife habitats • Biodiversity • Employment opportunities

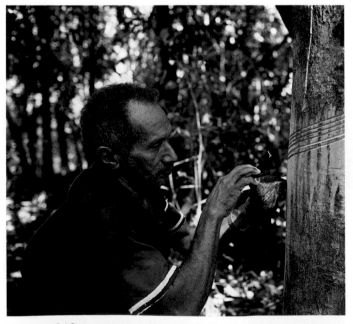

Figure 6.18 An indigenous person in Brazil collecting rubber

What goods are supplied by the tropical rainforest?

The plants of the tropical rainforest include many of the things we eat, such as cocoa, sugar and bananas. Cinnamon, vanilla and many other spices also come from the rainforest. Useful products like rubber, rope and baskets are made from rainforest plants. Some of the chemicals from rainforest leaves, flowers and seeds are used to make perfumes, soaps, polishes and chewing gum. Traditional subsistence farming is still very much about the harvesting of rainforest goods. The use of these forest products has been going on for centuries.

Finding new medicines

Today, however, we are beginning to realise that the forest has something more to offer. It is the stock of plants that pharmaceutical companies are finding which contain ingredients to help treat and cure diseases. Indigenous rainforest communities have a very long tradition of using parts (barks, resins, roots and leaves) of various plants for this purpose (Figure 6.18).

Currently, over 120 prescription drugs sold worldwide come from plant sources. About a quarter of the drugs used today in the developed world are derived from rainforest ingredients. Less than one per cent of the tropical rainforest trees and plants have been tested by scientists to find out whether they have any medicinal value. Twenty-five per cent of the active ingredients in today's cancer-fighting drugs come from organisms found only in the rainforest (Figure 6.19).

In 1980, there were no pharmaceutical companies researching possible new drugs and cures from plants. Today, there are well over a hundred. It is in the interests of global health care to protect the tropical rainforest and its stock of medicinal plants. It is vital that these plants are not over exploited. Either the wild plants are harvested in a sustainable way or they should be deliberately cultivated as crops, perhaps on deforested land.

What services are supplied by the tropical rainforest?

Some of the services listed in Figure 6.17, such as water and nutrient recycling (Section 6.1) and protection against soil erosion (Section 6.5), are services that benefit the environment and help to maintain the general health of the rainforest. On the other hand, the forest's biodiversity and wildlife habitats are benefits that people can enjoy, either as indigenous people or as tourists. However, it is the rainforest's air purification service that is perhaps of most value to both people and the environment – not just within the tropical rainforest, but globally.

Perhaps the single most important global issue today is global warming and climate change. Global warming will only be checked by:

- greatly reducing the burning of **fossil fuels** and so lowering emissions of carbon dioxide
- greatly reducing the rate of deforestation to make sure that as much of the Earth as possible is covered by trees to absorb the carbon dioxide in the atmosphere.

As one of the largest carbon sinks in the world, the tropical rainforest has a critical role to play. Protecting the remaining rainforest requires doing two things:

- making sure that much of it is left untouched, so that it stays in a pristine state, for example making large areas of rainforest into nature reserves or national parks
- allowing the resources of the rainforest, its goods and services, to be used, but only in a controlled and sustainable way (Section 6.7).

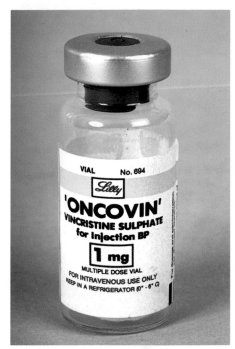

▲ Figure 6.19 Vincristine: an anti-cancer drug derived from a rainforest plant, periwinkle

→ Activities

1 What is the difference between the goods and the services of an ecosystem?
2 Study Figure 6.17.
   a) Identify which goods and services provided by the rainforest are important to us (even though we don't live there).
   b) For each good or service you have identified, explain why it is important.
3 Suggest what might be done to save the knowledge of the rainforest that indigenous communities have.
4 Describe the ways in which tropical rainforests are good for human health, for:
   a) people who live in the rainforest
   b) the wider human race.
5 Explain why the tropical rainforest is needed to fight global warming.

✪ KEY LEARNING

➤ Why sustainable management is needed
➤ International agreements
➤ Government intervention

Strategies for managing tropical rainforests sustainably

Why does the tropical rainforest need to be managed sustainably?

If the goods and services of the tropical rainforest are not protected, then they will soon become lost forever. Sustainable management means using goods and services in such a way that they are still available for the benefit of people in the future. If that does not happen, then the forest's stock of renewable resources will gradually become exhausted. Further large-scale deforestation has no place in any sustainable management of the rainforest.

What actions will bring about sustainable management of the tropical rainforest? Most actions are taken at three levels: international, national and local. Some topics, such as **conservation** and education, crop up in all three action levels. Other actions tend to occur at only one of these levels.

What can be done at an international level?

Three different types of action illustrate what is being done at this level.

Inter-government agreements on hardwoods and endangered species

The first of these involves agreements between governments aimed at protecting the biodiversity and resources of the rainforest. They include:

■ the International Tropical Timber Agreement (2006), which restricts the **trade** in hardwoods taken from the tropical rainforest. The very high prices paid for tropical hardwoods have encouraged a huge amount of illegal felling. This tends to take place in remote areas of the rainforest and so often goes on unnoticed by forestry officials. The 2006 Agreement restricts the trade in hardwood timber to timber that has been felled in sustainably managed forest (Section 6.8). All such timber has to be marked with a registration number (Figure 6.20).

■ the CITES (Convention on International Trade in Endangered Species, 1973) treaty blocks the illegal trade in rare and endangered animals and plants. The illegal trade is still worth millions of pounds.

Debt reduction by HICs

Most of the countries with tropical rainforest are newly emerging economies (NEEs) or low-income countries (LICs) . They may also have large debts, often resulting from overseas aid in the form of loans. Schemes known as debt-for-nature swaps are sometimes arranged. In 2010, for example, the USA signed an agreement to convert a Brazilian debt of £13.5 million into a fund to protect large areas of tropical rainforest. These swaps are all part of what is known more widely as **debt reduction**, where some high-income countries (HICs) agree to write off the debts of some low and middle-income countries (Section 19.6).

▲ Figure 6.20 Legally registered hardwood timber awaiting collection in Cameroon

Conservation and education by NGOs

The third type of international action is by **non-governmental organisations (NGOs)** such as the WWF, Fauna & Flora International, Birdlife International and the World Land Trust. All of them are charities that rely on volunteers and donations. As well as supporting tropical rainforests, they operate anywhere in the world where they think ecosystems are being seriously threatened. Such organisations:

■ promote the conservation message largely through education programmes in schools and colleges

■ provide training for conservation workers (another aspect of education)

■ provide practical help to make programmes more sustainable (see Section 6.8)

■ buy up threatened areas and create nature reserves.

Important initiatives about conservation can also be undertaken at both national (see below) and local levels (Section 6.8).

What should national governments do?

In terms of conservation and education, achieving a sustainable balance between protection and development in the tropical rainforest is the prime responsibility of government. All governments have the powers to pass laws to achieve this, for example by:

■ creating protected areas or reserves

■ stopping the abuse of the rainforest and other biomes by developers

■ making subjects, such as ecology or environmental studies, a compulsory part of the school curriculum.

Some governments have been more successful in protecting the natural environment. For example, in Costa Rica, a small country in Central America, 24 per cent of the land is protected as national parks and nature reserves. There are, however, some problems. For example:

■ Few governments are willing to do anything that might slow down the rate of economic development in their country. Citizens expect or want better living standards rather than new nature reserves.

▲ Figure 6.21 Teaching students in Brazil about forest conservation in an area where the land was devastated and has now been afforested

■ Governments seem unwilling to enforce and monitor laws aimed at protecting or conserving the rainforest.

■ There is a lot of corruption in the way rainforests are treated, for instance by illegal loggers and developers paying bribes.

Conservation is the prevention of wasteful use of resources. Natural resources such as timber can still be used, but must be used sustainably.

Environmental protection is the act of protecting the environment by individuals, organisations and governments, so ecosystems can remain balanced.

→ Activities

1 Describe the aims of sustainable management.

2 Design an educational leaflet aimed at people from LICs to explain why the conservation of the tropical rainforest in their countries is important.

3 Explain the conflict between economic development and conservation.

4 Suggest how corruption threatens the tropical rainforest.

5 Research one of the following NGOs and what it is doing to help the tropical rainforest:
   ■ Birdlife International
   ■ Fauna & Flora International
   ■ WWF (World Wide Fund for Nature)
   ■ World Land Trust.

Strategies for managing the tropical rainforests sustainably

Sustainable actions start at a local level. **Sustainability** emphasises the importance of local actions, such as:

■ respecting the environment and cultures of local people

■ using traditional skills and knowledge

■ giving people control over their land and lives

■ generating income for local people

■ using **appropriate technology** – machines and equipment that are cheap, easy to maintain and do not harm the environment.

What can be done at the local level?

In answering this question, aim to distinguish between two different situations:

■ areas with logging

■ areas still untouched by logging.

For the first, there are four possible actions.

Selective logging: this involves felling trees only when they are fully grown, and letting younger trees mature and continue protecting the ground from erosion. It involves a cycle lasting between 30 and 40 years.

Stopping illegal logging: given the remoteness of rainforest areas, illegal logging can easily go on unnoticed. It is still happening on a large scale. However, satellites and drones are now helping to monitor this.

Agroforestry: this involves combining crops and trees, by allowing crops to be grown in carefully controlled, cleared areas within the rainforest, and by growing rainforest trees on plantations outside the rainforest.

Replanting: a project in the Atlantic rainforest of Brazil (REGUA) has shown it is possible to recreate a forest cover almost like the original. This is done by collecting seeds from remaining patches of original forest, growing the seeds into saplings in nurseries and then planting the saplings back in the deforested areas (Figure 6.22).

It is amazing how quickly a new forest cover develops with almost the same gene bank as the original cover. No doubt what has been learnt in this small project will help Brazil fulfil its promise to reforest 12 million hectares, agreed in a new US–Brazil climate partnership in June 2015.

For the areas untouched by logging, **ecotourism** presents a type of sustainable action. Scenery, wildlife, remoteness and culture are the main attractions (Figure 6.24). It aims to educate visitors and increase their understanding and appreciation of nature and local cultures. It is small scale and local (controlled by local people, employing local people and using local produce). Its profits stay in the local community. It tries to minimise the consumption of **non-renewable resources** and the ecological impact. In this eco format, tourism becomes both a sustainable and a profitable activity.

➤ Figure 6.22 The REGUA nursery produces 70,000 saplings of 180 species a year

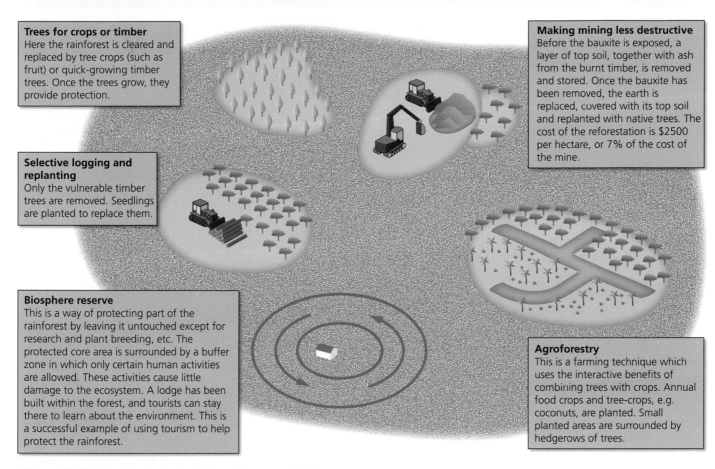

Trees for crops or timber
Here the rainforest is cleared and replaced by tree crops (such as fruit) or quick-growing timber trees. Once the trees grow, they provide protection.

Making mining less destructive
Before the bauxite is exposed, a layer of top soil, together with ash from the burnt timber, is removed and stored. Once the bauxite has been removed, the earth is replaced, covered with its top soil and replanted with native trees. The cost of the reforestation is $2500 per hectare, or 7% of the cost of the mine.

Selective logging and replanting
Only the vulnerable timber trees are removed. Seedlings are planted to replace them.

Biosphere reserve
This is a way of protecting part of the rainforest by leaving it untouched except for research and plant breeding, etc. The protected core area is surrounded by a buffer zone in which only certain human activities are allowed. These activities cause little damage to the ecosystem. A lodge has been built within the forest, and tourists can stay there to learn about the environment. This is a successful example of using tourism to help protect the rainforest.

Agroforestry
This is a farming technique which uses the interactive benefits of combining trees with crops. Annual food crops and tree-crops, e.g. coconuts, are planted. Small planted areas are surrounded by hedgerows of trees.

▲ Figure 6.23 Using the tropical rainforest in a sustainable way

For both types of area, local communities involved in projects such as replanting and ecotourism will gain a better understanding of the tropical rainforest. They will also help spread the message of sustainability to neighbouring communities. Community involvement represents yet another action falling under the broad heading of conservation and education.

Figure 6.23 shows different ways of using the tropical rainforest sustainably.

→ **Activities**

1 Write a campaign slogan to persuade local people to use tropical rainforests in a sustainable way.

2 Explain why it might be difficult to persuade locals that selective logging is a good idea.

3 Identify the resources needed for replanting in rainforests and explain why each one is needed.

4 Study Figure 6.23.

   a) Identify ways in which each sustainable activity helps the natural environment.

   b) Make a list of the goods being produced by these sustainable uses of the rainforest.

5 Study Figure 6.24. What makes this eco-lodge sustainable? Draw an annotated sketch.
Think about location, scale, resources and ecological impact.

▲ Figure 6.24 An eco-lodge in Costa Rica

7 Hot deserts

> ➤ The physical characteristics of hot desert climates

Hot desert environments

These are found in subtropical areas between around 20° and 30° north and south of the Equator. The Tropic of Cancer or the Tropic of Capricorn passes through most of the world's hot desert regions. In Chapter 3, we learned that this climate is characterised by hot and dry sinking air. The extremely arid conditions occur where less than 250 millimetres of rain falls annually (Figure 7.1).

What are the physical characteristics of a desert climate?

Large areas of the Earth's land surface are covered by **hot deserts**, including the Australian, Thar, Arabian and Kalahari deserts. Largest of all is the Sahara, which measures 9 million square kilometres. Its towering **sand dunes** can reach 150 metres. Despite its extreme climate, the Sahara manages to support 2 million people at its edges and in towns along caravan trails.

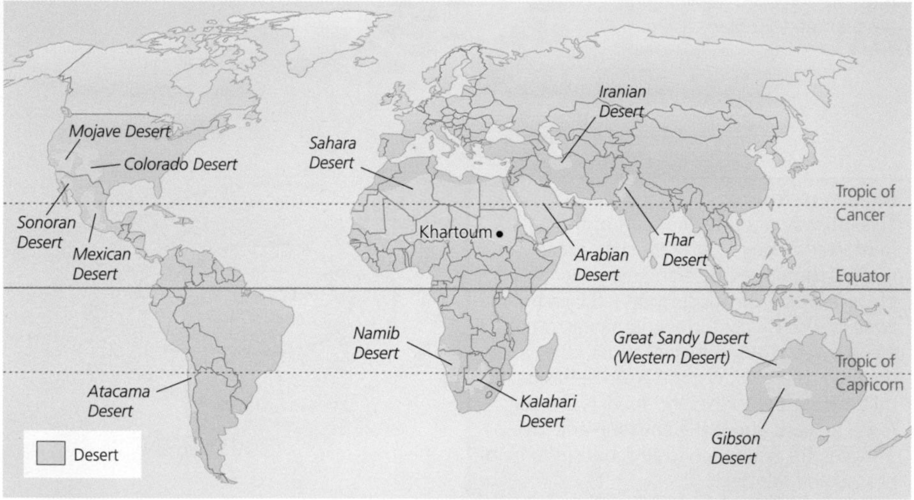

On the borders of hot deserts are the world's **semi-arid** areas, also called drylands or desert fringe areas (see Section 7.6). The Sahel is a long strip of semi-arid drylands that borders the south of the Sahara. It includes parts of Sudan, Chad, Burkina Faso and Niger. Water can sometimes be obtained in deserts from aquifers (see page 88), or rivers, like the Nile, which transport water from wet regions across much drier ones.

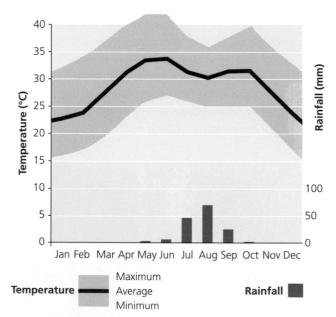

▲ Figure 7.1 The global distribution of the world's hot desert environments (semi-arid areas or drylands not shown)

Hot desert climates

Not only is annual precipitation low in hot desert environments, it is extremely unreliable too. Parts of Chile's Atacama desert have not seen rain for 400 years; others receive just one millimetre per year! More typically, hot desert regions experience around 100–200 millimetres of rainfall in most years. Figure 7.2 shows the annual climate graph for Khartoum in Sudan. Between June and October, Khartoum records on average six days with ten millimetres or more and 19 days with one millimetre or more of rainfall. Like other hot desert areas, rainfall can be unpredictable.

▲ Figure 7.2 The climate graph and monthly temperature data for Khartoum, Sudan

Figure 7.2 also shows the extreme high temperatures Khartoum experiences. An even higher temperature of 57.8 °C was once recorded in the northern Sahara at El Azizia in Libya.

Another characteristic is the extreme range of temperatures often experienced in a single day. The cloudless skies that allow high levels of insolation in the daytime also permit rapid heat loss at night. This can bring a drastic fall in temperature, and occasionally sub-zero temperatures. Figure 7.3 shows that the **diurnal temperature range** for a desert may exceed 35 °C.

Hot desert soils

In hot climates, soil-forming processes are limited by the shortage of water and vegetation. Over time, weathering creates deep deposits of sand and loose material. There may be little organic content due to the lack of vegetation growing there. These sandy, rocky soils are typically around one metre deep, although in some places, wind action builds tall dunes where deeper soils can potentially develop. Sand dunes should not be classified as soils if there is no organic matter present there at all.

Some desert soils are potentially very fertile because important nutrients for plant growth, such as calcium, have not been leached away over time. Once irrigated, the land can become highly productive for agriculture. Large desert regions of oil-rich Saudi Arabia and the southern states of the USA have benefited greatly from **irrigation**.

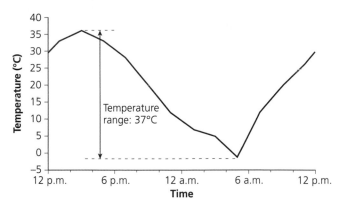
▲ Figure 7.3 Diurnal temperature range in a hot desert

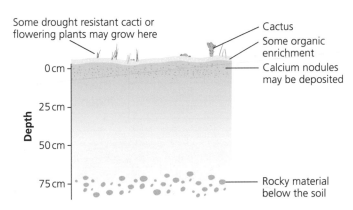
▲ Figure 7.4 A desert soil profile

→ Activities

1 Describe the global distribution of hot desert environments shown in Figure 7.1.
2 Suggest what is meant by 'maximum' and 'minimum' temperatures in Figure 7.2.
3 Use Figure 7.4 to describe the main characteristics of soils in hot desert environments.
4 This chapter reminds us that the climate of hot deserts 'is characterised by hot and dry sinking air'. Explain why this is the case. (Refer back to Chapter 3, pages 22–23 for the explanation.)

Geographical skills

Manipulating data sets

Study the climate graph for Khartoum (Figure 7.2).

1 Calculate the total annual rainfall.
2 Calculate the mean monthly rainfall.
3 Describe the annual pattern of rainfall. Are there wet or dry seasons? If so, when are they?
4 Calculate the annual temperature range (the difference in average temperature between the hottest and coldest times of the year).
5 Describe how temperature varies through the year. Does it show a distinctive seasonal pattern? If so, at what time of year are the hotter and cooler seasons?

➤ How plants have adapted to life in hot desert environments

➤ How animals have adapted to life in hot desert environments

➤ The interdependence of hot desert environments, ecosystems and people

Hot desert ecosystems and biodiversity issues

How have plants adapted to life in hot desert environments?

Desert biodiversity is far lower than in other global ecosystems you have studied. In Chapter 6, we learned that tropical rainforests have high levels of biodiversity (page 69). In contrast, far fewer species are supported by the extreme climate of hot deserts (Figure 7.5). Environment challenges include:

- dry conditions – plants that can survive in very dry conditions are called **xerophytes**. They use a range of adaptations, including thick, waxy cuticles and the shedding of leaves to reduce **transpiration**, to minimise water loss
- high temperatures – some plants have the bulk of their biomass below the ground surface where temperatures are cooler
- short periods of rainfall – deserts bloom suddenly after rainfall in order to complete their lifecycle quickly. This is an important issue to consider when discussing biodiversity.

The relatively small numbers of plants that survive are adapted in a range of ingenious ways (Figure 7.7).

▼ **Figure 7.5** Hot desert biodiversity (Grand Canyon National Park)

Group	Number of species
Mammals	56
Birds	400
Reptiles	36
Plant species	1,700
Mosses	60
Lichens	195

▼ **Figure 7.7** Plant adaptations that aid survival in hot desert environments

Drought-tolerant trees	Acacia trees have developed short, fat trunks that act as reservoirs for excess water. They are also fire-resistant. Their roots can penetrate 50 metres into the ground and reach out sideways to find as much water as possible.
Cacti	Cacti are 'succulents'; they store water in their tissues (Figure 7.6). The USA's saguaro cactus can grow up to 15 metres. The cacti's spikes deter consumer species, and their small, waxy leaves minimise transpiration losses.
Flowering plants	Desert flowers have seeds that only germinate after heavy rain and can lie dormant for years in between rains. They are 'ephemerals': they can complete their lifecycles in less than a month. Plants immediately produce brightly coloured flowers to attract insects.
Lichen	Lichen appears as a flaky crust on the ground, rocks and tree trunks. Lichens do not need soil to grow and for this reason are called **pioneer species**. They can grow on a bare rock surface in high temperatures. Lichen break down the rock chemically using their own organic acids. This helps them to extract nutrients they need.

▲ **Figure 7.6** The prickly pear is a type of cactus that grows in the Sonoran Desert

How have animals adapted to life in hot desert environments?

In the hottest desert regions, few animals have adapted to survive besides tough scorpions and small reptiles. In areas with a greater supply of water, the level of biodiversity rises as tough grasses, shrubs, cacti and hardy trees begin to form the basis of a larger food web that will include mammals (foxes, coyotes) and raptors (buzzards, hawks). Because deserts are found in most continents, different species have evolved to fill the niches in African, Asian and American deserts.

A simplified food web for the Mojave Desert in the USA is shown in Figure 7.8. This is the hottest and driest hot desert region in North America. The coyote and hawk are this environment's top carnivores (Figure 7.9). In desert regions around the world, consumer species have adapted to their environment:

- Kangaroo rats do not need to drink water; they get it from food. They live in burrows during the day to avoid extreme heat. They do not perspire and have highly efficient kidneys that produce very little urine.
- Desert foxes have thick fur on the soles of their feet, protecting them from the hot ground. The light-coloured fur on their bodies reflects sunlight and keeps them cool.

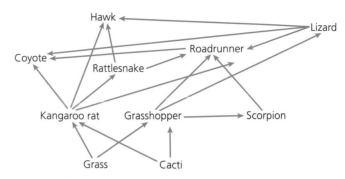

▲ Figure 7.8 A simplified food web for part of the Western Desert

▲ Figure 7.9 Coyote in the Western Desert

What interdependence exists in hot desert environments?

The biotic and abiotic components of an ecosystem are interdependent. Living or 'biotic' creatures play an important role in maintaining a healthy environment (the 'abiotic' parts) and vice versa. Abiotic components of hot deserts include the soil, underlying rocks and water supplies. The interdependency between different biotic and abiotic parts of hot deserts is shown by:

- links between different parts of the food web (animals eating plants that have gained nutrients from soils and water, for example)
- the role that vegetation roots play in stabilising sandy soils in semi-arid areas at the edges of deserts. The plants stop the soil from being blown away by the wind. From page 94, you will learn how the destruction of desert vegetation sometimes triggers a process called **desertification**
- the exchange of nutrients between soil and vegetation, and vice versa. Plants extract nutrients from the soil and return them to the soil when they die. Ultimately, the health of both soil and vegetation is dependent on each other.

→ Activities

1 State four ways in which hot desert animals are adapted to survive in an extreme environment. Mention both appearance and behaviour.
2 Using Figure 7.8 and your own understanding, explain the difference between primary consumers and secondary consumers in an ecosystem (see Chapter 5, page 61 to remind you).
3 Suggest why only a relatively small number of mammal species are supported by the hot desert environment (see Chapter 5, page 60 to remind you).
4 Outline one reason why a study of desert biodiversity, made immediately after rainfall, might have different findings compared with a study carried out during a long dry period.

Geographical skills

Quantitative skills

Study Figure 7.5.

1 Calculate the total biodiversity of the hot desert environment.
2 Calculate the percentage of all species that are mammals.

⭐ KEY LEARNING

➤ Migration to the USA's Western Desert
➤ Economic development opportunities of the Western Desert

Development opportunities in the USA's Western Desert

The USA's Western Desert region is actually made up of three different hot deserts. These are the Mojave Desert, part of the Sonoran Desert and part of the Chihuahuan Desert. In total, the Western Desert covers 200,000 square kilometres.

How has migration changed the Western Desert?

The Western Desert's indigenous people are made up of many different cultural groups. Arizona is home to the Navajo people, and the Havasupai people live by the Grand Canyon, which is a spiritually significant place for some indigenous peoples. The Sonoran Desert's Cocopah Tribe, also known as the River People, settled by the Colorado River centuries ago. People of European descent arrived and expelled the indigenous people in this region in the 1800s.

The Western Desert includes parts of several states: California, Nevada, Utah, Arizona and New Mexico. Population is low and nearly half live in the large cities of Phoenix and Tucson (both in Arizona), and Las Vegas and Henderson (both in Nevada). The majority of people in these cities are of European descent. In recent years, many people have retired here from cities like New York (retirement **migration**).

What are the Western Desert's development opportunities?

The Las Vegas region is home to around 2 million people, and Phoenix has 4.8 million residents. Urban residents can always find work in retailing and **service industries**. In the less populated, more inhospitable areas of the Western Desert, people earn their living from farming, mineral extraction, energy and tourism industries.

Farming

High temperatures and sunlight are generally favourable for agriculture, provided water can be found for irrigation. Two important sources of irrigation water are:

■ aquifers: large stores of water lie beneath some hot desert regions. Sometimes, a layer of permeable rock lies on top of impermeable rock. Rainwater and groundwater seep into the permeable layer and become trapped. This water can be brought to the surface by digging a well.

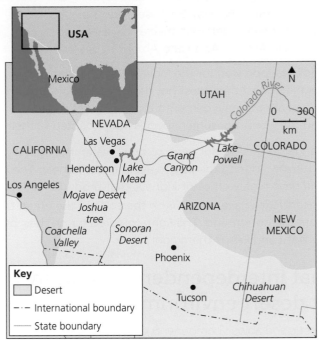

▲ Figure 7.10 The location of the Western Desert, USA

Despite being part of the Sonoran Desert, aquifer-based farming in California's Coachella Valley produces lush crops of vegetables, lemons, peppers and grapes (and in turn, a wine industry).

■ canals: most canals are used for large-scale industrialised agriculture. Farmers are allocated 80 per cent of Colorado water, even though they make up just 10 per cent of the economy.

▲ Figure 7.11 Irrigated agriculture in the Coachella Valley

Mineral extraction

The Western Desert states are rich in minerals, including copper, uranium, lead, zinc and coal. Not all these opportunities have been exploited, due to possible conflicts with other land uses, like tourism and farming. A plan for uranium mining near the Grand Canyon, Arizona was halted in 2012 due to a campaign by the Havasupai people. As uranium is used in **nuclear power** plants, they were concerned about the risk to wildlife and endangered species, and the contamination of water supplies.

Copper mining has taken place for centuries in the Sonoran Desert near Ajo, Arizona. The lack of water discouraged large-scale mining and settlement until underground water was found in an ancient lava flow north of Ajo. Today, opencast mining is carried out on a large scale (Figure 7.12).

▲ Figure 7.12 The Ajo copper mine, Arizona

Energy

The strong insolation in desert regions provides a fantastic opportunity for solar power. The entire Western Desert region is predicted to benefit from the construction of new solar power plants. The Sonoran Solar Project in Arizona is a solar power plant project, with 1.2 million panels, that will ultimately produce energy for 100,000 homes. It is planned to be the size of 330 football pitches and requires 360 workers to help build it.

Hydroelectric power (HEP) plants also supply Western Desert communities with some of their electricity. These are powered by water leaving Lake Mead. During the mid-1930s, 5000 people were employed to build the Hoover Dam.

Fossil fuels bring opportunities to the Western Desert too. People have been drilling for oil in Arizona since 1905. Today, there are 25 active oil production sites, all of which are on land owned by the Navajo people. Since 1998, the Navajo Nation Oil and Gas Company (NNOGC) has exploited this economic opportunity for the benefit of local Navajo people. More than 100 employees work to produce oil worth US$50 million annually. However, land ownership remains disputed and unemployment is high among some Navajo people.

Tourism

As US society has grown to have more money and leisure time, tourism has become the Western Desert's most important source of income:

- The national parks offer visitors a chance to experience a **wilderness area**. Important areas include the Grand Canyon and California's Joshua Tree National Park (named after the dominant plant type).
- The heritage and culture of Native Americans are celebrated at the Colorado Museum in Parker, Arizona.
- The entire economy of Las Vegas is built around entertainment, attracting around 40 million visitors in 2018.
- Two major lakes have been created as part of water management projects: Lake Mead and Lake Powell. Combined, they attract 2 million visitors a year and offer sailing, power boating, water-skiing and fishing.

→ Activities

1 Using Figure 7.10, describe how the Western Desert is distributed across different parts of the USA.

2 Using Figure 7.11 and your own understanding, explain how aquifers create important opportunities for people living in hot deserts.

3 Outline two reasons why mining opportunities in hot deserts are not always exploited.

4 To what extent do the negative environmental impacts of human activities in hot deserts outweigh the economic benefits they create? In your answer, you should:
- refer back to the opportunities you mentioned in 2 and 3
- describe how energy opportunities have benefited different groups of people
- write about tourism and the different people and interest groups who might visit a hot desert
- for each, explain the environmental costs and economic benefits
- weigh up arguments both for and against the statement before arriving at a final overall judgement.

⭐ KEY LEARNING
➤ The uneven development of the Western Desert
➤ Why accessibility is a challenge
➤ How people have adapted to the climate

Development challenges in the Western Desert

What explains the uneven development of the Western Desert?

Adapting to the hot desert environment of the Western Desert is a challenge, for both traditional Native American communities and more recent settlers. Figure 7.13 shows population distribution in the Western Desert. The pattern shown can be linked with variations in temperature (coasts can be cooler), water supply and accessibility.

Temperatures and water supply

There is no settled population in some areas, mainly due to very high temperatures. In the Mojave Desert's Death Valley, temperatures approach 50 °C in July. This is reaching the survival limit for plants. The absence of people in places like Death Valley reflects the low **carrying capacity** of the land (Figure 7.14).

The first Native Americans settled in areas where temperatures were at least tolerable, both for themselves and for their crops or animals. In all three desert areas (Mojave, Sonoran and Chihuahuan), indigenous communities formed near sources of water, either rivers or aquifers. Some groups developed an economy based on subsistence farming based on maize and corn, or alfalfa grown for livestock. Others hunted wild animals which thrive in areas with greater water supply and plant growth, such as rabbits.

Why is accessibility a challenge in the Western Desert?

The low population density of less than one person per square kilometre means that parts of the Western Desert lack surfaced roads. Accessibility is thus severely limited in areas of Nevada north of Las Vegas. Tourists and explorers must find their own way. Even where roads and rough tracks have been provided, the extreme temperatures make this a dangerous place if your car breaks down. In 2015, a elderly tourist died of dehydration in the Los Coyotes Reservation near the edge of the Mojave Desert. He had become lost after attempting to drive off-road. However, accessibility is less of a challenge than it used to be.

■ By the late 1800s, railroad developers moved in. Their choice of sites for stations influenced the growth of future key settlements. For instance, developers determined that the water-rich Las Vegas Valley would be a perfect location for a train station. Soon after, the first saloon bars, shops and hotels were built.

Key
Persons per km²
- <1
- 1–4
- 5–14
- 15–39
- >40

— State boundaries
— County boundaries
● Cities over 1,000,000
● Cities 350,000 to 1,000,000

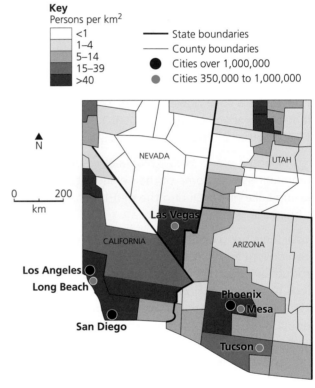

▲ Figure 7.13 Population distribution in the Western Desert

▲ Figure 7.14 Some parts of the Western Desert are uninhabitable, like Death Valley

- Better roads were laid in the 1900s. Soon people were driving through the desert in buses or their own cars (Figure 7.15).
- Major cities can now be reached directly by air. Las Vegas airport receives over 40 million people annually.

How have people adapted to the climate?

Traditional Native American housing was adapted to the extreme climate of hot deserts. Cocopah people lived in earth houses, made with a wooden frame packed with clay and thatched with grass. The thick earth walls kept the house cool in the daytime heat, and warm in the cold of night: perfect for life in a hot desert!

Nineteenth- and early twentieth-century migrants quickly adapted to the climate too. They wore wide-brimmed cowboy hats to prevent sunburn. Before the arrival of air conditioning and improved water supplies, their houses:

- had flat roofs to help collect rainwater
- were small, to reduce sunlight and keep temperatures low inside
- had whitewashed walls to reflect sunlight and keep buildings cool, a tradition which continues today (Figure 7.16).

Recent water shortage concerns have led people to change their behaviour accordingly. Some old sports pitches have been replaced with fake grass. More people have adopted drought-resistant 'desert landscaping' in their gardens.

▲ **Figure 7.15** Route 70 cuts through the Western Desert in Utah. Since 1926, Route 66 has connected Chicago to California via the Western Desert.

▲ **Figure 7.16** A whitewashed building in Taos, New Mexico

→ Activities

1 Using Figure 7.13, describe how population is distributed in the Western Desert.

2 Suggest how population distribution is influenced by:
   a) temperature
   b) water
   c) accessibility.

3 To what extent is it possible to overcome physical development challenges in the Western Desert? In your answer, you should:
   - use Figure 7.16 to explain how this urban landscape shows signs of adaptation to the hot desert environment
   - use Figure 7.15 to write about the risk motorists face crossing the desert and how they should look after themselves
   - suggest reasons why the carrying capacity of the land may vary, making challenges worse in some parts of the desert than other parts
   - weigh up the different arguments before arriving at a final overall judgement.

Case study

⭐ KEY LEARNING
➤ The costs and benefits of irrigating the Western Desert
➤ Future water supply and population growth issues

The Western Desert's water crisis

Until now, cities in the Western Desert have prospered thanks to massive **water transfers**. Vast volumes of water have been transferred from the Colorado River, but there are limits to what can be achieved. Further population growth may not be possible.

What are the costs and benefits of irrigating the Western Desert?

Twentieth-century migrants could see plenty of opportunities in the Western Desert's sunny skies. Farming and tourism would flourish if they could tackle the issue of water shortages. The solution was close at hand: the Colorado River. This massive, 2300 kilometre continental river brings meltwater from the Rockies and Wind River Mountains across the USA and down to Mexico.

On a small scale, for centuries, the Cocopah people had long been drawing Colorado water through canals to irrigate their fields. But while the snowmelt brings huge volumes of water in summer, the Colorado has a very low flow between September and April. In the most extreme years of the early 1900s, the Colorado's **discharge** was 13 times higher in mid-summer compared with winter.

In 1935, work began on the Hoover Dam, which stores the equivalent of two years' river flow in Lake Mead. The Glen Canyon Dam followed in 1963 (Figure 7.17). Together, the two dams and their reservoirs smooth out the Colorado's flow through the year and remove its flood peaks, and bring additional benefits to many other places and communities.

Reservoir water is piped along aqueducts, including the US$4 billion Central Arizona Project (Figure 7.19). Where required, it feeds the homes, farms and golf courses of the Western Desert. This water transfer has brought many benefits, but there are also costs, particularly to the Colorado River's ecosystem.

▲ Figure 7.17 Lake Powell and the Glen Canyon Dam

▼ Figure 7.18 Economic benefits and ecological costs of taking water to the Western Desert

Benefits	Costs
• Colorado's giant reservoirs bring water to cities throughout the Western Desert area, including Phoenix, Tucson, Albuquerque, San Diego, Las Vegas (see Figure 7.21) and Los Angeles. • The Colorado's aqueducts bring life-giving water to farms growing fruits and vegetables in places like Coachella. In total, more than 1.4 million acres of irrigated land throughout the Colorado River Basin produce about 15 per cent of the USA's crops and 13 per cent of its livestock. The total agricultural benefits are calculated to be US $1.5 billion per year.	• Silts and sands get trapped behind both dams. This makes the water that leaves the dam colder (silt heats up in sunlight, warming the water around it). As a result, the river's ecosystem has changed and many species have been lost. • Sandbanks along the sides of the river in its lower course have been starved of **sediment** and are smaller. Plants and animals that live on the sandbanks – such as the Kanab snail and the willow flycatcher – have also declined. The sandbanks that were once used for fishing and rafting have disappeared.

What are the issues around future water supply and population growth in the Western Desert?

Already, 30 million people in the southwest USA depend on water from the Colorado. Phoenix takes the maximum share of water it is allowed, but is predicted to double its population by 2050 (Figure 7.20). Between 2000 and 2020, several states experienced rapid population growth, more than double the national average: Nevada grew by 45 per cent, and Arizona and Utah both grew by around 35 per cent.

Yet while the cities of the Western Desert continue to grow, there is a physical limit to how much water can be taken from the Colorado. There is a political limit too, because of an international agreement which states that water must also be allowed to flow into Mexico (where the Colorado ends).

The region's **water security** is further threatened by climate change. Scientists have suggested that reduced rainfall could occur in places where water is already naturally scarce for part or all of the year. In 2016, Lake Mead reached a new record low level. The Western Desert region is projected to warm faster than the whole world in coming decades. By 2100, average annual temperatures in many areas could be five degrees higher than they were in the 1970s. The combination of a changing climate and the region's rapid population growth means that even greater water scarcity is expected in the future.

▲ Figure 7.19 Part of the Central Arizona Project

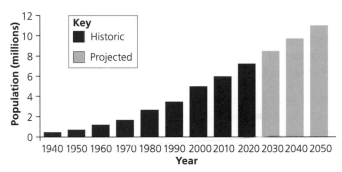

▲ Figure 7.20 Population growth in Arizona

→ Activities

1 State three economic ways in which Western Desert communities benefits from the Colorado River's water.

2 Outline three problems created by the management of the Colorado River.

3 a) State what is meant by 'water security'.
   b) Describe the population changes in Figure 7.20.
   c) Suggest what the implications of Figure 7.20 are for the future water security of people living in the Western Desert region.

4 To what extent can future water supply needs be met in the Western Desert?

In your answer, you should:

■ use Figures 7.17 and 7.19 to explain how technology increases water supply

■ suggest what may happen if new developments such as golf courses are allowed near cities like Las Vegas and Phoenix

■ mention physical geography (the region's climate)

■ weigh up the different arguments before arriving at a final overall judgement.

▲ Figure 7.21 Las Vegas

✪ KEY LEARNING
➤ The characteristics of desert fringe areas
➤ The link between desertification and natural climate change

Desert fringes and desertification

What are the characteristics of desert fringe areas?

Desert fringe areas have many names, including semi-deserts, semi-arid areas and drylands. In some places, rain falls in a fairly predictable pattern, making settled agriculture possible.

At the borders of hot deserts, desert fringe areas support greater biodiversity and larger plants. Grasses grow in higher rainfall desert fringe areas, such as the American Prairies, and there are more drought-resistant trees, such as the baobab tree in Africa's Sahel region, and Australia's eucalyptus tree.

Despite their higher rainfall, desert fringes are classified, alongside hot deserts, as **fragile environments**. Catastrophic ecological and environmental consequences can be caused by climate change or poor land management. As a result, desert fringes are at constant risk of desertification (Figure 7.22).

Causes of desertification

Desertification – the spread of desert conditions into semi-desert and grassland areas – is a major problem in many parts of the world, as Figure 7.23 shows. Around 1 billion people, or 15 per cent of the world's population, either experience or are threatened by habitable desert fringe areas turning into hot desert areas. This includes desert fringes in Australia, China, the USA and large parts of Africa.

The Sahel, including countries such as Burkina Faso, Niger and Chad (see page 84), is where the human risks created by desertification are greatest. This is because more than 100 million very poor and vulnerable people live there. There is little money and technology available in Sahel countries to help people adapt to the challenge.

Climate data for the Sahel suggest that desertification may have a physical cause: a long-term reduction in rainfall has taken place (see Figure 7.24). The most likely explanation for this is a natural rainfall cycle. The African continent has a long history of rainfall fluctuations of varying lengths, including cycles lasting decades.

▼ Figure 7.22 A sandstorm approaches the desert fringe in Niger

Long-term climate change

We do not know yet whether global warming caused by humans will create even greater rainfall deficiencies in the Sahel or other desert fringes. While many scientists are certain that global temperatures will rise this century, they are less confident when it comes to predicting how rainfall patterns will change. Some think there is a possibility of rain returning to the Sahel due to global warming, as the heating of oceans adds more water vapour to the atmosphere. This could, in turn, bring more rainfall that leads to the 'greening' of the Sahel. Equally, it may be that the climate becomes even drier, leading to the spread of sand dune systems across valuable crop land.

The climatic system is complex and there remains much uncertainty over what will happen in future.

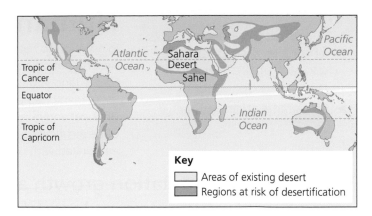

▲ Figure 7.23 World regions at risk of desertification

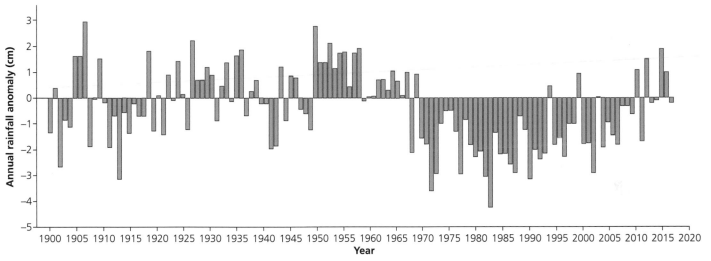

▲ Figure 7.24 Annual rainfall in Sahel countries, 1900–2017. Each bar shows how much the year's rainfall was above or below the long term average

→ Activities

1 Outline one physical and one human difference between desert and semi-desert areas. Physical characteristics include climate, vegetation and water supply. Human characteristics include population density and economic activities.

2 Using Figure 7.23, describe the global distribution of areas at risk of desertification. Refer to lines of latitude, names of continents and proximity (nearness) to desert areas.

3 a) State what is meant by a 'temperature anomaly'.

   b) Using Figure 7.24 and your own knowledge, suggest how and why the climate of the Sahel region is changing. First, suggest how the climate is changing (is it getting hotter and drier or wetter?); second, suggest why any changes are occurring.

⭐ KEY LEARNING

➤ The role of population growth and other local factors in desertification

➤ Desertification in Darfur

Human causes of desertification

In addition to natural pressures, desertification occurs when fragile land in desert fringe areas is overexploited by humans. Physical and human causes of desertification are often interlinked with one another.

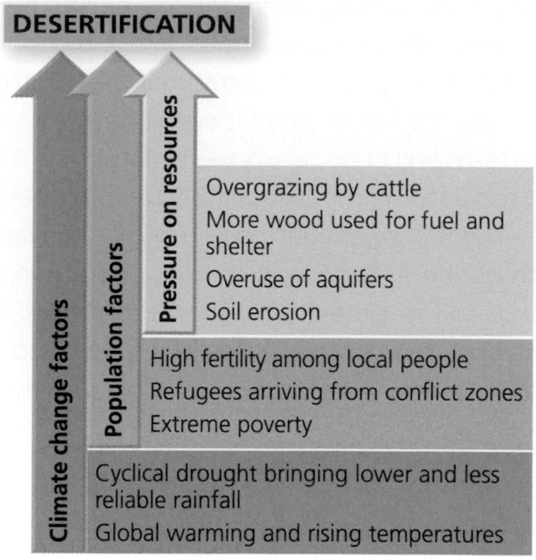

DESERTIFICATION

Pressure on resources
- Overgrazing by cattle
- More wood used for fuel and shelter
- Overuse of aquifers
- Soil erosion

Population factors
- High fertility among local people
- Refugees arriving from conflict zones
- Extreme poverty

Climate change factors
- Cyclical drought bringing lower and less reliable rainfall
- Global warming and rising temperatures

▲ Figure 7.26 The causes of desertification

What role do population growth and other local factors play in desertification?

The global issue of climate change plays a role in desertification (see page 94). Another contributing factor is that of local pressures, such as: population growth, removal of fuel wood, **overgrazing**, **over-cultivation** and soil erosion (see Figure 7.25). Figure 7.26 shows how these local pressures are often interlinked with global climate change, making desertification a particularly serious and difficult problem to solve.

▼ Figure 7.25 How local pressures lead to desertification

Population growth	There were just 30 million people living in the Sahel in 1950. Today, the figure is closer to half a billion, and by 2050, it is expected to reach one billion. This growth is due to: • High fertility and people living longer than they used to (Figure 7.27). • Migration – drought and desertification in one region will often displace people to another fragile environment. As the number of people increases, desertification can happen in the new environment too, so the problem gets spread from place to place. In addition to 'climate change refugees', millions of people have been forced to move into desert fringe areas by armed conflicts.
Removal of fuel wood	Wood and vegetation are still used for fuel in many countries: • Wood is a source of fuel for cooking and heating that many rural people depend on because no other energy supply is available. Even if the necessary **infrastructure** were provided, people may lack the money to pay for electricity or gas. • Fuel wood can be removed unsustainably – large areas are stripped of vegetation, leaving the soil unprotected and prone to wind or water erosion (when it rains), leading to desertification.
Over-cultivation	Over-cropping land can exhaust the soil's fertility: • As health has improved over time, more of the children born into farming families are surviving infancy, which leads to a rise in small-scale subsistence agriculture. More crops are planted and some aquifers have been drained dry. • Commercial farming worsens this situation. For example, some European companies use large areas of fragile land in Ghana to grow water-hungry cash crops such as jatropha (to make vegetable oil).
Overgrazing	If too many goats and cattle are grazed for too long on one site, all the vegetation is eaten and may be unable to regrow: • Nomadic (migratory) groups used to wander freely, following the rain wherever it fell. They would give vegetation a chance to recover. Now they cannot, due to new political boundaries drawn by European rulers or because large western companies have bought up land rights. • Civil war and political instability also force herders to stay too long in fragile places.
Soil erosion	Fuel wood removal, over-cultivation and overgrazing all lead in turn to soil erosion: • If vegetation has been burned, eaten by cattle or killed by drought, the exposed topsoil becomes baked hard by sunlight. • When it finally arrives, intense rain washes over the soil rather than soaking into the ground, and carries the topsoil away. Once the soil has eroded, it becomes impossible for the vegetation to grow back. A 'tipping point' is reached.

Why has desertification become a problem in Darfur?

You may have heard of Sudan's Darfur region, where 250,000 people have been killed since 2003 and an estimated 2.5 million were homeless in 2018. The environment was already under pressure before 2003:

- One year out of every five brings drought, crop failure and livestock loss to Sudan.
- In Sudan and neighbouring countries, food production has not kept pace with the growing population (Figure 7.28).

In 2003, nomadic cattle herders and settled farmers began to fight over water supplies and land. Herders were deliberately prevented from reaching water sources by farmers, which led to **overgrazing**. Once the vegetation was gone, their cattle died. In revenge, some herders chased farmers from their villages, and cut down their crops and trees.

▼ Figure 7.27 Population data for selected Sahel nations, 2018

Country	Pop. (millions)	Fertility rate (children per woman)	Pop. growth (% per year)	Pop. density (people per km²)
Burkina Faso	20.3	5.4	2.9	75
Chad	15.9	5.9	3.0	12
Mali	19.6	6.1	3.0	15
Niger	23.3	7.2	3.9	18
Senegal	16.3	4.8	2.8	80
Sudan	42.8	4.5	2.4	22

Millions of people fled their land and homes. They were housed in refugee camps, with help from the UN. But refugee camps create new environmental stress wherever they are located and cause desertification to spread (Figure 7.29). When people finally return home, further desertification is expected. One charity estimates that each returning family will need 30–40 trees to rebuild their houses and fences. The desert fringes of Sudan do not have enough trees left to support this.

▼ Figure 7.28 Population and land use trends in Darfur prior to conflict. In 2018, the population had grown to 9 million.

	1973	2003
Population	1.3 million	6.5 million
Type of land use	Percentage in 1970s	Percentage in 2000s
Bush and shrub	23.8	17.5
Rain-fed agriculture	22.7	34.4
Wooded grassland	11.8	7.1
Closed forest (natural)	10.7	7.9
Grazing/pastures	9.0	6.8

▲ Figure 7.29 Refugee camps greatly increase the population density of some fragile desert fringes

Geographical skills

Developing data analysis skills

Figure 7.27 is a large and complex table which can be used to help develop your data analysis skills

1. For each column of the table:
   a) describe the spread of data (say if the numbers vary a lot or a little)
   b) identify the maximum and minimum value in each case
   c) subtract the minimum from the maximum in order to calculate the data range.
2. Look for patterns which occur in the rows of the table (horizontally). For example, is there a country which is highest- or lowest-scoring in several categories?

→ Activities

1. Using Figure 7.28, describe how the population and environment in Darfur changed from 1973 to 2003.
2. Explain how the problems of overgrazing and soil erosion are linked.
3. To what extent does desertification occur mainly as a result of human factors? In your answer, you should:
   - explain how the population changes in Figure 7.27 may lead to negative environmental changes
   - explain possible physical causes of desertification in fragile desert fringe areas (unconnected to human activity)
   - weigh up the different arguments before arriving at a final overall judgement.

KEY LEARNING

- How better land management can help combat desertification
- How planting trees can help
- How appropriate technology can help

Tackling desertification

How can better land management help combat desertification?

The majority of people who live in the Sahel region still suffer from poverty. Niger is losing 250,000 hectares of farmland every year through desertification. Millet crops have failed and sand dunes are advancing. Women in some villages now walk as far as 25 kilometres a day to fetch water for their families.

However, a range of land management measures can help to preserve soil quality and water supplies, such as:

- tree-planting schemes to bind and protect the soil
- planting grass on slopes to help stabilise the topsoil
- building small rock dams to trap rainwater in gullies
- collecting rainwater on roofs by designing a flat roof with a surrounding lip
- building terraces (flattened sections with a retaining wall) on farmed slopes.

One successful strategy, introduced in the Sahel countries of Mali and Burkina Faso, is the construction of low stone walls called bunds (Figure 7.30). The stones are placed in lines across the slope gradient. They help to prevent soil erosion and slow down the flow of rainwater over the baked ground. When water pools behind a bund, instead of running fast over the land, it has time to soak into the ground.

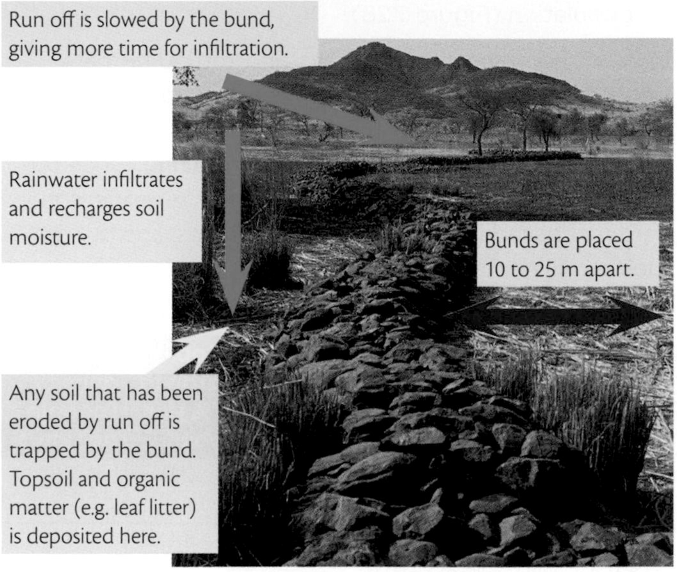

Run off is slowed by the bund, giving more time for infiltration.
Rainwater infiltrates and recharges soil moisture.
Bunds are placed 10 to 25 m apart.
Any soil that has been eroded by run off is trapped by the bund. Topsoil and organic matter (e.g. leaf litter) is deposited here.

Figure 7.30 How bunds work

How can planting trees help to stop desertification?

Tree roots help to stabilise soil, while their decomposing leaf litter adds valuable nutrients (see Section 5.1). This makes the planting of trees a practical way to tackle desertification. The African Union's proposed 'Green Wall' is a plan to plant a wall of trees across the entire Sahel region, running from the Atlantic Ocean in the west, all the way to the Indian Ocean in the east (Figure 7.31). It will be decades before the Green Wall reaches maturity, but it offers hope for **sustainable development** among communities. In addition to making the physical environment more secure, the project will also generate work for desperately poor communities in all the Sahel countries. Finally, it will help bring about political

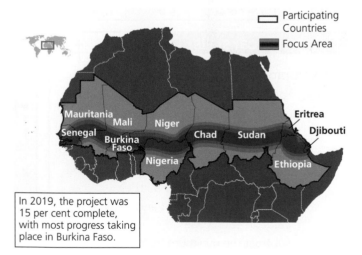

Participating Countries
Focus Area
Mauritania
Mali
Niger
Senegal
Burkina Faso
Chad
Sudan
Eritrea
Djibouti
Nigeria
Ethiopia
In 2019, the project was 15 per cent complete, with most progress taking place in Burkina Faso.

Figure 7.31 The proposed Great Green Wall, with a total distance of 7,775 km and a total area of 11.6 million hectares

co-operation in the region. This might reduce conflict and the number of refugee camps, which, unfortunately, contribute to desertification. However, climate change projections suggest increased aridity may threaten the survival of the trees in the long term.

How can appropriate technology be used to change the way people cook?

Removing trees for firewood is one of the most damaging human activities in desert fringe areas like the Sahel. For millions of years, people have been using wood as a cooking fuel. Yet population growth in the Sahel has meant that wood is a vanishing resource. When trees are cut down in large numbers, the resulting soil erosion effectively prohibits any future regrowth. This is unsustainable land management, but people in rural areas have no access to gas or electricity infrastructure.

Recently, however, an alternative way of cooking has begun to be adopted, using appropriate technology called 'efficient stoves'. One example is the Toyola stove in Ghana (see Figure 7.32), and another is the Upesi stove in Kenya. The stove designs are being distributed to rural desert fringe communities by charities like Practical Action and the Global Alliance for Clean Cookstoves. The key to their success is that the stoves can be made locally using more available materials like clay, and burn much smaller amounts of wood and charcoal.

Some stove designs also incorporate a thermocouple, which generates sufficient electricity from the heat to charge a mobile phone – which growing numbers of Sahel farmers own. In turn, mobile phone access is helping farmers to gain access to weather forecasts, which can help them prepare for drought or rain.

Another important energy development for desert fringe areas is the move towards solar power (see page 52). As well as providing energy for cooking and other needs, earnings from solar power could provide Sahel nations with money to tackle desertification even more effectively. Hot deserts and their fringes may eventually be seen as the world's most resource-rich places.

▲ Figure 7.32 Cooking on a stove that uses far less fuel than a traditional fire in Tanzania

→ Activities

1. a) Using Figure 7.30, describe the characteristics of a bund.
   b) Using Figure 7.30 and your own understanding, suggest how bunds can help tackle the problem of desertification.

2. Use Figure 7.31 to help you assess possible challenges for the African Union as it carries out the Great Green Wall project. Think about different physical and human challenges. Suggest possible solutions for the problems that may develop.

3. a) State what is meant by appropriate technology.
   b) To what extent can the use of technology help to protect the environment in desert fringe areas? In your answer, you should:
   - explain how the process of desertification, once it has begun, can become an irreversible problem
   - outline how the use of more energy efficient technologies, such as efficient stoves, can mean there is less vegetation removal and desertification
   - think about the many pressures which other human activities put on the environment in desert areas
   - weigh up the different arguments before arriving at a final overall judgement.

8 Cold environments

⭐ KEY LEARNING

➤ The characteristics of polar and tundra climates

➤ How permafrost affects cold environments

Polar and tundra environments

Covering one quarter of the Earth's land surface, the world's cold environments are high-latitude world regions where cold, sinking air generates freezing winds and sunlight is weak. At the highest latitudes, the Sun does not even rise for several months of the year. Few people want to live in such extreme conditions.

Where are polar and tundra environments located?

There are two types of cold climate environment: **polar** environments and **tundra** environments.

Polar environments

- These are found in inland areas, far from the warming influence of the sea. They include Greenland, Northern Canada, Northern Russia (Siberia) and Antarctica.
- Most polar regions are partly or completely covered with ice caps (see Figure 8.1).
- The average monthly temperature is always below freezing (see Figure 8.2). This allows snow and ice to accumulate over time.

Tundra environments

- These are found south of the ice caps in the northern hemisphere.

- They occupy around one-fifth of the Earth's land surface, including enormous areas of Russia and Canada. These places lack permanent ice cover, but experience very cold weather for most of the year.
- Most of the ground is permanently frozen. The treeless tundra ecosystem is composed of low-lying shrubs and mosses.

What are the characteristics of polar and tundra climates?

Polar climates

Figure 8.2 shows the annual climate graph for Antarctica's McMurdo research station. You can also see the **temperature range** for the summer months. In some years, daytime temperatures very occasionally reach 10 °C – but the average temperature usually remains below freezing in all polar climates.

Precipitation in a polar climate falls mostly as snow. Overall, there is very little precipitation, the same as in **hot deserts**. This is because cold air cannot hold much water vapour.

Tundra climates

Figure 8.3 shows the climate graph for the town of Barrow, Alaska. It has a short, cool summer, although the Sun does shine 24 hours a day! In December, however, there is complete darkness. In some years, temperatures have fallen below −40 °C. The tundra climate is very harsh in winter. In the coldest tundra environments, the average monthly temperature nudges above freezing for just one month of the year. The total annual amount of precipitation is low, due to the cold air temperature.

▲ Figure 8.1 Ice caps in a polar environment

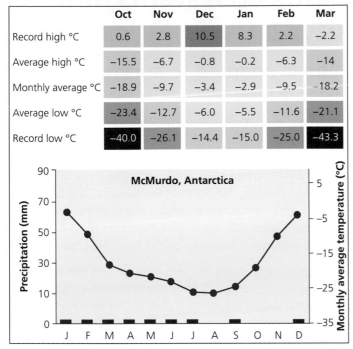

	Oct	Nov	Dec	Jan	Feb	Mar
Record high °C	0.6	2.8	10.5	8.3	2.2	-2.2
Average high °C	-15.5	-6.7	-0.8	-0.2	-6.3	-14
Monthly average °C	-18.9	-9.7	-3.4	-2.9	-9.5	-18.2
Average low °C	-23.4	-12.7	-6.0	-5.5	-11.6	-21.1
Record low °C	-40.0	-26.1	-14.4	-15.0	-25.0	-43.3

▲ Figure 8.2 Climate graph and monthly temperatures for McMurdo, Antarctica (a southern hemisphere, polar climate)

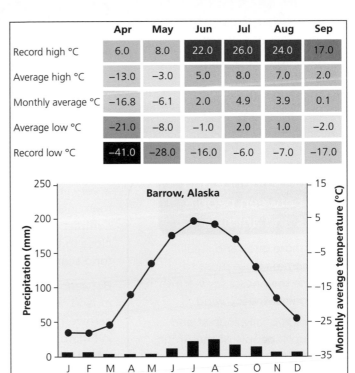

	Apr	May	Jun	Jul	Aug	Sep
Record high °C	6.0	8.0	22.0	26.0	24.0	17.0
Average high °C	-13.0	-3.0	5.0	8.0	7.0	2.0
Monthly average °C	-16.8	-6.1	2.0	4.9	3.9	0.1
Average low °C	-21.0	-8.0	-1.0	2.0	1.0	-2.0
Record low °C	-41.0	-28.0	-16.0	-6.0	-7.0	-17.0

▲ Figure 8.3 Climate graph and monthly temperatures for Barrow, Alaska (a northern hemisphere, tundra climate)

Figure 8.3 shows that daytime temperatures in Barrow occasionally rise above 20 °C in the summer months. However, the monthly average temperature is always much lower, due to colder temperatures at night.

Permafrost

In cold climates, most of the ground is permanently frozen in a state of **permafrost**. Around one-quarter of the Earth's surface is affected by permafrost, including tundra, polar and some mountain regions. In tundra regions, ice in the uppermost **active layer** of the soil thaws for one or two months of the year during the brief summer, but there is still ice below the active layer. The ice acts as an impermeable barrier to the downward movement of melted water in the soil, resulting in waterlogged conditions.

→ Activities

1 Outline the main landscape and climatic characteristics of polar and tundra environments.
2 Using Figures 8.2 and 8.3 and your own understanding, explain one reason why annual temperature patterns differ for Antarctica and Alaska.

Geographical skills

Manipulating data sets

Study Figure 8.3.

1 Calculate the total annual precipitation (use the blue bars).
2 Calculate the mean monthly precipitation.
3 Describe the annual pattern of precipitation.
4 Calculate the annual temperature range (use the red line).

Cold climate ecosystems and biodiversity

How have animals adapted to survive in cold environments?

Many different animal species live in cold climate regions. The polar bear is perhaps the most well known. Much of the polar bear's time is spent hunting for seals along the northern edges of the Arctic Ocean, in places where there is very little plant life (Figure 8.4). When seals are scarce, the polar bears roam inland into the tundra, in the very far north of Russia and Canada. In contrast, brown bears graze and hunt at the southern edges of the tundra, close to the forest boundary (Figure 8.6). To escape freezing conditions and food scarcity in winter, they hibernate, insulated by fat from the cold.

Between the northern and southern edges of the tundra lie vast expanses of land, where the Arctic fox and tundra wolf are the environment's top carnivores (Figure 8.5). A simplified food web for part of this region is shown in Figure 8.7. The consumer species are adapted to their environment in ways which aid survival:

■ Snowshoe rabbits have white fur. This means they cannot be seen easily against the winter snow.
■ Caribou and musk ox have two layers of fur to help them survive the bitter cold. They also have large hooves to help them travel over soggy ground and break through ice to find drinking water during the winter months.

Some species, like the musk ox, are permanent tundra residents. Many birds and mammals are not, however. They use the tundra as their summer home and migrate elsewhere in winter. Biodiversity is relatively low overall (Figure 8.9).

Interdependence and links between people, plants and animals

■ Tundra birds and small mammals use moss to line their nests for warmth against the icy wind.
■ Traditionally, Inupiat and Yup'ik people of the Arctic Circle depended on animal skin and feathers for their clothing (see page 104).
■ Historically, indigenous people in coastal areas have depended on marine species (including fish, sharks and whales) for food and other uses (see page 104).

▲ Figure 8.4 The northern edge of the tundra, where polar environments take over

▲ Figure 8.5 An Alaskan tundra wolf

▲ Figure 8.6 Looking south from the edge of the tundra towards the treeline

How are plants adapted to tundra environments?

At the base of the tundra food web are many producer species (Figure 8.7). These include low-lying shrubs, mosses and lichens. Tundra plants are adapted in a range of ingenious ways to help them overcome environment challenges (Figure 8.8). In turn, the biotic (plants) and abiotic (soils) components of ecosystems have become interdependent. Low-lying vegetation helps to protect the soil from wind erosion, for instance.

▼ Figure 8.7 A simplified tundra food web

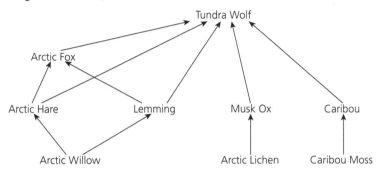

▼ Figure 8.8 Tundra plant adaptations

Permafrost	● The permafrost chills the ground and is a barrier to root growth. Only plants with shallow root systems can survive, including mosses, lichens, some flowering plants and low-growing shrubs. Trees, which rely on deep roots for stability in the wind, cannot survive here (the word 'tundra' means 'treeless' in Finnish).
Poor drainage	● In the flat, low-lying tundra regions, summer melting of the active layer leaves many areas waterlogged. This favours hardy organisms like mosses that can tolerate both extremely dry and wet conditions. This allows them to cope with the seasonal change from very dry to wet conditions, as the active layer melts.
Low insolation	● Due to the high latitude, light is weak even with long summer days. Snow covers plants for many months of the year. They have therefore adapted in ways that maximise photosynthesis during the short growing season. The snow buttercup and Arctic poppy have adapted to the short growing season by producing flowers very quickly, while the snow is still melting. They have cup-shaped flowers that face up towards the Sun so the Sun's weak rays are directed towards the centre.
Strong wind	● High air pressure over the North Pole generates strong, cold winds that blast tiny particles of ice southwards over the tundra. Plants have adapted in ways that keep them warm and minimise transpiration loss to the wind. They grow close together and near ground level (few species reach 40 centimetres in height). This allows plants to trap pockets of warmer air. Their leaves are small and fringed with tiny hairs to capture heat.

▼ Figure 8.9 Tundra biodiversity

Group	Number of species
Mammals	75
Birds	240
Insects	3,300
Flowering plants and shrubs	1,700
Mosses	600
Lichens	2,000

→ Activities

1 State three ways in which polar and tundra animals are adapted to survive in an extreme environment. Mention both appearance and behaviour.

2 Using Figure 8.7 and your own understanding, explain the difference between primary consumers and secondary consumers in an ecosystem. (See Chapter 5, page 61 to support your answer.)

3 Using Figure 8.8 and your own understanding, suggest why only a relatively small number of mammal species are supported by the tundra environment. (See Chapter 5, page 61 to support your answer.)

Geographical skills

Quantitative skills

Study Figure 8.8.

1 Calculate the total biodiversity of the tundra environment.

2 Calculate the percentage of all species that are mammals.

⊘ KEY LEARNING

➤ Migration to Alaska
➤ The economic development opportunities of Alaska's environment

Alaska's development opportunities

How have migrants contributed to the development of Alaskan environments?

Covering nearly 2 million square kilometres, the US state of Alaska borders Canada and the Arctic Ocean (see Figure 8.10). Like many of the Earth's coldest regions, Alaska has been settled for thousands of years, despite challenging environmental conditions.

- Alaska's indigenous peoples include the Aleut, Yupik and Inuit (Inupiat).

- During the last Ice Age, Arctic indigenous peoples spread widely throughout the Arctic Circle into Canada, Alaska, Russia and Scandinavia. Back then, the USA and Russia were joined by a land bridge.

- Since the 1800s, European settlers have come to Alaska, taking over indigenous territory, changing the indigenous peoples' way of life and increasing the population of Alaska from 100,000 to almost 750,000.

- Alaska is also home to **economic migrants** travelling north temporarily to work for oil and mining companies.

▲ Figure 8.10 The location of Alaska, the 49th state of the USA

What are Alaska's development opportunities?

Given its great size, Alaska is one of the most **sparsely populated** places in the world. This reflects the lack of economic opportunities in this extreme environment. Nearly half of the residents live in the city of Anchorage. Some urban residents work in retailing or education, health and government services. Others, along with much of Alaska's rural population, earn a living from fishing, mineral extraction, tourism or energy industries.

The fishing industry

The 3,000 rivers, 3 million lakes and 10,686 kilometres of Alaskan coastline provide many economic opportunities linked with fishing.

There are two main sectors of the industry:

- *Commercial fishing.* Since the 1870s, the sector has grown to employ one in ten Alaskans. Some of the biggest salmon, crab, and whitefish fisheries in the world are in Alaska. They provide around 80,000 jobs and add as much as US$6 billion to the state economy annually. Some jobs are only seasonal, however.

- *Subsistence fishing.* Native American communities remain dependent on fish for several uses. Fish provide food, oil (for fuel) and bones (to help make clothing and tools).

Alaska's fisheries are widely viewed as a successful example of sustainable management.

▼ Figure 8.11 Fishing opportunities in different Alaskan regions

Arctic-Yukon	Subsistence fishing: many Inuit fishing villages rely on salmon and herring for their survival.
Central	The world's most important commercial sockeye salmon fisheries (Bristol Bay and Copper River). Inuit communities depend on shrimp and scallop fishing in Prince William Sound.
Southeast	Commercial fishing: species include salmon, herring and red king crab.
Westward	The largest crab and Pacific cod fisheries in the state.

Mineral extraction

In the late 1800s, Alaska was known as 'the gold rush state'. Today, one-fifth of the state's mining wealth still comes from gold (although silver, zinc and lead mining are also very important).

Large gold mines must be managed carefully to minimise **environmental impacts**. Humans and ecosystems can be harmed by the toxic chemicals used to process gold ore (such as mercury, cyanide and nitric acid).

Mining development has sometimes been halted due to environmental campaigns. In 2013, the Pebble Mine gold project was closed down. It would have been North America's largest open-pit operation. Native American communities ran an effective 'No Dirty Gold' campaign. Fifty businesses supported the campaign by saying they would not buy Pebble Mine gold (see Figure 8.12). Anglo American, one of the world's biggest mining companies, walked away from a half-billion dollar investment due to the scale of opposition. In 2019, however, US environmental law-makers stated they would not object to any future development. Pebble Mine may yet be exploited.

Tourism

Tourism attracts between 1 and 2 million visitors each summer making the sector one of Alaska's biggest employers, although some work is seasonal and poorly paid. Fishing, whale watching and kayaking are all popular. Around 60 per cent of summer visitors are cruise ship passengers.

Hiking, skiing, rock climbing and sightseeing by helicopter are also available. The state has numerous national parks, preserves, refuges and monuments. There are historical sites for those interested in the Inupiat and Yup'ik heritage. These tourists arrive mostly by air.

Energy

Energy production is another big employer, especially the oil industry (see page 109).

■ More than 50 hydroelectric power (HEP) plants supply Alaskan communities with one-fifth of their electricity. Previously glaciated U-shaped valleys in Alaska are a perfect site for HEP generation.
■ **Geothermal energy** is also being harnessed in tectonically active parts of the state. Alaska's coastline is part of the Pacific 'Ring of Fire' (see page 5). A tourist resort at Chena Hot Springs near Fairbanks is now powered entirely by geothermal power.

▲ Figure 8.12 The Pebble Mine generated opposition from the Bristol Bay fishing industries and communities

→ Activities

1   Using Figure 8.10, describe the location of Alaska in relation to the rest of the USA.
2   Using Figure 8.11 and your own understanding, explain how fishing creates important opportunities for people in Alaskan regions.
3   Outline one reason why mining opportunities in Alaska are not always exploited.
4   To what extent do the negative environmental impacts of human activities in Alaska outweigh the economic benefits they create? In your answer, you should:

■   refer to your answers for Activities 2 and 3
■   describe the benefits of energy opportunities
■   write about tourism and the different kinds of people and interest groups who might visit Alaska
■   explain possible environmental costs and economic benefits of the activities
■   weigh up the different arguments before arriving at a final overall judgement.

> ⭐ KEY LEARNING
> ➤ Extreme temperature challenges and development in Alaska
> ➤ Why accessibility is a challenge in Alaska
> ➤ Protecting buildings and infrastructure

Alaska's development challenges

How have extreme temperatures affected the development of Alaska?

Adapting to the cold tundra environment of Alaska has always been a development challenge, for both traditional Inuit communities and more recent settlers. There is virtually no settled population in the northern interior (see Figure 8.13), due to perilously low temperatures and months without sunlight. The absence of people reflects the low **carrying capacity** of the land there: frozen ground due to the presence of permafrost (Figure 8.14) and the short **thermal growing season** rules out crop production. Instead, the first indigenous people who settled along Alaska's coastal margins developed a subsistence economy based on fishing and seal hunting.

Life was still very challenging for the first settlers. Traditionally, Inupiat and Yup'ik people coped with extreme cold temperatures by making coats from caribou skin and sealskin boots. Goose down was used as a lining. Over time, however, they have increasingly adopted the use of modern textiles like Gore-tex.

Why is accessibility a challenge in Alaska?

The low average population density of less than one person per square kilometre means that most of Alaska lacks surfaced roads. Hunters, miners and explorers must make their own way across the tundra. Even where roads and rough tracks have been provided, physical processes make their use difficult and dangerous.

- Snow and ice make some roads and tracks unusable for months of the year.
- A process called **solifluction** takes place in summer. On slopes, the soil's active layer starts to flow downhill. The thawed soil slides easily over the impermeable frozen layer below. Large amounts of soil and mud can collect at the base of slopes, covering highways that run along valley floors, cutting places off for months.

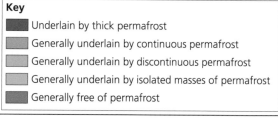

Key
- ⬛ Underlain by thick permafrost
- ⬜ Generally underlain by continuous permafrost
- ⬜ Generally underlain by discontinuous permafrost
- ⬜ Generally underlain by isolated masses of permafrost
- ⬜ Generally free of permafrost

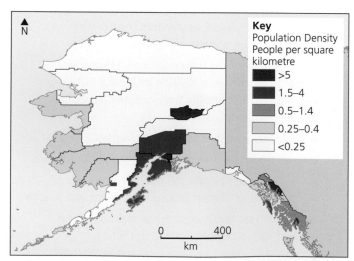

Key
Population Density
People per square kilometre
- ⬛ >5
- ⬛ 1.5–4
- ⬛ 0.5–1.4
- ⬜ 0.25–0.4
- ⬜ <0.25

▲ Figure 8.13 Variations in population density in Alaska show how this is an unevenly developed place

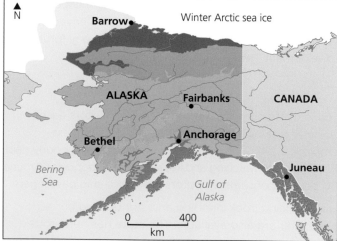

▲ Figure 8.14 The distribution of permafrost caused by cold temperatures helps to explain Alaska's uneven development

- Permafrost underlies most of Alaska (Figure 8.14). The seasonal thawing of the active layer means that off-road travel cannot take place during summer.
- Over time, the seasonal thawing and re-freezing of the active layer results in great expanses of uneven ground surface called **thermokarst** (Figure 8.15) making travel impossible in some places.
- Frost heave – where pebbles and stones slowly rise upwards to the ground – can make tracks dangerous.

What can be done to protect buildings and infrastructure?

Indigenous people and newcomers alike use high-pitched steep roofs for their homes so snow can slide off. Triple-glazed windows help to keep the cold at bay.

The active layer thawing causes the most serious challenge for building. The heat that buildings and settlements create – known as the 'urban heat island' effect – can make this worse. Many buildings erected by early European settlers in the 1880s soon became unusable (Figure 8.16). Escaping heat from the underside of properties led to the thawing of the frozen icy ground beneath. As ice loses volume when it turns to water, the land under many homes subsided.

Millions of kilometres of permafrost have been damaged due to insufficient attention being paid to the sensitivity of soil conditions in Alaska. Today, new buildings are always raised on piles to prevent thawing. These piles can lift a structure several metres above the surface and are sunk deep into the land, well below the lower limit of the active layer. Increasingly, telescopic piles are used. These expand or contract if there is any residual ground movement. The same principles are applied to the vital **infrastructure** that connects places together.

- Roads are now built on gravel pads one to two metres deep that stop heat transfer from taking place.
- Utilities such as water, sewerage and gas cannot be buried underground or they would freeze too. Instead, they are carried by utility corridors or 'utilidors'.
- Airport runways are painted white to reflect sunlight and stop them from warming up too much on sunny days.

▲ **Figure 8.15** A thermokarst land surface in Alaska

▲ **Figure 8.16** House destroyed by permafrost thawing, North America

→ Activities

1 Using Figure 8.13, describe the population distribution of Alaska. Refer to data in Figure 8.13 to support your answer.

2 Explain how solifluction, permafrost thawing and frost heave all create development challenges in Alaska.

3 Look at Figure 8.14. Suggest why some parts of Alaska remain free of permafrost.

4 Suggest why climate change could mean that measures taken to protect roads and buildings in Alaska may prove to be insufficient in coming years.

⊛ KEY LEARNING

➤ The benefits and costs of Alaska's onshore oil fields
➤ The issues surrounding future development of offshore oil fields

The challenges and opportunities of Alaskan oil

Oil and gas contribute a large part of the state's annual earnings, although the amount earned in 2018 (US$3 billion) was much lower than in 2012 (US$10 billion) due to falling oil prices. The industry provides 100,000 jobs, employing one in seven Alaskans.

What are the benefits and costs of Alaska's onshore oil fields?

In 1968, vast onshore oilfields were discovered near Alaska's north coast. Oil production began at Prudhoe Bay in 1977. In the early days, almost 2 million barrels a day were produced. The 800-kilometre trans-Alaskan oil pipeline was built to transport the oil to the southern coast port of Valdez (see Figure 8.17). This was made necessary by the challenge of ice in the northern seas, which meant that oil tankers could not be used.

The pipeline took five years and US$8 billion to build. This was seen as a price worth paying in the 1970s. Rising oil prices and political problems in the Middle East had left the USA desperate to improve its own **energy security**.

Overcoming challenges

Engineers were careful to modify the pipeline to help overcome Alaska's environmental challenges:

■ The pipeline was raised off the ground on stilts (they are eleven metres deep and cost US$3,000 each to build back in the 1970s).
■ Pipeline suspension bridges were used to cross the state's major rivers, including the 700-metre wide Yukon.
■ The pipeline zigzags in some places (see Figure 8.18). This means that it is flexible and can adjust to ground movement from **earthquakes** (which are a risk in this part of the world).

Over time, Prudhoe Bay's oil production has declined. It peaked in the 1980s, when Alaska produced a quarter of all US oil. Today, there are fierce ongoing political battles over the potential costs and benefits of drilling for new oil in neighbouring areas (Figure 8.19). Between 6 billion and 16 billion barrels of oil lie beneath the 80,000 square kilometre Arctic National Wildlife Refuge (ANWR). But this area is home to rare animals such as polar bears, wolverines

and snow geese. In 2005, the US Senate voted to block a proposal to begin drilling for oil there. However, after his election in 2016, President Trump announced he would be happy to allow drilling. The ANWR therefore faces an uncertain future.

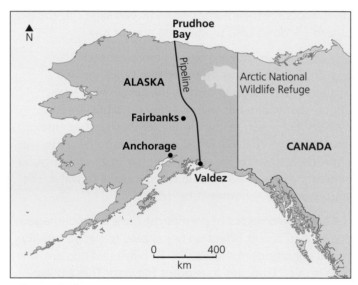

▲ Figure 8.17 The trans-Alaskan oil pipeline and Arctic National Wildlife Refuge

▲ Figure 8.18 A zig-zagging Alaskan oil pipeline. This design increases resilience against earthquakes

▼ Figure 8.19 Costs and benefits of oil production in Alaska

Benefits	Costs
Many working Alaskans rely on the oil and gas industries for their income. More than 90 per cent of taxes raised by the Alaskan state come from this sector, so it pays for education, health, policing and important community services.	Migrant workers take the majority of jobs created by the oil industry, spend little locally and often only have short-term contracts. In Prudhoe Bay, locals take just 400 of the 2,000 available jobs.
In some places along its route, the trans-Alaskan pipeline passes underground so that it does not disturb the **migration** routes of the tundra caribou. The pipeline is thickly insulated to protect it from freezing and stop the permafrost from melting.	In 1989, an oil tanker, Exxon Valdez, ran aground on the southern Alaskan coast. Only 15 per cent of the 1.2 million spilled barrels was ever recovered. Around 5,000 sea otters and many seals and eagles were killed. A broken pipeline spilled 1 million litres of oil in the fragile North Slope region in 2006.

What are the issues surrounding future development of offshore oil fields?

In addition to the dispute over the ANWR, Alaska's offshore waters are also a source of controversy. There are believed to be 30 billion barrels of recoverable oil, and many trillions of cubic metres of gas beneath the Beaufort Sea and Arctic Ocean (Figure 8.20).

No drilling is allowed off the coast of Kaktovik and Barrow. These waters are home to bowhead whales, and Inupiat residents do not want the whales to be disturbed by outsiders. Inupiat people, however, are allowed to conduct carefully controlled whale hunts. Sharing whale meat among families is an enduring tribal tradition. There is also a drilling ban in Bristol Bay, on the south coast of Alaska, to protect the area's sockeye salmon fishery.

Most of the Beaufort and Chukchi seas remain open to exploration. Big companies like Shell, Chevron and Statoil began exploratory undersea drilling there in 2012. For the Inupiat, supporting offshore drilling is a tough decision. Like many in Alaska, they are dependent on the oil industry for jobs.

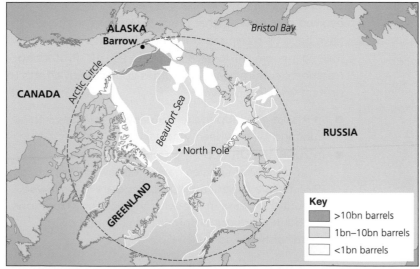

▲ Figure 8.20 Oil and gas fields under the Arctic Ocean

→ Activities

1 Outline two reasons why is it potentially difficult and/or expensive to make use of Alaska's oil and gas reserves.
2 Using Figure 8.19 and your own understanding, suggest why wilderness areas like Alaska face an increasing threat from economic development.
3 a) Outline one physical risk which Alaska's cold environment creates for the trans-Alaskan oil pipeline.
   b) Explain one way in which this risk is being managed.
4 To what extent should the exploitation of new oil resources be allowed in Alaska? In your answer, you should:
   ■ use Figure 8.18 to explain how technology is used to minimise the risk of oil spillages
   ■ refer to your answer to Activity 2
   ■ suggest what may happen if new offshore oil resources are exploited
   ■ weigh up the different arguments before arriving at a final overall judgement.

Geographical skills

Map analysis and interpretation
Study Figure 8.20.

1 Identify countries with offshore oil and gas fields near their coastline.
2 Use evidence to explain why these offshore oil and gas deposits could become a source of political conflict between the USA (Alaska) and other countries.

⭐ KEY LEARNING

➤ The need to protect wilderness environments

➤ Why wilderness cultures are under threat

➤ How the Antarctic wilderness is managed

Wilderness environments

Why is there a need to protect wilderness environments?

Wilderness areas are unspoilt and remote regions of the world. Extreme climate and inaccessibility used to keep mass tourism and economic development at bay. Now, modern transport gives easy access to previously inaccessible areas. In our 'shrinking world', wilderness areas have been opened up to tourism and businesses. Travel companies are keen to market new 'exotic' locations, while energy and mining companies hope to discover new **natural resources**.

There are several reasons for preserving wilderness areas and protecting them from development:

- There is a need to maintain the 'gene pool' of wild organisms, to make sure that genetic diversity is maintained over time.
- Scientists need to have access to undisturbed animal and plant communities for their studies.
- There should always be some places left in their natural state, so we can understand how much developed places have changed.

Wilderness areas perform vital **ecosystem services** that the world relies on. The white snow and ice in polar regions reflects sunlight and helps to regulate Earth's temperatures. Permafrost keeps enormous volumes of methane, a potent greenhouse gas, locked in ice (if released, it would contribute significantly to global warming).

The tundra is a **fragile environment** too, and thus in need of special protection. Extreme climates give rise to physical environments that are extremely sensitive to change. The slow growth of tundra plants means that it takes many years for the ecosystem to regrow after damage. It can take 50 years for tyre tracks left by off-road vehicles to disappear.

Why are wilderness cultures under threat?

As well as the physical environment being threatened, 'cultural erosion' has taken place in Alaska with the arrival of Europeans, both as colonists and tourists. Their influence has caused the loss of unique characteristics from the indigenous cultures, such as language. In the past, 20 native languages were spoken in Alaska, but English has been imposed on indigenous people and is now spoken by the younger generation. Sometimes, English names have replaced people's original names. In the 1970s, American schooling insisted on classroom use of English. Now, some languages such as Eyak have lost their last speaker, while others are on the verge of dying out.

Other cold countries have their own highly distinctive cultures too. Life in a challenging environment has evolved over time so that practices born in hardship have become treasured traditions. For instance, in Iceland, people still love to eat 'rotten shark'. Protecting traditional cultures can sometimes conflict with efforts to protect wildlife, however.

- In Alaska, the Inupiat are still allowed by US law to hunt and kill bowhead whales.
- In 2008, the US Supreme Court made polar bears an endangered species having outlawed hunting in 1972. Native American Inuit people are still allowed to hunt the bears in line with their cultural traditions, although there are different opinions on how sustainable this is.

▲ Figure 8.21 A polar bear cub relaxes with her mother

How is the Antarctic wilderness managed?

Antarctica remained unexplored until just over a century ago, although large-scale slaughter of seals and whales in the Southern Ocean dates back to the 1700s. Public awareness first began to develop when the Shackleton expedition of 1914–17 brought back moving black-and-white film images of spectacular glaciers and wildlife.

Concerned that Antarctica could be spoiled by unchecked commercial exploitation, several nations signed the Antarctic Treaty in 1961. The more recent 1998 Protocol on Environmental Protection to the Antarctic Treaty is one of the toughest sets of rules in the world. Under the agreement, no new activities are allowed in Antarctica until their potential impacts on the environment have been assessed and minimised. Tourist boat operators have to follow incredibly strict guidelines.

Even so, there has been a tremendous growth in tourism over the last two decades. Between 2000 and 2008, the number of tourists tripled (see Figure 8.22). Just 100 years after the very first explorers set foot there, Western tour operators now charge up to £10,000 per person. Some tourists merely gaze at the coast from ships, while more active individuals go mountaineering or shoot wildlife photography (Figure 8.23). For US$50,000 it is even possible to take a guided ski trek to the South Pole.

▲ Figure 8.23 A tourist photographs emperor penguins in Antarctica

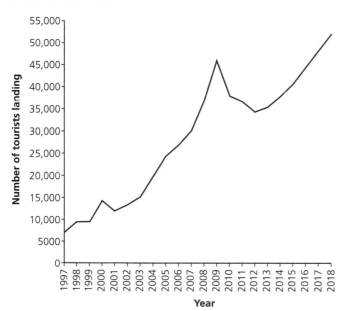
▲ Figure 8.22 Tourist growth in Antarctica

→ Activities

1 Give three characteristics of wilderness areas.
2 Using Figure 8.23 and your own understanding, suggest why tourism is a potentially valuable economic development strategy in cold environments.
3 a) Using Figure 8.22, describe the rise in visitor numbers over time in Antarctica.
   b) Suggest three reasons for this rise.
4 To what extent are wilderness areas in cold environments being managed successfully? In your answer, you should:

■ explore the management of the landscape, ecosystems and also people's culture
■ outline examples of successful management
■ write about times when management has not worked well or has been lacking
■ identify future pressures and challenges which could be hard to manage
■ weigh up the different arguments before arriving at a final overall judgement.

➤ How technology can help development of Arctic wilderness communities

➤ Why global technology companies are relocating to cold environments in Scandinavia

Then	Now
Animal skins and hides were used as clothing. People had to hunt animals and make clothes.	Modern materials such as polyester are used. People buy clothes online.
Traditional Inuit games were played outside, depending on the season.	Today's younger Inuit generation often play computer games and watch TV.
People travelled by walking, kayaking or using dogsleds (some still do).	The Inuit now use modern transport (see Figure 8.25).

▲ Figure 8.24 Technology has changed Inuit lifestyles

Balancing development and conservation using technology

Economic development in cold environments has often had disastrous impacts for their ecosystems. Hunting, whaling and fur trapping have pushed some land and marine species close to extinction. **Pollution** from oil and mining industries has added further stress. Indigenous communities struggle to maintain their traditions, and out-migration of the young in search of education and employment threatens the sustainability of some communities. Yet some technological developments have helped make life easier (see Figure 8.24) and have encouraged the young to stay, particularly since the arrival of information and communications technology (ICT). Digital technology offers exciting new development opportunities which may limit environmental damage and cultural erosion.

▲ Figure 8.25 Inuit people have adopted modern technology including the snowmobile

How can ICT help Arctic wilderness communities to develop?

The Scandinavian countries of northern Europe were among the first to see the potential of ICT for community survival. By the late 1980s, isolated villages in Finland, Sweden and Norway had been provided with shared computer and internet facilities by their governments. For these isolated communities inside the Arctic Circle, the internet became a survival lifeline.

In recent years, Inuit communities in Alaska and Canada have also discovered the value of ICT for their sustainable development.

Things first began to change in the 1990s when wireless radio and satellite links were provided for all the main populated areas, although coverage remains very uneven. As a result, many isolated parts of Alaska are 'leapfrogging' forwards in their access to technology. People can now receive a mobile phone call in places where landlines are still absent. The tiny Inuit village of Little Diomede is home to 120 people. Located on an island in the Bering Strait, with Alaska to the east and Russia to the west, mobile phone service is available despite the fact that the mail only arrives once a week.

■ In some places, two-way video conferencing is transforming how communities can access education and health services. In Canada, local government has collaborated with the company Cisco to allow remote Inuit schools to be taught in real time by teachers in other schools. They also collaborate with students of the same age throughout Canada.

■ The University of Alaska offers a range of degrees and courses which may be completed entirely online by students in isolated areas, provided they have internet access.

■ The Alaska Native Knowledge Network is an online database that collects and preserves Inuit culture.

More progress is around the corner. Soon, even the most remote Alaskan and Canadian villages will have access to high-speed broadband networks. As the Arctic ice thins, it is becoming easier to lay fibre optic cable. Several plans are in place to bring thousands of kilometres of fibre optic cable to the Arctic coastline. Thanks to the efforts of companies like Anchorage-based Quintillion, some of the world's remotest communities are now experiencing **globalisation**.

Why are global technology companies relocating to cold environments?

Some of the world's largest internet companies have relocated their data centres to cold environments. These offices have a low environmental impact and also offer employment to people in some of the world's most remote places. The global relocation of Facebook and Google to northern Scandinavia provides a good example of how development and conservation can be balanced.

Facebook users uploaded 350 million new photos each day in 2019, and generated 4 million 'likes' every minute. All this information must be stored and backed up in the company's data centres. As a result, data centres are very expensive to run. Energy is used to power the computers and also to cool them down with fans or air conditioning.

Consequently, more companies are placing their data centres in Arctic areas because the climate cools down the machinery. Google has a data centre in Finland. Facebook has two giant data centres in Luleå, Sweden – a long way from its desert headquarters in California! Each site covers 30,000 m² (the size of ten football pitches) and cost around US$750 million to build. Luleå is at the same latitude as Fairbanks, Alaska, USA and is part of Sweden's coldest region. Physically, the site of Luleå offers Facebook:

- a cold climate – located on the edge of the Arctic Circle, Luleå has short, mild summers and long, cold and snowy winters. Winter temperatures are well below freezing. For eight months of the year, the high-power computer equipment will cool itself at no cost
- low HEP costs – Luleå lies near hydroelectric power stations at the mouth of the River Lule. This provides cheap electricity for lighting (the heat from the computers is used to warm the Facebook staff's offices)
- flat land – Luleå is built on a flat, glacially eroded valley floor.

Facebook is not the only technology company in Luleå. The town is now home to 2,000 employees working for several large technology industries that have clustered there.

▲ Figure 8.26 The location of Luleå

▲ Figure 8.27 The first Facebook centre in Luleå was built in 2013. A second data centre was added soon after, with a third planned for 2021.

→ Activities

1 Outline how the introduction of ICT could have:
   a) one negative impact on Inuit culture
   b) one positive impact on the development of Inuit communities.

2 Using Figure 8.27 and your own understanding, explain why global companies have relocated parts of their businesses to cold environments.

3 To what extent is the use of technology always helpful for communities in cold environments? In your answer, you should:
   - explain ways in which different technologies have helped Inuit communities
   - outline how the use of communications technology can help development
   - outline ways in which new technologies can also harm local communities and cultures
   - weigh up the different arguments before arriving at a final overall judgement.

⊛ KEY LEARNING

➤ Protective actions at different scales

➤ The balance between conservation and development in the Arctic

➤ The Arctic and climate change

Managing cold environments

What actions at different scales can protect cold environments?

Managing the balance between development and conservation is rarely easy to achieve. A range of management actions have been taken at global, national and local scales.

■ **International agreements** and treaties can influence what happens to cold environments and their ecosystems. For instance, the number of bowhead whales in Arctic waters has been growing at three per cent per year since the 1970s (see Figure 8.28). Population size has recovered since a global ban on commercial whale hunting was introduced by the International Whaling Convention in 1986. The Antarctic Treaty is another international success story (see page 111). Countries in the Arctic Circle have created a similar organisation called the Arctic Council, which wants to deliver sustainable development throughout the entire Arctic region.

■ **National governments** sometimes struggle to manage their own regions because they are expected to support the conflicting interests of many different groups, from indigenous people to big businesses. In Alaska, some politicians want to increase oil production to increase Alaska's income, but former US President Barack Obama (2009–16) banned oil exploration from taking place in 12 million acres of the Arctic National Wildlife Refuge (ANWR). In contrast, President Trump came to office in 2017 saying he would allow drilling on protected Arctic land.

■ **Non-governmental organisations (NGOs)** support the interests of groups of people who struggle to be heard.

The Inuit Circumpolar Council (ICC) is a non-profit organisation that represents indigenous people from Nunavut and other northern regions. Inuit NGOs have frequently taken action to save the environment. The campaign of Alaskans against Pebble Mine is one example (see page 105). In 2013, UK based NGO Greenpeace sent campaigners to Russia's Arctic Ocean to protest against oil exploration. The Russian government arrested and imprisoned them. In a later international court case, Russia paid Greenpeace US$3 million in compensation.

▲ **Figure 8.28** A bowhead whale

Can conservation and development be balanced in the Arctic?

Figure 8.29 shows how international organisations, national governments and NGOs have varying views about how the Arctic should be managed.

▼ **Figure 8.29** Three Arctic management approaches

The Arctic Council (an international organisation)	National governments	Greenpeace (an NGO)
Represents eight countries and the indigenous people of the Arctic. Established in 1996 to promote co-operation. Priorities: sustainable development, environmental protection and limiting harm to the economies of indigenous people. In the future, the Arctic Council could become an organisation with legal powers, potentially setting fishing or hunting quotas.	Currently entirely responsible for managing their own ecosystems. Some have given areas protection as national parks (see Figure 8.30). Just over ten per cent of all Arctic land now has some level of special protection. Each country has its own laws controlling mining and oil pollution. International agreements and laws are not always followed: for example, Norway continues to hunt whales.	Called for a 'global sanctuary' in the Arctic, stating that: 'The best way is to make its resources off limits. That's why we're campaigning for a global sanctuary and a ban on oil drilling and industrial fishing.' But would this approach limit indigenous people's freedom to use Arctic resources for their own economic development?

Does climate change mean Arctic management strategies may fail?

Despite efforts to manage them sustainably, polar and tundra regions face an uncertain future. Climate change may already be causing permanent harm. One estimate shows the size of the tundra climate zone and ecosystem has reduced by 20 per cent since 1980. For complex reasons, global warming is felt most in polar regions.

- Temperatures in Newtok, Alaska, have risen by 4°C since the 1960s, and by as much as 10°C in winter months. Permafrost is thawing at an even faster rate, causing more buildings to subside, tilt and sink.

- Traditional life for the Yup'ik population is under threat from global warming – the ice pack on the Bering Sea is now too thin for them to fish and hunt safely.

It may be possible to adapt to these changes, of course. During building construction, Yup'ik workers now push building piles four metres into the ground to guarantee stability (three metres used to be enough). However, Arctic temperatures may rise by another 7°C by 2100, potentially causing major environmental changes and the extinction of species. This means that conservation of the Arctic's natural environment will be increasingly hard to achieve – even if strong international, national and local agreements are put in place to protect the region from economic development.

> ### → Activities
>
> 1 a) Name two international agreements that aim to protect cold environments.
>   b) Explain one strength and one weakness of the Arctic Council.
> 2 State three climate change impacts in Arctic regions.
> 3 Using Figure 8.30, describe the distribution of protected areas in the Arctic.
>
>   Think carefully about how to describe different locations. Pay attention to where the North Pole is positioned.
>
> 4 To what extent can Arctic environments be protected in the future? In your answer, you should:
>
>   - explain the main threats to Arctic landscapes and Arctic waters
>   - outline how far climate change and other threats can be prevented
>   - think critically about the relative importance of local and global actions and management strategies
>   - weigh up the different arguments before arriving at a final overall judgement.
>
> 5 You can carry out your own up-to-date research for this topic. The intergovernmental Panel on Climate Change (IPCC) has published reports which include predictions about how cold environments may be affected by climate change by 2100. A report you can examine is:
>
>   - Polar regions (Arctic region and Antarctic) – http://archive.ipcc.ch/pdf/assessment-report/ar4/wg2/ar4-wg2-chapter15.pdf

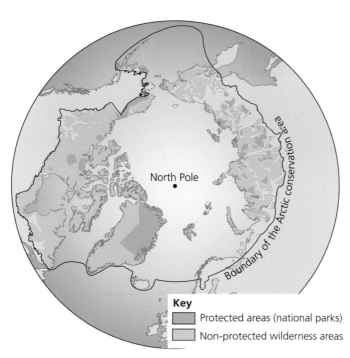

Key

■ Protected areas (national parks)
■ Non-protected wilderness areas

▲ Figure 8.30 Protected areas already make up ten per cent of Arctic wilderness areas

Question practice: Paper 1 Section B

2.1 Study the data in Figure 1. It shows population changes for three animal species in an area of deciduous forest in Scotland.

Which species shows the greatest proportional change between 2014 and 2016?

Species	2010	2012	2014	2016
Fox (carnivore)	55	55	40	50
Rabbit (small herbivore)	4,000	2,500	2,300	3,000
Deer (large herbivore)	80	80	100	95

Figure 1 Population changes for three animal species

[1 mark]

2.2 Outline **one** natural reason why the deer population in Figure 1 rose between 2012 and 2014. [2 marks]

The word 'natural' appears in the question, meaning that human management cannot be suggested as a possible answer.

2.3 Which **one** of the following describes the mean monthly temperature range for a deciduous forest? Select **one** letter only.

A 18 °C in July and 5 °C in January
B 28 °C in July and 10 °C in January
C 18 °C in July and –5 °C in January
D 32 °C in July and 14 °C in January

In multiple choice questions, eliminate the wrong answers rather than looking for the right one.

[1 mark]

2.4 Which **one** of the following statements best describes the soils of a tropical rainforest? Select **one** answer only.

A Deep with well-developed soil horizons
B Infertile and rapidly recycles nutrients
C Alkaline, thin and fertile
D Moderately deep and slowly recycles nutrients

[1 mark]

2.5 Using Figure 2 and your own understanding explain the pattern of deforestation in the Amazon rainforest. [6 marks]

'Using Figure 2 and your own understanding, explain the pattern' means you that you should try to give reasons why the deforestation has occurred in the places shown on the map using both information you can see and what you have learnt while studying the Amazon rainforest.

▲ Figure 2 Deforestation in the Amazon basin

2.6 Study Figure 3. Calculate the difference in the total rainfall between the wettest and driest months. [1 mark]

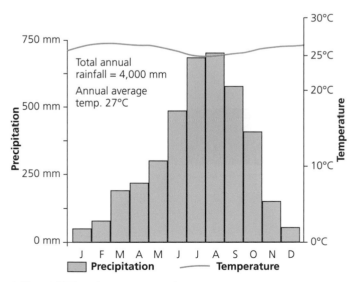

▲ Figure 3 The climate at Douala (Cameroon)

2.7 Outline **one** social impact of deforestation of the Amazon rainforest. [2 marks]

2.8 Suggest **one** way of making use of the tropical rainforest more sustainable. [2 marks]

2.9 For a hot desert environment **or** a cold environment that you have studied, to what extent has human activity threatened that environment? [9 marks]

2.10 Figure 4 shows a diagram of the nutrient cycle in a tropical rainforest. The circles represent the stores of nutrients and the arrows represent the flow or transfer of nutrients from one store to another.

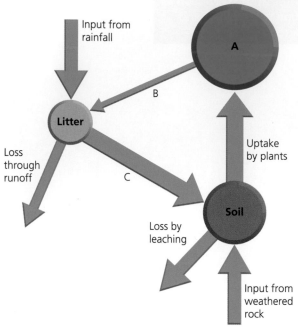

> With a diagram such as this take plenty of time to look carefully and make sure you understand what it is showing before you try to answer the question.

▲ Figure 4 A nutrient cycle in a tropical rainforest

Give the name of the store represented by the letter A in Figure 4. [1 mark]

2.11 Give **one** reason why the arrows are different widths in Figure 4. [1 mark]

2.12 Describe the flows shown by the letters B and C in Figure 4. [2 marks]

> Be very careful to make sure you are referring to the correct letters on the diagram.

2.13 Using Figure 5 and your own understanding, explain how plants adapt to the climate in the tropical rainforest.

[4 marks]

Make sure you use the photo by referring directly to something you can see such as the tree trunk. Then use your own understanding to explain how it is an adaptation to the climate.

▲ **Figure 5** Vegetation in a tropical rainforest

2.14 'The increasing demand for beef is the main cause of deforestation in the tropical rainforest.' Using an example you have studied discuss this statement.

'Discuss' means you need to mention both sides of the argument. So, in this case, give reasons why you think this may be true and reasons why you think it may not be true.

Make sure you use an example of a real place you have studied such as the Amazon rainforest.

[6 marks]

2.15 Outline **one** strategy which could be used at an international level to manage the tropical rainforest in a more sustainable way.

[2 marks]

2.16 For a hot desert environment **or** cold environment that you have studied, to what extent does that environment provide both opportunities and challenges for development?

[9 marks]

[+ 3 marks SPaG]

There are three important things you must include in this answer:
- opportunities for development
- challenges for development
- all must be in real places you have studied.

119

Weathering and mass movement

How does weathering weaken a cliff face?

Weathering is the breaking down of rock *in situ* (where it is). It is caused by day-to-day changes in the atmosphere, such as extremes of temperature and **precipitation**.

Chemical weathering

Chemical weathering is caused by a chemical reaction when rainwater hits rock and decomposes it or eats it away.

■ Carbonation is when carbonic acid in rainwater reacts with calcium carbonate in limestone to form calcium bicarbonate. This is soluble, so limestone is carried away in **solution**.

■ Hydrolysis is when acidic rainwater breaks down the rock, causing it to rot. For example, when rainwater changes granite into clay.

■ Oxidation is when rocks are broken down by oxygen and water, often giving iron-rich rocks a rust-coloured surface (iron oxide).

Mechanical (physical) weathering

Mechanical weathering results in rocks being disintegrated rather than decomposed. It is usually associated with extremes of temperature:

■ **Freeze-thaw weathering** (frost shattering) happens when water enters cracks. When the temperature falls below freezing, this water freezes and increases in volume by nine per cent, putting pressure on the rock around the crack. If the temperature rises above freezing, the ice will thaw and relieve the pressure. Constant repetition of this freezing and thawing cycle causes angular rock fragments to break away and collect as scree at the base of the **cliff**.

■ Salt weathering is when salt spray from the sea gets into a crack in a rock. It may evaporate and crystallise, putting pressure on the surrounding rock and weakening the structure.

How does mass movement happen?

Mass movement is the downslope movement of rock, soil or mud under the influence of gravity. Heavy rainfall is usually the trigger, but the scale of movement is determined by the extent of weathering on the slope.

Sliding

Landslide is the generic term for the downhill movement of a large amount of rock, soil and mud. **Sliding** occurs on steep cliffs previously weakened by weathering. Heavy rain infiltrates the soil and percolates down into the rock. The now heavier, saturated mass falls away along a distinct slip plane, which is a line of weakness such as a fault or bedding plane. A slide happens quickly. It starts by tearing away the vegetation on the top edge of the cliff. Once the slide has begun, its descent is aided by lubrication from the wet rocks below. The cliffs in the Jurassic Coast near Durdle Door in Dorset have suffered from large landslides in recent years.

A rock slide is where a large amount of rock slides down a cliff. This happens along a fairly straight slip plane, where rock falls as a block which maintains contact with the cliff. The leading edge of the slide collects as a pile of rocks on the beach or in the sea.

Mud slides are usually wet, rapid and tend to occur where slopes are steep (over ten degrees). Monmouth Beach at Lyme Regis, Dorset is prone to mudslides. They usually occur where vegetation cover is sparse so cannot hold the soil in place. They happen after a period of heavy rain. At the base of a mudslide, the saturated soil spreads out to make a lobe (see Figure 10.5).

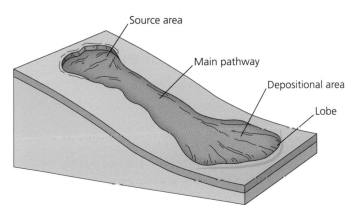

▲ Figure 10.5 Mud slides

Rock falls

Bare, well-jointed rocks are prone to freeze-thaw weathering, which results in falling rocks losing contact with the cliff face. At the bottom of the cliff they fan out to form a scree slope. Rock falls are common on vertical cliffs such as at Burton Bradstock, Dorset, where 400 tonnes of rock fell from the 49-metre vertical cliff in July 2012 (see page 132).

Slumping

While a slide takes a fairly straight path down a cliff, a slump has a concave slip plane, so material is rotated backwards into the cliff face as it slips. At Barton on Sea, Hampshire, the cliff is **slumping** at up to 30 centimetres a day (Figures 10.6 and 10.7).

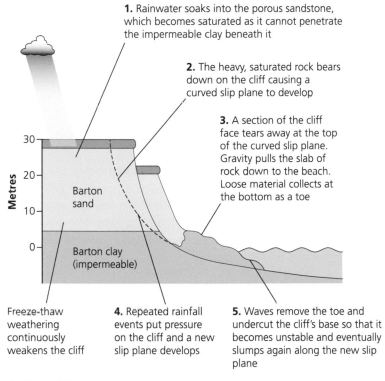

1. Rainwater soaks into the porous sandstone, which becomes saturated as it cannot penetrate the impermeable clay beneath it

2. The heavy, saturated rock bears down on the cliff causing a curved slip plane to develop

3. A section of the cliff face tears away at the top of the curved slip plane. Gravity pulls the slab of rock down to the beach. Loose material collects at the bottom as a toe

Metres

Barton sand

Barton clay (impermeable)

Freeze-thaw weathering continuously weakens the cliff

4. Repeated rainfall events put pressure on the cliff and a new slip plane develops

5. Waves remove the toe and undercut the cliff's base so that it becomes unstable and eventually slumps again along the new slip plane

▲ Figure 10.6 Slumping

▲ Figure 10.7 Rotational slumping at Barton on Sea, Hampshire

→ Activities

1 Draw a series of three or four labelled diagrams to show how freeze-thaw weathering widens cracks in a rock.

2 a) Define the terms weathering and mass movement.
   b) Explain the link between weathering and mass movement.

3 With your own diagrams, explain the difference between a rock slide and a rock fall.

4 Draw a sketch of the cliffs at Barton on Sea (Figure 10.7). Annotate it to describe and explain what is happening to the cliff.

➤ The processes of coastal erosion
➤ How material is transported by waves
➤ Causes of deposition

Marine processes

What are the processes of coastal erosion?

Marine erosion is the removal of material by waves. Waves erode by hydraulic power, abrasion and attrition.

Hydraulic power

Hydraulic power is the relentless force of destructive waves pounding the base of cliffs. This causes repeated changes in air pressure as water is forced in and out of joints, faults and bedding planes. The forward surge of water compresses air in these cracks and, when the wave retreats, there is an explosive effect as pressure is suddenly released. This onslaught is aided further by the weakening effect of weathering. Material breaks off cliffs, sometimes in huge chunks.

Abrasion (corrasion)

Destructive waves have enough energy to hurl sand and shingle at a cliff. The resulting scratching and scraping of the rock surface is called **abrasion**. This is concentrated between the high and low watermarks and is particularly effective in high-energy storm conditions.

Attrition

Attrition is the grinding down of **load** particles. During transport, pebbles collide with each other. Over time, this wears away jagged edges to make smooth, rounded pebbles. Some collisions may cause a pebble to smash into several smaller pieces, each of which will be further smoothed and rounded.

▲ **Figure 10.8** Seven Sisters cliffs in East Sussex, UK. Note the rounded pebbles due to attrition

The rate of erosion will be higher where:

- a rock has many joints
- the coastline is exposed to a large fetch, such as the Needles on the Isle of Wight (Hampshire), which have an 8,000 kilometre fetch across the Atlantic Ocean
- strong winds blow for a long time and create destructive waves. These conditions are common in winter
- an area has no beach to act as a buffer between the sea and the cliffs

- a headland juts out into the sea. Waves converge on a headland (wave refraction) and gain height and erosive energy
- there are soft rocks. The average annual rate of erosion of the unconsolidated, soft, boulder clay rocks of the North Norfolk coast is five metres a year. Contrast this with the hard granite rocks of South West England which erode at 0.001 metres a year.

How is material transported by waves?

Load is transported material. Most marine load originates from river deposits, from eroded headlands and from the seabed. Load varies from fine silt to large rocks.

Transport onto the beach

Load is transported by waves. The larger and heavier the load particle, the greater is the velocity needed to transport it. The lightest load is carried in **suspension** or saltated, while heavier load is moved onto a beach by traction. These processes are outlined in Figure 11.3 on page 153.

Transport along the beach parallel to the shore

Load is transported along the shore by **longshore** (littoral) **drift**. The direction is determined by the **prevailing wind**. Along the Dorset coast, the prevailing southwest wind causes a drift in an easterly direction, making the swash surge up the beach at an oblique angle. In response to gravity, the backwash goes back down the beach at right angles to the shore. Suspended load is therefore carried easterly in a zig-zag manner.

What conditions cause deposition?

Deposition is when waves drop and leave behind the load they were transporting. The deposited load is called **sediment**. Deposition results in more sediment staying on the beach than is taken away by the backwash. This will happen:

- in low energy, sheltered bays, where constructive waves are dominant
- if there is a large source of sediment updrift, such as a rapidly eroding headland
- where there are large expanses of flat beach so the swash spreads out over a large area. This weakens the wave so that its backwash is not strong enough to transport the sand back out so sea
- when, on an outgoing tide, tidal material is trapped behind a **spit** (see page 140)
- where engineered structures like groynes trap sediment on the updrift side (see Figure 10.10).

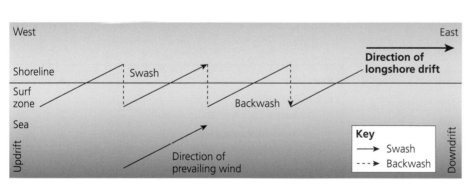

▲ Figure 10.9 Longshore drift

▲ Figure 10.10 Groynes trap sediment, slowing down longshore drift

→ Activities

1 Give three types of coastal erosion. For each, write its definition in your own words.
2 Draw an annotated diagram to explain how hydraulic power causes erosion.
3 Draw a series of diagrams to explain the effect of attrition on pebbles.
4 Create a fully annotated diagram to show longshore drift where the prevailing wind is from the east.
5 Suggest why local citizens in a seaside resort may be concerned about longshore drift in their area.

 Fieldwork: Get out there!

Using two markers, such as rulers, a tape measure, a floating object and a digital watch, devise a method for measuring the speed of longshore drift.

Example

⭐ KEY LEARNING

➤ How landforms are affected by hardness of rock

➤ How rock structure affects landforms

➤ The location of major landforms of the Dorset coast

Geology and rock structure on the Dorset coast

Throughout this chapter, examples are used from the Dorset coast where differing **geology**, or rock types, and rock structures have led to a variety of coastal landforms.

The Dorset coast is sometimes known as the 'Jurassic coast'. The name comes from the age of the rocks, dating from the Triassic, Jurassic and Cretaceous periods, spanning almost 200 million years. The coastline is made up of layers of different sedimentary rock, each of which give rise to distinctive landscapes.

How do hard and soft rocks affect landforms?

As hard rocks are less easily eroded than soft rocks, they project into the sea as headlands (see page 130) and form high cliffs. In the Isle of Purbeck, the hard Portland limestone forms steep cliffs while the softer Bagshot beds, Wealden clays and sandstone form low-lying bays. Hard chalk rocks have produced arches and stacks (see page 134).

How does rock structure affect landforms?

Rock structure includes:

■ how rocks are aligned in relation to the coast (concordant and discordant coasts)

■ how rocks dip down to the sea as a result of folding.

Concordant and discordant coasts

Along the east coast of the Isle of Purbeck, the alternating layers of hard and soft rock run at right angles to the shore, giving rise to headland and **bay** formation (see Figure 10.11 and page 130). This is typical of a discordant coastline. In contrast, the southern coast is fairly smooth in shape where the rock is uniform. This is typical of a concordant coastline, with alternating layers of hard and soft rock running parallel to the coast. This can be seen in particular in the Kimmeridge clays and Lower Purbeck (limestone) rocks. However, at Lulworth, a cove has formed where waves have broken through a weakness in the hard Portland limestone and scooped out the softer rocks behind it (Figure 10.12).

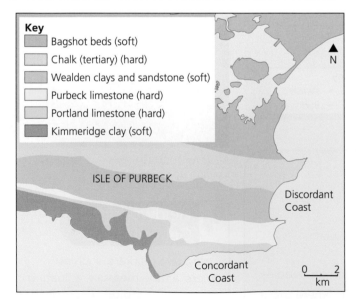

▲ Figure 10.11 Concordant and discordant coast, Isle of Purbeck, Dorset

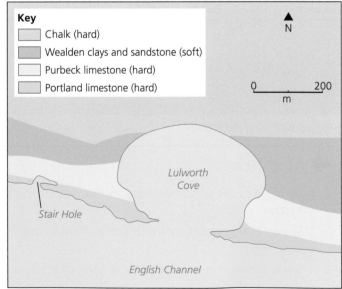

▲ Figure 10.12 Geology map of Lulworth Cove, Dorset, a concordant coastline

Rocks and angle of dip

Sedimentary rocks are formed on the sea bed and are raised by mountain building processes, which folded the rocks over millions of years. In places along the Dorset coast, an up-folded area called an anticline can be seen in a headland. This can be seen on the eastern edge of Stair Hole near Lulworth Cove (Figure 10.13). From left to right, there is a slight anticline, then a syncline (or downfold) moving into a clear anticline nearest the sea. The rising angle of dip at the coast results in a steep cliff profile.

▲ **Figure 10.13** Folding at Stair Hole with the cove at Lulworth behind

What are the major landforms at the Dorset coast?

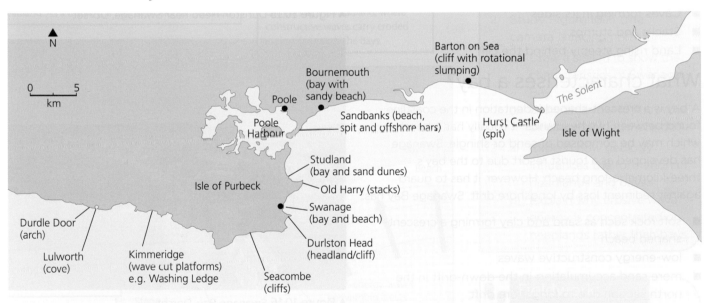

▲ **Figure 10.14** The location of major landforms on the Dorset/Hampshire coast

→ Activities

1 Define the terms geology and rock structure.

2 Outline the difference between a concordant coast and a discordant coast.

3 Use Figure 10.14 to compile a table for coastal erosion landforms and coastal deposition landforms along the Dorset/Hampshire coast.

Geographical skills

Draw a geological sketch of Figure 10.13. Annotate it to show how Lulworth Cove was formed. Mention:

- soft rock
- erosion
- hard rock
- folded rock.

➤ How caves, arches and stacks are formed
➤ The characteristics of caves, arches and stacks

Caves, arches and stacks

Section 10.5 showed how headlands project into the sea and how they are eroded by destructive waves associated with wave refraction. So, despite the hardness of rock, headlands are constantly being re-shaped by the waves. Landforms located at headlands include **caves**, sea **arches**, **stacks** and **stumps**.

How are caves, arches and stacks formed?

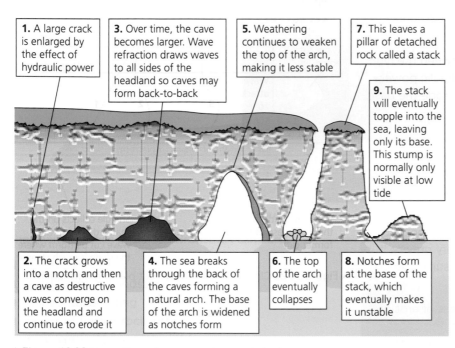

1. A large crack is enlarged by the effect of hydraulic power

3. Over time, the cave becomes larger. Wave refraction draws waves to all sides of the headland so caves may form back-to-back

5. Weathering continues to weaken the top of the arch, making it less stable

7. This leaves a pillar of detached rock called a stack

9. The stack will eventually topple into the sea, leaving only its base. This stump is normally only visible at low tide

2. The crack grows into a notch and then a cave as destructive waves converge on the headland and continue to erode it

4. The sea breaks through the back of the caves forming a natural arch. The base of the arch is widened as notches form

6. The top of the arch eventually collapses

8. Notches form at the base of the stack, which eventually makes it unstable

▲ **Figure 10.22** Formation of caves, arches and stacks

What characterises a cave?

Caves at a headland may be several metres high at their entrance and taper back a long way (Figure 10.22). The Dorset coast has caves in the limestone cliffs at Durlston Head, in the chalk stack at Ballard Down and also in some of the chalk stacks at the Foreland (Figure 10.25).

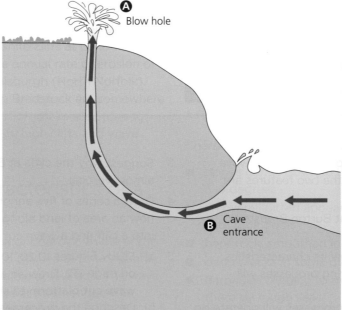

A blow hole (gloup) may form in the roof of the cave towards the back.

Pressure from waves may push water up the blow hole so that it emerges on the cliff above

Ⓐ Blow hole

Ⓑ Cave entrance

The widest part of the cave is at its entrance

▲ **Figure 10.23** Characteristics of a cave

What characterises a sea arch?

- Unsupported top of the arch
- Wave-cut notches at the base of the arch, so wide at the base
- Water going right through the gap
- The arch is an extension of the headland

What characterises stacks?

- Detached blocks or pillars of rock located off a headland
- Some may be pinnacle shaped like the needles of the Isle of Wight and some stacks off Ballard Down
- Often several metres high
- Hard rock
- Wave-cut notches at the base

▲ Figure 10.24 Characteristics of Durdle Door, Dorset – a sea arch

What characterises stumps?

Some headlands also have stumps, which are the bases of collapsed stacks. They can only be seen at low tide. Old Harry's Wife is a stump that lies just beyond Old Harry. The stump is submerged.

→ Activities

1. a) Describe a cave.
   b) Draw a series of three to four diagrams to explain how a cave is formed.
   c) Explain why caves may form on either side of a headland.
2. a) Imagine you were to take a photo of the Old Harry Rocks in 200 years' time. Sketch what you think this headland would look like.
   b) Explain the changes you have made to the original photo (Figure 10.25).
3. a) Draw a diagram to show the characteristics of a stump.
   b) Explain how a stump is formed.

▲ Figure 10.25 Old Harry Rocks – sea stacks at the Foreland, off Ballard Down, Dorset

Geographical skills

1. Draw a sketch of Durdle Door arch (Figure 10.24).
   Label the arch to show its characteristics.
2. Draw a sketch of Figure 10.25. Annotate with the following labels:
   a broad stack composed of hard chalk, a new small arch, the steep pillar-shaped stack of Old Harry, rock debris from the previous collapse of an arch, the headland.

KEY LEARNING

- How beaches are formed
- The characteristics of beaches
- Why beach profiles change

Beaches

A beach is a landform of coastal deposition that lies between the high- and low-tide levels. Most beaches are formed of sand, shingle or pebbles, as well as mud and silt. A beach that forms in a bay is crescent-shaped (see Figure 10.26A), but its shape is distorted by longshore drift, so the beach is narrower updrift than downdrift.

How are beaches formed?

Section 10.3 showed how waves transport materials from the sea to the shore. Two distinct types of beaches can be formed.

Sandy beach

In sheltered bays, low-energy constructive waves transport material onto the shore (see Section 10.1). The swash is stronger than the backwash, so sediment is slowly but constantly moved up the beach. Once the tide has gone out, there is more material on the beach than before. An example is Sandbanks Beach, Poole, Dorset (Figure 10.26 A).

Pebble beach

Exposed beaches such as West Bay, part of Chesil Beach in Dorset (Figure 10.26 B), sometimes have a large fetch. The plunging nature of destructive waves (see Section 10.1), along with their stronger backwash, means that pebbles are not moved far up the beach, which makes the **beach profile** steep. A storm beach may form when there is wild, stormy weather and waves hurl boulders and large pebbles to the back of a beach.

What characterises a beach?

Characteristics of a sandy beach	Characteristics of a pebble beach
Gradient: Generally shallow, almost flat	**Gradient:** Generally steep
Dominant waves: Constructive	**Dominant waves:** Destructive
Distance stretches inland: A long way	**Distance stretches inland:** Not far
Back of beach: Sand dunes (sometimes)	**Back of beach:** Storm beach with large pebbles
Other characteristics: At low tide, small water-filled depressions called runnels form. These are separated by small sandy ridges running parallel to the shore. The wet sand may have a rippled appearance	**Other characteristics:** Pebbles increase in size towards the back of the beach
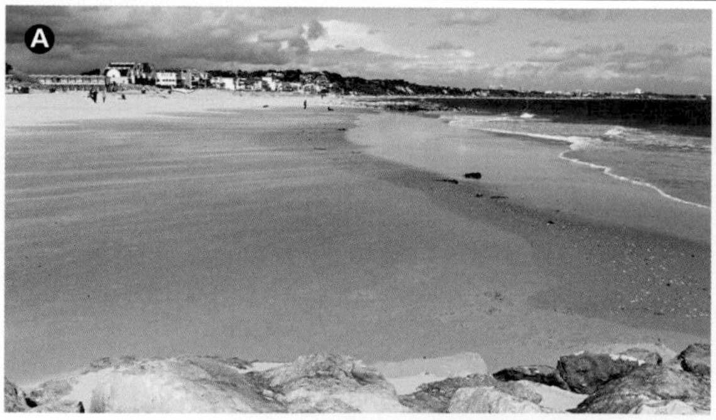	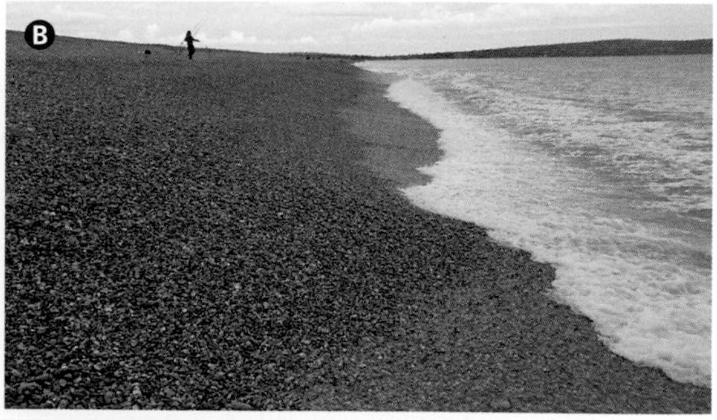

Figure 10.26 Contrasting Dorset beaches

What is a beach profile?

A beach profile shows the gradient from the back of the beach to the sea. A sandy beach generally has a gentle, fairly flat profile, whereas a pebble beach usually has a steep, stepped profile.

Why do beach profiles change?

A berm is a terrace on a beach that has formed in the backshore, above the water level at high tide. On broad beaches there may be three or more subparallel berms, each formed under different wave conditions. Berms are formed in calm weather when constructive waves transport material onto the beach. While an existing berm is moved up the beach by storms and spring tides, a new berm may develop and change the beach profile.

- In winter, berms, and sometimes the sand dunes at the back of the beach, are eroded by destructive waves which drag beach deposits offshore to create an offshore **bar**. This lowers the height of a beach.
- In late spring and summer, so long as longshore drift is not depleting the beach of sand, constructive waves will rebuild the beach. The offshore bar is transported by the waves to rebuild the berms, and dunes are replenished by saltation by the wind (see Section 10.9).
- Destructive waves often result in winter profiles that are narrower and steeper than those in summer.

▲ Figure 10.27 A sandy beach at low tide

Labels on Figure 10.27: Water-filled trough called a runnel; Sand dunes; Berm; Dry backshore; Ridge; Ripples in the sand made by waves; Wide sandy beach shows a large **tidal range**

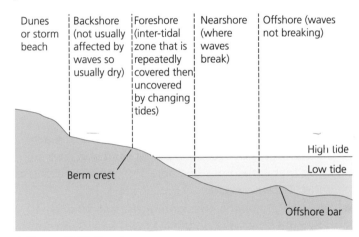

▲ Figure 10.28 Beach profile in summer

Labels on Figure 10.28: Dunes or storm beach; Backshore (not usually affected by waves so usually dry); Foreshore (inter-tidal zone that is repeatedly covered then uncovered by changing tides); Nearshore (where waves break); Offshore (waves not breaking); High tide; Low tide; Berm crest; Offshore bar

Fieldwork: Get out there!

Research how to measure the gradient (profile) of a beach using a clinometer.

→ Activities

1   Draw labelled sketches of photos A and B in Figure 10.26 to show the different characteristics of a sand beach and a pebble beach.

2   Explain why pebble beaches are steep.

3   The following descriptions are mixed up. Can you correct them?

Nearshore	The inter-tidal zone repeatedly covered, then uncovered by changing tides
Backshore	The breaker zone where waves break
Foreshore	Fairly far out to sea where the waves do not break
Offshore	An area that is not usually affected by waves, so the sand is usually dry

4   a) Define the term beach profile.
    b) Draw the shape of a beach profile in summer (Figure 10.28).
    c) On the same diagram, use a different colour to draw how you would expect this profile to look in winter.
    d) Explain the change you have made to the original profile.

5   a) Explain how a sandy beach is formed.
    b) Study Figure 10.27. Describe the changing nature of this sandy beach if you were to walk from the sand dunes towards the sea.

➤ How sand dunes are formed
➤ The characteristics of sand dunes
➤ Dune succession

Sand dunes

Sand dunes (large heaps of sand) form on the dry backshore of a sandy beach.

How are sand dunes formed?

For a sand dune to form, it needs:

- a large flat beach
- a large supply of sand
- a large tidal range, so there is time for the sand to dry
- an onshore wind to move sand to the back of the beach
- an obstacle such as drift wood for the dune to form against.

Wind moves sand in three ways (Figure 10.29):

- When there are obstacles, such as driftwood, the heaviest grains of sand will settle against the obstacle to form a small ridge. Lighter grains may be transported and will settle on the other side of the obstacle.
- Eventually, the area facing the wind begins to reach a crest. This is because the pile of sand becomes so steep that it becomes unstable and begins to collapse under its own weight.
- When this happens, the lighter grains of sand fall down the other side on the lee (slip) face. Sand stops slipping once a stable angle has been reached at 30–34 degrees.
- The repeated cycle of wind blowing up the *windward* side and slipping down the *leeward* side causes a sand dune to migrate inland over time.
- A sand dune itself becomes an obstacle, so more dunes may form in front of it. The height of dunes depends on the strength of the wind. Stronger winds create higher dunes.

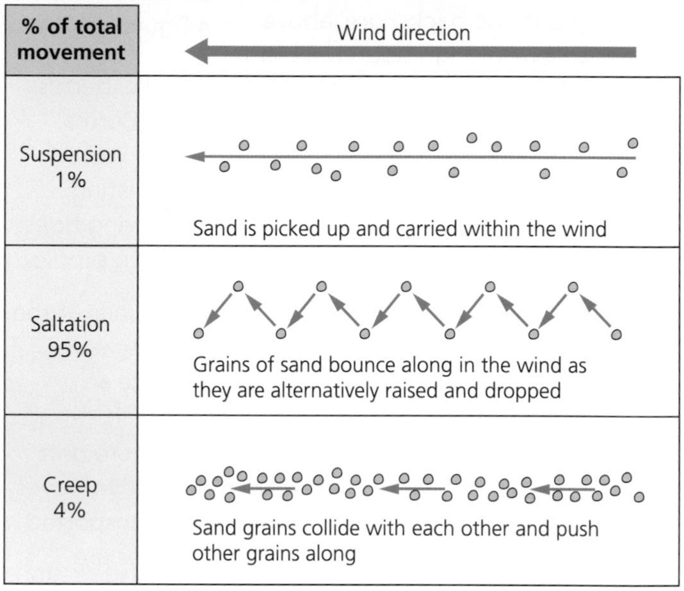

% of total movement	Wind direction
Suspension 1%	Sand is picked up and carried within the wind
Saltation 95%	Grains of sand bounce along in the wind as they are alternatively raised and dropped
Creep 4%	Sand grains collide with each other and push other grains along

▲ Figure 10.29 Wind transport of sand

What characterises sand dunes?

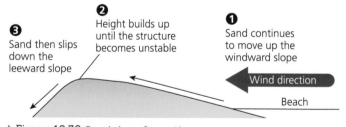

❸ Sand then slips down the leeward slope
❷ Height builds up until the structure becomes unstable
❶ Sand continues to move up the windward slope
Wind direction
Beach

▲ Figure 10.30 Sand dune formation

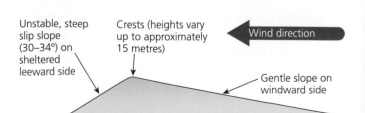

Unstable, steep slip slope (30–34°) on sheltered leeward side
Crests (heights vary up to approximately 15 metres)
Wind direction
Gentle slope on windward side

▲ Figure 10.31 Characteristics of a sand dune

▲ Figure 10.32 A sand dune near Studland, Dorset

How do dunes change inland?

Several lines of dunes may run parallel to the shore (Figure 10.33). The change in vegetation with increased distance inland is known as a dune succession:

- Dunes grow taller. Embryo dunes are only a few metres high whereas mature dunes may be up to 15 metres high
- Size increases inland as long-rooted marram grass and other vegetation bind the sand together, thereby preventing further migration. Marram grass grows quickly and aids sand accretion. Its long roots bind the sand and help build up the height of the dunes.
- Dunes closest to the beach have a yellow, sandy colour and not much vegetation. Dunes further back look grey and less sand-like.

- Inland, the dunes become increasingly colonised by vegetation.
- Each line of dunes is separated by a trough called a slack. Slacks are formed by the ongoing removal of sediment from the leeward base of one line of dunes and up the windward side of the next dune line. Sometimes, slacks are eroded so much that they reach down as far as the **water table**, resulting in the formation of salty ponds.
- Occasionally, a dune may develop a huge depression called a blowout, when strong winds remove sand from an area that has lost its protective vegetation cover.

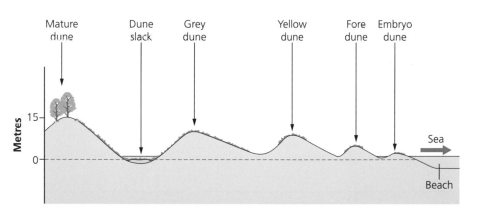

▲ Figure 10.33 How sand dunes change inland

Fieldwork: Get out there!

1 Describe how you might use a quadrat (a square frame) and a tape measure to carry out systematic sampling of vegetation over a line of sand dunes, going inland from the sea.

2 How would you expect the density of vegetation to change going inland?

3 Suggest a reason for the change.

→ Activities

1 The following definitions have been mixed up. Can you match the correct term and definition?

Embryo dune	The slope that faces away from the wind
Marram grass	The upper horizontal limit of wet sand
Saltation	Where there is a trough or low point in a line of dunes
Crest	A newly formed sand dune closest to the sea
Water table	The slope that faces the wind
Dune slack	A plant found in sand dunes that has long, binding roots
Leeward slope	How sand is bounced along by the wind
Windward slope	The top of a sand dune

2 Explain why a large tidal range aids the formation of sand dunes.

3 Which form of wind transport is dominant in sand dune formation? Can you suggest why?

4 Draw an annotated diagram to show how sand initially builds up behind an obstacle on the beach.

5 Explain why sand dunes are steeper on one side.

KEY LEARNING

➤ How spits are formed
➤ The characteristics of spits
➤ How bars are formed
➤ The characteristics of bars

Spits and bars

If unchecked, longshore drift (see page 127) can deplete updrift beaches of their sand and create new landforms downdrift. Two landforms associated with longshore drift are spits and bars.

How is a spit formed?

A spit is a sand or shingle beach that is joined to the land but projects downdrift into the sea. Spits form where the coastline suddenly changes shape or at a river **estuary**.

A spit is an unstable landform. It will continue to grow until the water becomes too deep or until material is removed faster than it is deposited. Hurst Castle spit in Hampshire (Figure 10.35) is growing in length, but losing shingle from its main ridge. Sandbanks Spit in Poole, Dorset, is not growing, due to tidal currents and dredging.

What characterises a spit?

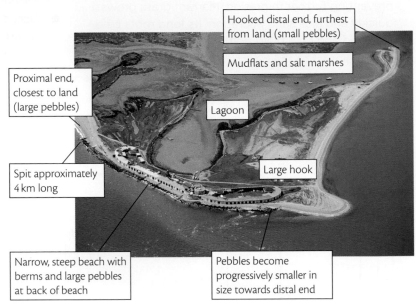

Hooked distal end, furthest from land (small pebbles)

Mudflats and salt marshes

Proximal end, closest to land (large pebbles)

Lagoon

Spit approximately 4 km long

Large hook

Narrow, steep beach with berms and large pebbles at back of beach

Pebbles become progressively smaller in size towards distal end

▲ Figure 10.35 Characteristics of Hurst Castle spit

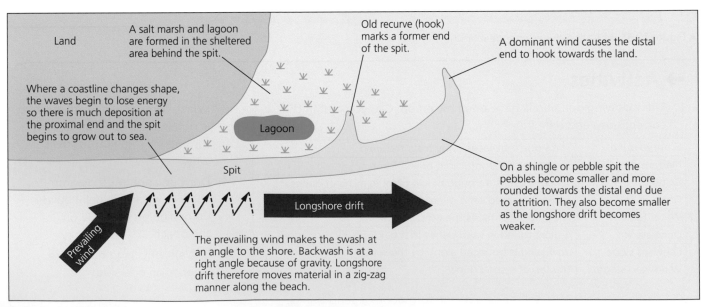

Land

A salt marsh and lagoon are formed in the sheltered area behind the spit.

Old recurve (hook) marks a former end of the spit.

A dominant wind causes the distal end to hook towards the land.

Where a coastline changes shape, the waves begin to lose energy so there is much deposition at the proximal end and the spit begins to grow out to sea.

Lagoon

Spit

Longshore drift

Prevailing wind

The prevailing wind makes the swash at an angle to the shore. Backwash is at a right angle because of gravity. Longshore drift therefore moves material in a zig-zag manner along the beach.

On a shingle or pebble spit the pebbles become smaller and more rounded towards the distal end due to attrition. They also become smaller as the longshore drift becomes weaker.

▲ Figure 10.34 How longshore drift forms a spit

How are bars formed?

A bay bar (or barrier beach bar) is a ridge of sand or shingle that stretches from one side of a bay to the other, forming a lagoon behind it. Barrier beach bar formation is due to longshore drift transporting sediment from one side of a bay to the other (see Figure 10.34 and Figure 10.36).

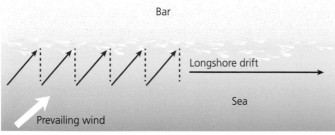

▲ Figure 10.36 Formation of a barrier beach bar

A submerged offshore bar is a raised area of the seabed that lies offshore. Submerged bars form in shallow waters by the **transportation** of sediment off and then back onto a beach. In stormy weather, destructive waves drag beach material out to sea to form an offshore bar. When it is calm again, constructive waves steadily transport bar sediment back towards the shore.

Barrier islands are visible offshore bars that form parallel to the coast, often in chains. They may be:

- offshore bars formed by waves churning up the sand to make a vertically forming submarine bar
- due to waves breaching a spit, for example Scolt Head, North Norfolk, which is detached from and aligned to the shore, with sand dunes, salt marshes and mudflats
- coastal submergence along a discordant coast, which leaves higher land as islands.

Bars can also form out to sea where tidal currents result in a build-up of sediment. Rising sea levels due to ice melting have driven some offshore bars onshore.

What characterises bars?

Figure 10.38 shows a submerged beach bar. For bay bars or barrier beach bars, at least some part of the bar is visible at all times.

▲ Figure 10.37 Slapton Sands bay bar, Devon

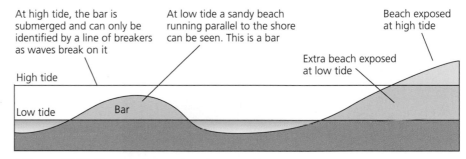

▲ Figure 10.38 Characteristics of a submerged bar

→ Activities

1 Give a definition of a) a spit b) a bar. For each, suggest where on the coastline you would expect them to be formed.

2 Explain the role of longshore drift in the formation of a spit.

3 Draw a sketch of the bar at Slapton Sands (see Figure 10.37). Annotate the sketch to show the characteristics of the bar.

4 Use annotated diagrams to explain why a submerged bar is usually only a temporary feature.

5 State two differences between a submerged bar and a barrier beach bar.

6 a) Use an atlas or OS map to locate the nearest spit to your home. Use the internet to obtain a photo of that spit.

   b) Label the photo to show the spit's characteristics.

   c) Add arrows to show the direction of longshore drift.

Coastal management: hard engineering

Exposed coastal areas take a battering from erosive, destructive waves. If the coastline contains high-value buildings, then the local authority tries to protect the coastline from erosion.

Hard engineering is when expensive artificial structures are used for protection. They are effective, but do not blend in well with the natural environment. Hard engineering includes **sea walls**, **groynes**, **rock armour** and **gabions**.

What are sea walls?

A sea wall provides a barrier between waves and the land. It is placed along the back of a beach. Recurved sea walls (Figure 10.39) are more expensive than flat sea walls, but are more effective in reflecting waves and reducing overtopping. Steps are often added to the base to give extra stability.

The recurved face rotates the wave backwards so that some of its energy is reflected back out to sea. This impedes the next wave and reduces its energy, reducing its erosive power.

▲ Figure 10.39 Sea wall at Highcliffe, Hampshire

What are groynes?

Groynes are wooden or stone structures built in the foreshore; they look like fences or walls. They are built at right angles to the beach and are spaced at regular intervals, approximately 50 metres apart. Traditionally, groynes were made of hardwood timber, but stone groynes are now more popular.

Groynes trap sediment transported by longshore drift. This builds up the beach on the updrift side of a groyne (Figure 10.40). A larger beach provides a more effective buffer against coastal erosion as it absorbs the waves' energy, and reduces the impact of waves on the sea wall. Groynes are particularly effective when used in conjunction with **beach nourishment** (see page 146).

▲ Figure 10.40 A wooden groyne traps sediment

What is rock armour?

Rock armour (rip rap) is made up of thousands of tonnes of huge boulders of hard rock like granite, to act as a barrier between the sea and the land (Figure 10.41). Boulders are generally big enough not to be moved by storm waves. As water enters gaps between boulders, pressure is released and this reduces the waves' energy, so there is little scouring of the base.

What are gabions?

Gabions are steel wire-mesh cages filled with pebbles or rocks. They are placed at the back of a sandy beach to create a low, wall-like structure (see Figure 10.42). Water enters the cages and this absorbs and dissipates some of the waves' energy, thus reducing the rate of erosion. Gabions may also be placed in front of a cliff, where they reduce the risk of landslides.

→ Activities

1 Study Figure 10.43, a map of a stretch of coastline.
   a) In pairs, put forward an argument to protect this coastline.
   b) One person then writes a proposal for building groynes, while the other person writes a proposal for a sea wall. (Include social, economic and environmental aspects.)
   c) Present the proposals to another pair of students and gain their opinions on the strength of the proposals.
   d) Agree on a solution.
2 Study Figure 10.40. Assuming the camera was pointing south, in which direction is longshore drift travelling? Explain your answer.
3 Draw an annotated sketch of Figure 10.40 to show how a groyne reduces coastal erosion.
4 Draw an annotated diagram to show how either rock armour or gabions reduce the rate of coastal erosion.

▲ Figure 10.41 Rock armour at the back of a beach

▲ Figure 10.42 Gabions near Hengistbury Head, Hampshire

Key

▢ Beach		▢ Hotel	
⊥⊥⊥⊥ Cliff		⚑ Golf course	
▭ Road		▢ Farmland	
▢ Housing		♜ Historic castle	
▢ Shops			

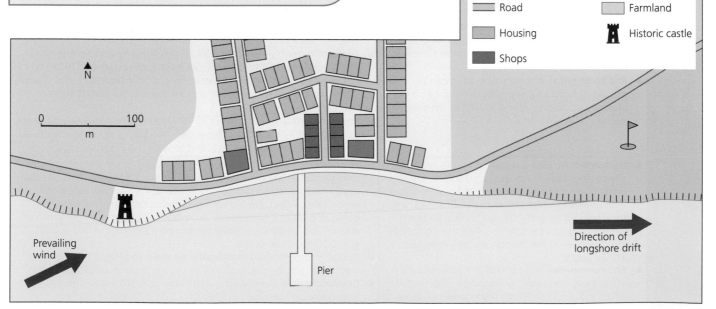

▲ Figure 10.43 A stretch of coastline

KEY LEARNING

➤ The benefits and costs of hard engineering

Benefits and costs of hard engineering

What are the benefits and costs of sea walls, groynes, rock armour and gabions?

▼ Figure 10.44 Benefits and costs of hard engineering

	Benefits	Costs
Sea walls	• **Social**: A sea wall gives people a sense of security. It often has a promenade on top of it, which doubles up as cycle route outside peak walking periods. Steps at the base of a wall act as seating areas for beach users. • **Economic**: If well maintained, sea walls can last for many years. • **Environmental**: Sea walls do not impede the movement of sediment downdrift, so they do not disadvantage other areas.	• **Social**: They restrict people's access to the beach and if waves break over the sea wall (overtopping), coastal flooding may occur. • **Economic**: At about £5,000 per linear metre, sea walls are expensive to build. Repairs are also expensive. Reflected waves scour the beach in front of a sea wall and this undermines its foundations. If damage is not repaired quickly, the result may be devastating. In Dawlish, Devon (February 2014), the sea wall carrying the main south coast railway collapsed causing £35 million repairs to the wall and track (see Figure 10.45). To reduce scouring, rock armour (page 143) and beach nourishment (page 146) may be needed. This adds to the cost. • **Environmental**: From the beach, a wall of concrete is ugly to look at. Sea walls can also destroy habitats.
Groynes	• **Social**: Rock groynes at Sandbanks, Poole, have concrete crests for people to walk along to reach a viewing or fishing point. Groynes also act as windbreaks. • **Economic**: At £5,000 each, groynes are relatively cheap and, if well maintained, can last up to 40 years. A larger beach, with more space for activities, attracts more tourists, which boosts the local economy.	• **Social**: Groynes are barriers, which impede walking along a beach. They are also dangerous, as they have deep water on one side and shallow water on the other. This is a particular hazard to children who find it hard to resist climbing on them. Groynes may also be a danger to wind surfers, who may collide with them. • **Economic**: By trapping sediment, groynes restrict the supply of sediment down-drift. For example, the new groynes at Poole restrict sediment movement towards Bournemouth. The problem is merely passed on to incur more cost. Groynes are ineffective in stormy conditions and need regular maintenance so they do not rot. • **Environmental**: Groynes may be considered unattractive, especially degraded ones.
Rock armour	• **Economic**: It is relatively cheap. Rock armour costs £1,000–3,000 a metre, compared to £5,000 a metre for a sea wall. The structure is quick to build and easy to maintain. It can be built in weeks rather than the months it takes to make a sea wall. If well maintained, rock armour lasts a long time. • **Environmental**: It is versatile, as it can be placed in front of a sea wall to lengthen its lifespan or used to stabilise slopes on sand dunes.	• **Social**: Rock armour makes access to the beach difficult, as people have to clamber over it or make long detours. People may have accidents when clambering over it as rocks may be unstable and, if rocks are regularly covered by the tide, they may collect slippery seaweed, which accentuates the hazard. • **Economic**: Highly resistant rocks from Norway and Sweden are often used in preference to rocks from local quarries. This may cause resentment and it inflates the cost considerably. Also, heavy storm waves will move rocks and so the armour needs regular maintenance. • **Environmental**: Rock armour is ugly and it often covers vast areas of a beach. Driftwood and litter become trapped in the structure and imported rocks do not blend in with the local geology.
Gabions	• **Economic**: At £110 a metre, they are relatively cheap and easy to construct. Gabions are often constructed on site using local pebbles. This makes them much cheaper than sea walls, rock armour or groynes. It also makes them ideal as a quick-fix solution. For the cost, they are good value for money, as they may last 20–25 years. • **Environmental**: They blend in better than other hard engineering methods, especially when sand is blown into them or when they are covered by vegetation.	• **Economic**: The use of gabions is restricted to sandy beaches, as shingle hurled at them would quickly degrade them. Gabions are easily destroyed, so regular maintenance is needed. Repair of embedded, vegetation-covered gabions can be expensive. The gabions built at Thorpeness, Suffolk in 1976 had their covering of topsoil and vegetation washed away by storms in 2010. It cost £30,000 to repair them. • **Social**: In a damaged state, gabions are dangerous. People may trip over them or cut themselves on the broken steel wire mesh. • **Environmental**: Damaged gabions are unsightly and sea birds may damage their feet in them.

There are many examples of hard engineering to protect the coastline along the south coast of England.

In Dawlish, Devon, the main south coast railway follows the coastline protected by a sea wall (Figure 10.45). In February 2014, severe storms led to the sea wall being undermined, causing the railway track to collapse.

▲ Figure 10.45 Overtopping of the sea wall at Dawlish, Devon

Further east along the coast, the seaside resort of Lyme Regis is heavily protected by hard engineering structures. They help maintain the beach. The consequences of not protecting the coast can be seen in the background of Figure 10.46, where cliffs have collapsed due to weathering and marine erosion.

➤ Figure 10.46 Hard engineering structures at Lyme Regis, Dorset

→ Activities

1 Study Figure 10.44. Which type of hard engineering might be preferred by:
   - a coastal resident?
   - a tourist?
   - an environmentalist?
   - a government minister responsible for the coastal protection budget?

   Give reasons for each preference.

2 a) Suggest what damage is being done by overtopping the sea wall at Dawlish (Figure 10.45).
   b) Suggest why the sea wall is not effective.

3 Study Figure 10.46.
   a) Write a report describing and commenting on the level of coastal protection given by these structures in Lyne Regis.
   b) Explain why coastal protection was needed.

Coastal management: soft engineering

Soft engineering works more in sympathy with nature than hard engineering. It is generally less expensive, but often less effective.

What is beach nourishment?

Beach nourishment is a broad term for the replacement of lost sediment. A nourished beach means fewer waves reach the back of a beach. As more wave energy is absorbed and dissipated by the beach, the rate of erosion is reduced. The following techniques show how beaches are nourished.

Beach recharge

This is where sediment is taken from a bay and placed on a beach that is losing sand. This happens every summer at Pevensey (East Sussex), where longshore drift removes 20,000 cubic metres of beach sediment a year. A dredger collects shingle from the seabed and, on the high tide, comes in twice daily to pump out the sand. At Sandbanks in Poole (Dorset), recharge takes place every ten years. Bulldozers are often used to spread out the sand.

Beach recycling

This is the removal of sand from a down-drift area, which is building up sand and returning it up-drift. At some beaches, for example at Seaford, East Suffolk, trucks move around 100,000 cubic metres of shingle twice every year.

What is beach reprofiling?

Beach reprofiling is the artificial re-shaping of a beach using existing beach material. In winter, a beach is lowered by destructive waves (see page 123). After winter storms, bulldozers move shingle back up the beach (see Figure 10.48). Like beach nourishment, reprofiling ensures that the beach is large enough to be an effective buffer between land and sea.

What is sand dune regeneration?

Sand **dune regeneration** is the artificial creation of new sand dunes or the restoration of existing dunes. Sand dunes act as a physical barrier between the sea and the land. They absorb wave energy and water. In this way, they protect the land from the sea.

Sand dunes near Studland, Dorset are often regenerated.

▲ Figure 10.47 Beach recharge at Poole

▲ Figure 10.48 Reprofiling

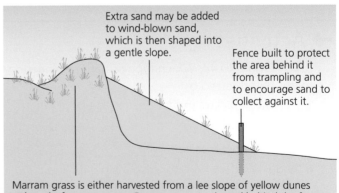

Extra sand may be added to wind-blown sand, which is then shaped into a gentle slope.

Fence built to protect the area behind it from trampling and to encourage sand to collect against it.

Marram grass is either harvested from a lee slope of yellow dunes or bought from a nursery. Grasses are transplanted behind the fence. Long-rooted marram grass helps to stabilise storm-damaged dunes when planted on their eroded windward faces.

▲ Figure 10.49 Sand dune regeneration

What are the benefits and costs of these soft engineering strategies?

▼ Figure 10.50 Benefits and costs of soft engineering

	Benefits	Costs
Beach nourishment	**Social**: A wider beach means more room for beach users. People living along the seafront are more protected from coastal flooding.	**Social**: During re-nourishment, access to the beach is restricted for several weeks. Beach recycling may cause resentment from residents living close to the donor area.
	Economic: At Sandbanks, the wider, nourished beach protects very expensive properties. The buffer of a widened beach reduces sea wall maintenance costs. A broader beach may also attract more tourists.	**Economic**: Although cheaper than hard engineering options, this has high overheads as it costs around £300,000 to hire a dredger. The 137,000 m³ of nourishment at Sandbanks in 2014 cost £1.95 million.
	Environmental: A nourished beach is natural and blends in with the environment.	
Beach reprofiling	**Social**: At Pevensey, the residential area behind the beach is now protected so residents feel safe.	**Social**: Bulldozers restrict access to Pevensey's beach, especially in winter.
	Economic: If the shingle ridge at Pevensey is breached, the estimated repair cost would be about £125 million, whereas the combined cost for nourishment and frequent reprofiling is £30 million over 25 years.	**Economic**: Major reprofiling costs can be expensive. Further west along the coast at Selsey (West Sussex), £200,000 a year was paid to realign the beach prior to the Medmerry Scheme (see Section 10.15).
	Environmental: In preventing a breach, the Pevensey Levels has been protected and the beach still looks reasonably natural.	**Environmental**: A steep, high-crested beach may look unnatural and uninviting to tourists.
Sand dune regeneration	**Social**: Sand dunes protect land uses behind them. Once established, they are popular as picnic and walking areas.	**Social**: While becoming established, regenerated sand dunes are fenced off and signs tell people to keep out. This may deter tourists.
	Economic: Small planting projects often use volunteer labour and local grass for transplants so costs are minimal.	**Economic**: • Dune regeneration has to be checked twice a year and have fertilisers applied. • Expensive systems have to be put in place to protect planted areas from trampling. Studland beach receives up to 25,000 people a day in summer. To reduce the risk of damage, boardwalks have been built through the dunes, fire warnings and fire beaters put in place, and all tourist facilities are contained around a solitary car park.
	Environmental: At Studland sand dune, regeneration has helped maintain a habitat for rare Dartford warblers, nightjars and chiffchaffs. Six species of reptiles, including adders and sand lizards, inhabit the dunes, along with damselflies and dragonflies, which hover in the slacks.	**Environmental**: It must also be remembered that sand dunes are a dynamic environment. Once regenerated, there is no guarantee that they will be stable. The grass may soon be damaged by storms, and even in favourable conditions it will take two to three years before grasses become established and begin to spread. If the coast is highly exposed, then additional, less attractive engineering such as rock armour is needed.

➜ Activities

1 Outline the difference between beach nourishment and beach reprofiling.

2 Devise an instruction manual for creating a new sand dune. Each stage needs a cartoon-type diagram.

  a) State two reasons why regenerated sand dunes are fenced off.

  b) Explain why there may be public opposition to fencing off the dunes.

3 Study the information in Figure 10.50. Compare it with Figure 10.44 on page 144. Which type of engineering – hard or soft – might be preferred by each of the following:

  ■ a coastal resident?

  ■ a tourist?

  ■ an environmentalist?

  ■ a government minister responsible for the coastal protection budget?

  Give reasons for each preference.

⭐ **KEY LEARNING**

➤ Managed retreat
➤ The benefits and costs of managed retreat

Coastal realignment

Creating an engineered new position of a coastline is called **coastal realignment**. In the context of managing coastal flooding in the UK, this involves moving the boundary inland.

What is managed retreat?

Managed retreat is when a decision is made to no longer follow a 'hold the line' strategy for managing coastal flooding and erosion. People are moved out, buildings are demolished then a breach is made in the existing sea defences so the sea can inundate the land and create new intertidal habitats. If the area is low-lying then, prior to the breach, an inland embankment is built to protect the area.

At Wallasea Island, Essex (Figure 10.52), over 600 hectares of mudflats, salt marsh and saline lagoons have been created. This photo was taken six years after the breach was made, allowing the River Crouch to inundate the land.

Coastal realignment offers a sustainable long-term solution to the prospect of rising sea levels associated with global warming. It is better to be pro-active and manage a planned inundation of water than to keep reacting to breaches. Defences may be installed further inland where they can operate more effectively.

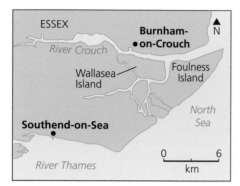

▲ Figure 10.51 Location of Wallasea Island, Essex

▲ Figure 10.52 Allfleet's Marsh Managed Realignment on Wallasea Island, Thames Crouch Estuary

What are the benefits and costs of managed retreat?

Benefits

- *Social*: It may help take the pressure off areas further along the coast and reduce their risk of flooding.
- *Economic*: It is often cheaper in the long term to use managed realignment than to continue to maintain hard engineering defences.
- *Environmental*: Managed realignment is designed to conserve or enhance the natural environment. It creates new intertidal habitats that compensate for those lost through coastal squeeze. At Wallasea, 38 species of bird have been recorded on the new mud flats, some in considerable numbers. These include ringed plover, dunlin, Brent geese and shellduck (Figure 10.53).

Costs

- *Social*: Relocation of people to new homes causes disruption and distress. If the long-term plan for the realignment of 40 square kilometres of the North Norfolk coast goes ahead, this will destroy six villages (Figure 10.54). Hundreds of people will have to be re-housed and whole communities will be split up. This could be a reality in 20–50 years' time. People feel 'let down' by managed retreat and feel the battle against the sea should continue.
- *Economic*: Short-term costs may be high, as relocation costs have to be paid. The recent Medmerry realignment scheme, in West Sussex cost £28 million, when it only cost £0.2 million a year to profile the shingle beach (Section 10.15).
- *Environmental*: Large areas of agricultural land are lost. Habitats of coastal birds such as bitterns, cranes and marsh harriers would be affected, so bird numbers would initially decline. It may take a long time to reach their previous numbers.

▲ Figure 10.53 Birds feeding on mudflats

Norfolk coastal erosion

Millions of people live close to the UK's coast and many will become increasingly vulnerable to rising seas as our climate changes. In northeast Norfolk, people are facing the possibility that they may be among the country's first climate change refugees. Environmental scientists suggest that one option to tackle erosion along a nine-mile stretch of coast is not to build more concrete sea defences but begin a 'managed retreat' from the sea. A section of the low-lying Broads, including six villages, could be lost to the waves if the plans were approved. But shocked residents are rallying to appeal for more money to protect their homes.

◄ Figure 10.54 An article from *The Guardian*, 17 April 2008

→ Activities

1 Define the term coastal realignment.
2 Draw a sketch of Figure 10.52. Label the position of the new coastline, the position of the old coastline, a lagoon, salt marshes, mudflats, and fields growing oilseed rape (indicated by yellow patches).
3 In pairs, conduct a 'for and against' argument for managed retreat. One of you could act as a resident in one of the threatened villages in north Norfolk. The other could act as a government minister with responsibility for the coastal protection budget.

Example

⭐ KEY LEARNING
➤ Why the scheme was needed
➤ What the strategy was
➤ What the positive effects and conflicts were

Coastal realignment in Medmerry

Medmerry in West Sussex is the largest managed coastal realignment scheme in Europe.

Why was the scheme needed?

The **Environment Agency (EA)** considered the area to the west of Selsey (West Sussex) to be the area of South East England most at risk of flooding due to **climate change**. A shingle ridge was the only protection from the sea, and from the 1990s beach reprofiling (page 146) took place every winter, at an annual cost of £200,000. This was becoming unsustainable. If breached, then 348 properties in Selsey, a water treatment plant and the main road between Chichester and Selsey would be flooded, along with many holiday homes and rental cottages. The last breach, in 2008, caused £5 million of damage.

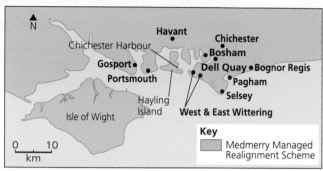

▲ Figure 10.55 Map of the south coast

What strategy was used?

Following public consultation, work to realign the coast began in 2011 and was completed in 2014. Managed retreat was achieved by the following:

- Building a new embankment, up to two kilometres inland from the shore, using clay from within the area. This embankment enclosed the future intertidal area and protected the properties behind it.
- Behind the embankment, a channel was built along its whole length to collect draining water. Four outfall structures were built into the embankment to take the water into the intertidal area.
- Rock armour was then placed on the seaward edges of the embankment, where it linked up with the remaining ridge. This used 60,000 tonnes of hard rock from Norway.
- Once the embankment and rock armour were in place, a 110-metre breach was made in the shingle bank to allow the sea to flood the land behind to create a new intertidal area.

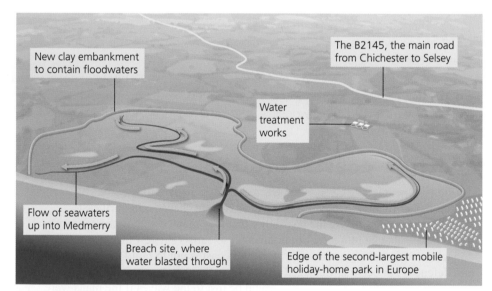

▲ Figure 10.56 Medmerry realignment

What happened as a result?

Positive effects

Social:

- Selsey now has a 1 in 1,000 chance of coastal flooding which provides the best level of protection in the UK. Disruption from a possible breach during the 2013 winter storms was avoided.
- A maintenance access track behind the embankment provides a cycle route and footpath. Today, there are ten kilometres of footpaths, seven kilometres of new bike paths and five kilometres of new bridleways in an area that previously only had two short footpaths.

Economic:

- Tourism, a main contributor to the local economy, is expected to increase. Two new car parks and four viewing points give easy access. An increase in visitor numbers to the nearby holiday village is expected as the area gains a reputation with birdwatchers and people interested in nature.
- The newly flooded area is expected to become an important fishing nursery that will boost the local fishing industry in Selsey. The salt-marsh vegetation will also be used for extensive cattle farming, to produce expensive salt-marsh beef.

Environmental:

- By carrying out a detailed environmental assessment prior to flooding, designers were able to take measures to protect existing species, such as water voles, crested newts and badgers.
- 300 hectares of new intertidal habitats are forming seaward of the embankment. Mudflats, salt marshes and transitional grasses have already attracted large numbers of ducks and lapwings. The area is turning into a huge nature reserve managed by the RSPB.

Controversy and conflicts

Social:

- Some local residents still feel that the EA should not have given up the land so easily and insist they should have looked into other options, such as offshore reefs or continued beach reprofiling.
- Some opponents of the scheme came from outside the area; they resented such an expenditure in a **sparsely populated** area. Would the money not have been better spent draining the Somerset Levels, for example? The need to compensate for coastal squeeze would have been a strong consideration at Medmerry.

Economic:

- At a cost of £28 million, the scheme was very expensive. Can this be justified if it only cost £0.2 million a year to maintain the shingle beach? However, with rising sea levels, continued reprofiling was not a viable long-term option.
- For the realignment scheme to take place, three farms growing oilseed rape and winter wheat had to be abandoned. Losing good agricultural land was regarded by some people as being wasteful, and this raises questions regarding the priority for protecting buildings over agricultural land. Is this a short-sighted approach?

Environmental:

- Despite planning, habitats of existing species, such as badgers, would have been disturbed.

→ Activities

1   In groups of four, engage in a role-play of a meeting of local residents when they first heard of the scheme. Possible roles: farmer, local fisherman, RSPB member, holiday homes manager. Together, decide whether you are for or against the scheme.

2   Explain why the newly created environment may increase tourism in the area.

3   Place tracing paper over Figure 10.56 to draw a frame for your own map. Trace the new coastline, the embankment, the waterworks, the B2145 road and outline of the area of holiday homes. Annotate your map to describe and explain the stages of the realignment scheme.

11 River landscapes

⭐ KEY LEARNING

➤ How a river erodes
➤ How a river transports its load
➤ How and why a river deposits its load

Rivers

A river carries excess water on land, mainly from **precipitation**, to the sea. Along its journey to the sea, **fluvial** (river) **processes** of **erosion**, transport and deposition take place, helping to shape the river's channel and its valley. The channel is the groove through which a river flows, and consists of its banks and bed.

How does a river erode?

Fluvial erosion is the process by which a river wears away the land. The ability of a river to erode depends on its velocity. Erosion takes place in four ways: hydraulic power, abrasion, solution and attrition.

Hydraulic action

This is when the sheer force of fast-flowing water hits the river banks and river bed and forces water into cracks. This compresses air in the cracks. Repeated changes in air pressure weaken the channel. **Hydraulic action** is responsible for **vertical erosion** in the upper course of a river making the valley deeper. In the lower course, it contributes to **lateral erosion** of the banks, especially when fast-flowing water hits the outside bend of a **meander**, changing the course of the river.

Figure 11.1 shows how erosion affects both the bed and banks of a river channel, causing vertical and lateral erosion.

■ Vertical erosion is the deepening of the river bed, mostly by hydraulic action. It is most evident in the upper course of a river. Here, what little energy the river has left over after overcoming friction is used to deepen its channel.
■ Lateral erosion is 'sideways' erosion. It wears away the banks of the river. This is most evident in the lower course of a river.

Abrasion

This is also called corrasion. Small boulders and stones may scratch and scrape their way down a river during transport, thereby wearing down the river banks and bed. Stones which have fallen into the channel quite recently will be angular and have sharp, jagged edges. These are particularly effective tools of **abrasion**. Ongoing abrasion is responsible for both vertical erosion and lateral erosion of the channel.

Solution

This is also called corrosion. **Solution** refers to the dissolving of rocks such as chalk and limestone. Rivers travelling over these rocks will erode them in this way.

Attrition

Attrition affects a river's **load**. When stones first enter a river, they will be jagged and angular. As they are transported downstream, stones collide with each other and also with the river banks and bed. This gradually knocks off the stones' jagged edges so they become smooth and more rounded (Figure 11.2). Some collisions may cause a stone to smash into several smaller stones. These re-sized stones will be further smoothed and rounded on their journey to the sea.

▲ Figure 11.2 Rounded boulders on a riverbed

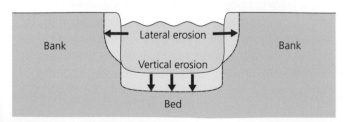

▲ Figure 11.1 Vertical and lateral erosion

How does a river transport its load?

Fluvial transport is the process by which a river carries its load (Figure 11.3). Load differs in size, from large, angular boulders in the upper course, to fine, suspended silt in the lower course. Load mostly comes from material that has weathered and tumbled down the hillside, though some also comes from eroded river banks.

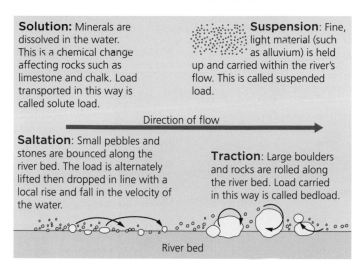

Solution: Minerals are dissolved in the water. This is a chemical change affecting rocks such as limestone and chalk. Load transported in this way is called solute load.

Suspension: Fine, light material (such as alluvium) is held up and carried within the river's flow. This is called suspended load.

Direction of flow

Saltation: Small pebbles and stones are bounced along the river bed. The load is alternately lifted then dropped in line with a local rise and fall in the velocity of the water.

Traction: Large boulders and rocks are rolled along the river bed. Load carried in this way is called bedload.

River bed

▲ Figure 11.3 River transport processes

Why do rivers deposit sediment?

Fluvial deposition is how a river drops its load. The material deposited is called **sediment**. The bigger the load particle, the greater the velocity needed to keep it moving. When velocity falls, large boulders are the first to be deposited. The finest particles are deposited last. This explains why mountain streams have boulders along their bed, while close to the river's mouth there is only fine silt. Along its course, a river will deposit its load wherever the velocity falls, where the river enters a sea or lake. It will also deposit more of its load during drought, when less water flows. The Hjulstrom Curve (Figure 11.4) shows the different critical velocities at which erosion, transport and deposition occur.

- The blue area shows that large 10 mm-diameter particles are transported between around 85 and 110 cm/sec. When velocity falls below the critical velocity of 85 cm/sec (and enters the green section), these 10 mm particles are deposited.

- A 0.1 mm diameter particle will be transported between around 0.9 and 35 cm/sec. When velocity falls below 0.9, these 0.1 mm particles are deposited.

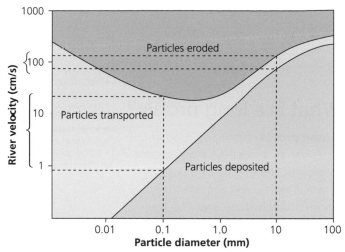

▲ Figure 11.4 The Hjulstrom Curve

→ Activities

1 Give four types of fluvial erosion. For each, write a sentence in your own words to describe it.

2 Explain how fluvial processes cause vertical and lateral erosion.

3 a) Draw three annotated diagrams to describe how a rock fragment that has fallen into a river will change as it moves downstream. Use appropriate adjectives, such as angular, jagged, rounded, rough and smooth.

  b) In your annotations, explain the changes you have shown. Include the following terms: attrition, transported, abrasion, traction, collide, banks and bed.

4 a) Define the term 'load'.

  b) State two origins of load found in a river.

5 Study Figure 11.4.

  a) At what range of velocities will a 1.0 mm particle be transported?

  b) Below what velocity will a 1.0 mm particle be deposited?

6 State the general relationship between particle size and the velocity at which a particle is deposited.

⊛ KEY LEARNING

➤ A river's long profile
➤ How and why the long profile changes
➤ Why discharge and velocity increase downstream

The long profile of a river

A river is nature's way of removing excess water from the land. As it does so, it changes the **landscape** and creates landforms. The area drained by a river and its tributaries is called a river basin or a **drainage basin**.

What is a long profile?

A **long profile** shows the gradient of a river as it journeys from source to mouth. The source of a river is where it starts, and the mouth is where it reaches the sea. The River Severn travels 354 kilometres from its source in the Plynlimon Hills in the Cambrian Mountains to its mouth in the Bristol Channel (Figure 11.5).

A river tries to achieve a smooth curve in order to reach its base level at the sea. This is called a graded long profile. Figure 11.7 shows the long profile of the River Severn. Notice how the gradient falls steeply at first, then becomes concave and then almost flat.

1. Upper course (above approx. 160 m)
River travelling through mountains. Has a very steep V-shaped valley
• Very steep gradient
• Vertical erosion
• Narrow shallow channel
• Low velocity
• Small discharge
• Rough river bed
• Large angular rocks on the bed

2. Middle course (approx. 60–160 m)
Travelling through hills. Has a broad valley
• Gradient less steep
• Lateral erosion more than vertical erosion
• Channel is wider and fairly deep
• Fairly high velocity
• Quite high discharge
• River bed less rough than in upper course
• River bed sediments smaller and smoother

3. Lower course (under approx. 60 m) Travelling through a large area of low-lying land. Has a very broad valley
• Very gentle gradient
• Very wide valley
• Mostly deposition
• Very wide and deep river channel
• High velocity
• High discharge
• Smooth river bed covered with alluvium

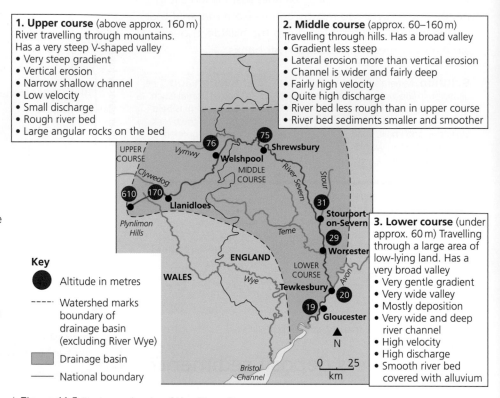

Key

⬤ Altitude in metres

---- Watershed marks boundary of drainage basin (excluding River Wye)

▨ Drainage basin

— National boundary

▲ Figure 11.5 Drainage basin of the River Severn

How and why does the long profile change?

Figure 11.7 shows how the long profile can be split into three courses based on gradient. Figure 11.6 describes the processes that happen in each of the three courses.

	Upper course	Middle course	Lower course
Erosion	Mostly vertical erosion by hydraulic action	Less vertical erosion, more lateral erosion. Much attrition and abrasion, some solution	Very little vertical erosion, only lateral erosion
Transport	Mostly traction. Large boulders moved	Mostly suspension, less traction. Load becomes smaller and less angular	Mostly suspension and solution. Very small particles of load. Great quantity of load
Deposition	Large boulders deposited	More deposition, especially on the inside bend of meanders	Deposition now the main fluvial process. Fine material is now deposited
How the long profile changes	The upper course is set in a landscape of high **relief**. The long profile starts at its source. Trickles begin to merge to form a single channel which flows down a steep gradient. The steep descent gives the river more potential energy. In places, there may be waterfalls and rapids.	The middle course is further downstream in an area of hilly rather than mountainous relief. The channel is deeper, and the volume of water has been increased by the many tributaries that have joined the main river. As vertical erosion reduces, the gradient of the long profile becomes concave.	The lower course is the section closest to the river mouth, where the surrounding land is low-lying. Erosion is now confined to lateral erosion at meanders. Lack of vertical erosion means the gradient is almost flat.

▲ Figure 11.6 Processes operating in the three courses

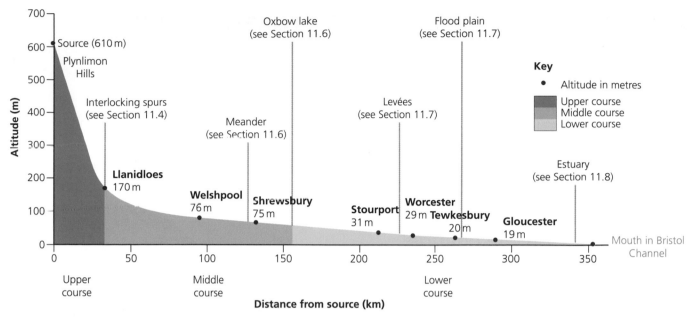

▲ Figure 11.7 Long profile of the River Severn, Wales and England

Why do discharge and velocity increase downstream?

Discharge is the volume of water passing through a given point on the river. It is measured in cubic metres per second (cumecs). Discharge = velocity × cross-sectional area. Discharge increases downstream as tributary streams join the main river and add their volume of water to it.

The average velocity of a river increases along its course. Despite the steep descent in the upper course, it is the lower course which has the greatest velocity. Velocity depends on how much water comes into contact with the channel's banks and bed. In the upper course, a small channel means there is much friction. So, despite the steep slope, velocity is low. Conversely, in the broad, deep channels of the lower course, less water is in contact

with the bed and banks, so velocity is much higher. This is because speed is boosted by the additional discharge from all the tributaries.

 Fieldwork: Get out there!

1 Assume you have been given the following equipment: a stop watch, two one-metre rules, a tape measure and an orange. Explain how you could use this equipment to work out the velocity of a river in its upper course.

2 If you repeated the experiment downstream, how would you expect the velocity to change? Explain your answer.

→ Activities

1 Study Figure 11.5.

a) Name a town in the upper, middle and lower course of the River Severn.

b) From the list below, select the area that best describes the size of the drainage basin of the River Severn (use the scale on the map):

■ 5,000 square kilometres

■ 11,000 square kilometres

■ 21,000 square kilometres.

2 Study Figure 11.7.

Draw an annotated long profile of a river to describe the three stages along its course. For each stage, mention the:

■ river gradient

■ channel slope

■ valley slope

■ velocity and discharge.

3 a) Describe how the following things change along the course of a river: river width, depth, type of erosion, transport and deposition.

b) Explain why the changes occur.

➤ A river's cross-profile
➤ How and why a channel's cross-profile changes downstream
➤ How and why a valley's cross-profile changes downstream

Changing cross-profiles of a river

What is a cross-profile?

A **cross-profile** is a section taken sideways across a river channel and/or a valley.

■ A channel cross-profile only includes the river.
■ A valley cross-profile includes the channel, the valley floor and the slopes up the sides of the valley (Figure 11.8).

How and why does a channel cross-profile change downstream?

How does it change?

Changes downstream in the channel cross-profile are summarised in Figure 11.9.

Why does it change?

■ In the upper course, the river erodes its bed by hydraulic and abrasive action. As the river travels downstream, it is joined by a number of tributaries. These increase the volume of water which gives the river kinetic energy, a higher velocity and thus more erosive power. This allows it to cut a much deeper channel with increased distance downstream.
■ The channel becomes wider downstream because, as the gradient becomes less steep, there is less vertical erosion. By the time the river is in the middle course, lateral (sideways) erosion is dominant. This erodes the river banks, which makes the channel wider.

Steep V-shaped cross-profile

Interlocking spurs

River takes up most of the valley floor

▲ Figure 11.8 Valley cross-profile of the upper course of the River Severn

▼ Figure 11.9 How a channel cross section changes downstream. Note: The diagrams are for a straight section of a river. At a meander, the cross-profile will become asymmetrical (see page 162)

Upper course	Middle course	Lower course
 River	 River	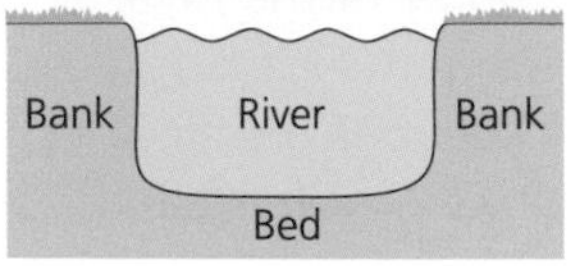 Bank River Bank / Bed
The channel is very narrow (only a few metres wide) and very shallow (less than 0.25 metres deep).	The channel becomes wider. It may be several metres wide. For many rivers, it will be over a metre deep.	The channel becomes wider still. A small river may only be 5–10 metres wide, whereas the River Severn is 3.2 kilometres wide at its mouth by the old Severn Bridge crossing. The channel is much deeper.

How and why does a valley cross-profile change downstream?

How does it change?

▼ Figure 11.10 How the valley cross-profile changes downstream

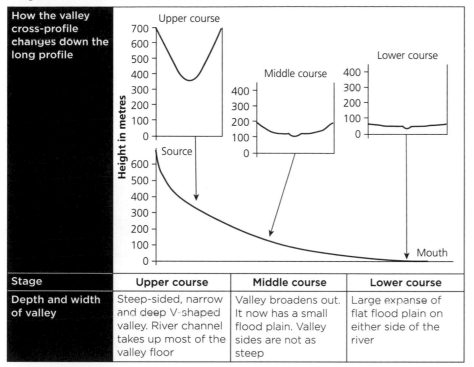

Stage	Upper course	Middle course	Lower course
Depth and width of valley	Steep-sided, narrow and deep V-shaped valley. River channel takes up most of the valley floor	Valley broadens out. It now has a small flood plain. Valley sides are not as steep	Large expanse of flat flood plain on either side of the river

Why does it change?

In the upper course, there is a steep, V-shaped cross-profile. Vertical erosion by the river is the dominant process operating in the valley. This creates a slope that weathered material from the valley sides can fall down. On reaching the river, this material is removed. Rivers tend to have their source in upland areas, which means the rock is harder. The valley sides are therefore not broadened out much by weathering and erosion so slopes remain steep.

In the middle course, the river is flowing through lower country. The gradient is less steep, so the river begins to bend and erode laterally (sideways) into the valley sides. This broadens out the valley. In addition to this, the rate of weathering increases on the softer rocks of the valley sides. As the river uses more energy in lateral erosion, it is not able to remove all of the weathered material, so this builds up the valley floor to give it a more gentle profile.

In the lower course, the river is passing through low-lying country. Deposition from **floods** builds up the **flood plain**, and widens the valley.

Geographical skills

Using a grid like this, draw a valley cross-section of the Exe valley in Figure 11.11. Draw the cross-section west to east along the northing 33.

(Graph: Height in metres (180–280), axis labelled W — River Exe — E)

→ Activities

1   a) Draw a labelled sketch of Figure 11.8. Add to the labels to describe, in detail, a valley cross-profile in its upper course.
    b) Describe how this valley cross-profile would change in its middle course.
    c) Explain these changes. Include the following terms in your answer: vertical erosion, lateral erosion, deposition, and weathering.
2   a) Draw a labelled diagram to show the channel cross-profile in the upper course.
    b) Describe how the channel cross-profile changes in the middle course.
    c) Explain the changes. Include the following terms: vertical erosion, lateral erosion and velocity.

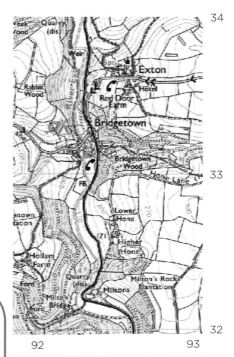

▲ Figure 11.11 OS map extract showing the River Exe at Exton. © Crown copyright and database rights 2020 Hodder Education under licence to OS

Interlocking spurs and rapids

Erosion processes result in the formation of **interlocking spurs** and rapids, on the upper course of a river.

What are interlocking spurs?

Interlocking spurs are projections of high land that alternate from either side of a valley and project into the valley floor, formed by fluvial erosion (Figure 11.12). They are found in the upper course of a river where rocks are hard, like in the Afon Dulas Valley, a tributary valley of the River Severn. Notice the zip-like interlocking nature of the hillsides, the very narrow valley floor with its winding river, and how the river takes up most of the valley floor. Figure 11.14 shows the OS map extract of the same area.

Characteristics of interlocking spurs:

- a steep gradient
- convex slopes
- project from alternate sides of the valley
- separated by a narrow valley floor mostly taken up by the river channel
- sometimes wooded
- may have scree slopes.

Figure 11.13 describes stages in the formation of interlocking spurs.

▲ **Figure 11.12** Interlocking spurs

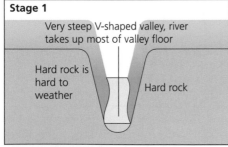

Stage 1

Very steep V-shaped valley, river takes up most of valley floor

Hard rock is hard to weather

Hard rock

In the upper course of a river, the river's water volume and discharge are low. The river uses most of its energy overcoming friction with the channel. What energy it has left over is used by hydraulic action to deepen the channel (vertical erosion).

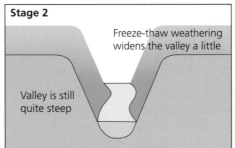

Stage 2

Freeze-thaw weathering widens the valley a little

Valley is still quite steep

In upland areas, the geology is composed of hard rock such as granite or slate. However, freeze-thaw weathering (see Section 10.2, pages 124–25) gradually broadens it out, giving the valley a steep, V-shaped cross profile. Repeated weathering weakens the rock so fragments break loose and tumble down the hillside as scree, which the river then removes.

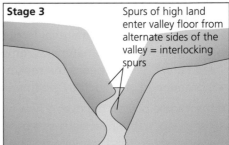

Stage 3

Spurs of high land enter valley floor from alternate sides of the valley = interlocking spurs

The winding path taken by the river is due to obstacles of harder rock in its path. The river takes the easiest route over the land. This results in projections of high land entering the valley from alternate sides. These projections are the interlocking spurs.

▲ **Figure 11.13** Stages in the formation of interlocking spurs

What are rapids?

Rapids are fast-flowing, turbulent sections of a river where the river bed has a relatively steep gradient, usually in the upper course. The ability of a river to erode its bed depends on the river's energy.

It also depends on the degree of hardness of the bedrock. In some cases, there may be vertical bedding, whereby alternate bands of hard and soft rock cross the channel (Figure 11.15). Differential erosion will occur, as soft rock is more easily eroded than hard rock. This makes the river bed uneven and the river's flow becomes turbulent, resulting in 'white water' sections typical of rapids. The River Severn has rapids just south of the Ironbridge Gorge.

Key

- Soft rock
- Hard rock

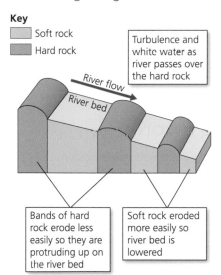

▲ Figure 11.15 Formation of rapids

Turbulence and white water as river passes over the hard rock

River flow

River bed

Bands of hard rock erode less easily so they are protruding up on the river bed

Soft rock eroded more easily so river bed is lowered

Fieldwork: Get out there!

Imagine you were visiting Jackfield Rapids (Figure 11.16) to carry out fieldwork to determine the velocity of the river.

1 Identify two potential hazards of the site.

2 Suggest precautions that may be taken to reduce the risk.

Notice the brown contour lines that show the height and slope of the land. They show high land projecting into the Afon Dulas Valley from alternating sides of the valley. The contours are close together, showing that the slopes of the interlocking spurs are steep.

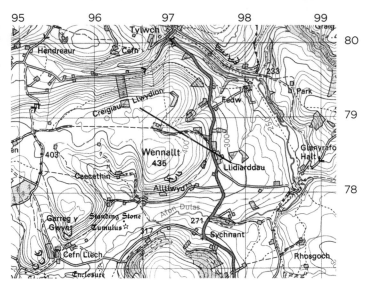

Scale 1 : 50 000

▲ Figure 11.14 OS map extract of the Afon Dulas Valley near Llanidloes. © Crown copyright and database rights 2020 Hodder Education under licence to OS

▲ Figure 11.16 Jackfield Rapids in River Llugwy, Wales

→ Activities

1 Draw a sketch of Figure 11.12 and add labels to describe the characteristics of the interlocking spurs and the nature of the valley.

2 Suggest three reasons why few people live in the area in Figure 11.14.

Geographical skills

1 Study Figure 11.14, the OS map extract.

a) State the four-figure reference for Garreg y Gwynt.

b) What is the maximum height shown on the map? Give a six-figure reference for its location.

c) Draw an annotated sketch cross-profile from north to south along eastings 96. Label the Afon Dulas River and interlocking spurs.

✪ KEY LEARNING

➤ The characteristics of a waterfall

➤ How a waterfall is formed

➤ The characteristics of a gorge

➤ How a waterfall creates a gorge

Waterfalls and gorges

Waterfalls and **gorges** are channel landforms found in the upper course of a river.

What are the characteristics of a waterfall?

A waterfall is where water falls down a vertical drop in the channel, usually from a considerable height (Figure 11.18). The River Severn itself does not have a waterfall, although the Water Break-its-Neck waterfall on one of its tributaries is sometimes referred to as the River Severn Waterfall (Figure 11.20).

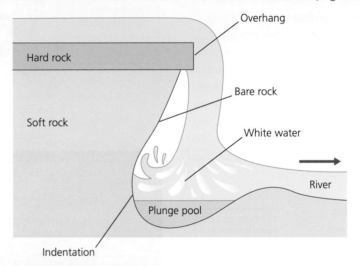

▲ Figure 11.18 Characteristics of a waterfall

What characterises a gorge?

A gorge is a narrow, steep-sided valley, with bare, rocky walls. A gorge of recession is a gorge found immediately downstream of a waterfall (see Figure 11.19). Gorges may be classed as either channel or valley landforms. Although the River Severn does have a gorge at Ironbridge, which displays some of these characteristics, it was primarily formed as a glacial overflow channel, so cannot be used as an example of fluvial (river) erosion.

Characteristics:

- very narrow valley
- very steep, high valley sides
- located immediately downstream of a waterfall
- river channel takes up most, if not all, of the valley floor
- turbulent, fast flowing white water
- many areas of bare rock on valley sides
- boulders litter the river bed.

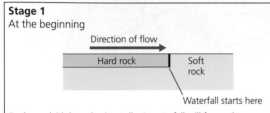

Stage 1
At the beginning

Rocks are laid down horizontally. A waterfall will form where there is a junction between a hard rock capping upstream and soft rock downstream. Differential erosion means the river erodes the softer rock and the water falls vertically from the hard rock to the soft rock below.

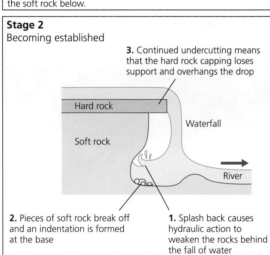

Stage 2
Becoming established

3. Continued undercutting means that the hard rock capping loses support and overhangs the drop

2. Pieces of soft rock break off and an indentation is formed at the base

1. Splash back causes hydraulic action to weaken the rocks behind the fall of water

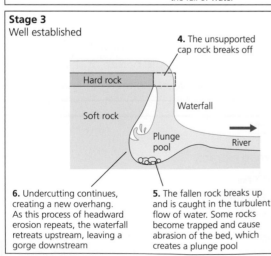

Stage 3
Well established

4. The unsupported cap rock breaks off

6. Undercutting continues, creating a new overhang. As this process of headward erosion repeats, the waterfall retreats upstream, leaving a gorge downstream

5. The fallen rock breaks up and is caught in the turbulent flow of water. Some rocks become trapped and cause abrasion of the bed, which creates a plunge pool

▲ Figure 11.17 Stages in the formation of a waterfall

How does a waterfall create a gorge?

By the end of stage 3 in the formation of a waterfall (Figure 11.17), the scene is set for a gorge to be formed. As the waterfall retreats upstream it leaves a steep-sided valley downstream which is called a gorge. Every time the overhanging cap rock breaks off, the waterfall retreats and the gorge grows longer.

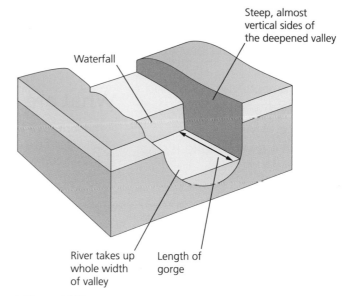

▲ Figure 11.19 Block diagram of a gorge

→ Activities

▲ Figure 11.20 Water Break-its-Neck waterfall, Radnor Forest, Wales, on a tributary on the River Severn

▲ Figure 11.21 Fairy Glen, a gorge on the Conwy River near Betws-y-Coed, North Wales

1 Draw a sketch of the waterfall in Figure 11.20. Label the following features:

Waterfall; Layer of hard rock overlying waterfall; Layer of softer rock at base of waterfall; Broken fragments of rock; Plunge pool created by erosion; Gorge (formed downstream of waterfall).

2 Draw a sketch of the gorge in Figure 11.21. Annotate it to explain how the gorge might have been formed, as the waterfall retreats upstream. Use the following terms:

Overhang; Headward erosion; Soft rock; Hard rock capping; Undercutting; Hydraulic power; Steep-sided.

3 a) Suggest what it is about landscapes with waterfalls and gorges that attracts tourists.
   b) Based on what you know about river processes, and their information from these pages, write a safety leaflet for people visiting a waterfall or gorge.

Meanders and oxbow lakes

A combination of fluvial erosion and deposition leads to the formation of meanders and oxbow lakes, mainly in the middle and lower courses of a river.

What are the characteristics of a meander?

A meander is a bend in a river. The characteristics of a meander are shown in Figure 11.22. The River Severn has many meanders some of which have carved out huge loops. The centre of Shrewsbury is inside one of these large loops.

Inside bank
Slip-off slopes on River Severn near Welshpool

- Curved, beach-like feature on the inside bank
- Very gentle, convex slope
- Sediment consists of sand, gravel and pebbles that are smoothed and rounded by attrition
- Vegetation begins to grow furthest from the water
- Slow flowing water

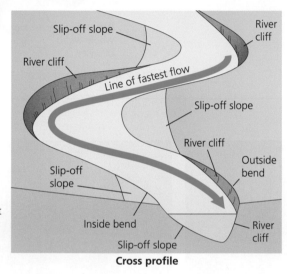

Cross profile

Outside bank
River cliff on River Severn near Buildwas

- A steep drop down into the river on the outside bend
- Can be several metres high
- Composed mostly of bare earth
- Unconsolidated material at the base
- Fast flowing water

▲ Figure 11.22 Characteristics of a meander

How is a meander formed?

In the early stages of meander formation, water flows slowly over shallow areas (riffles) in the riverbed and faster through deeper sections (pools). This eventually sets in motion a helicoidal flow that corkscrews across from one bank to another. This starts the erosion and deposition processes which continuously shape a meander.

▲ Figure 11.23 A meandering river and meander scars, which mark former positions of a meandering river

- Fast-flowing water on the outside bank causes lateral erosion through abrasion and hydraulic action, which undercuts the bank and forms a **river cliff**. (The point of maximum erosion is slightly downstream of the mid-point of the loop.)
- Helicoidal flow is a corkscrew movement. The top part of the flow hits the outside bank and erodes it. The flow then 'corkscrews' down to the next inside bend, where it deposits its load as friction slows the flow.
- Fast flow causes vertical erosion on the outside bend. This deepens the river bed, resulting in an asymmetrical cross-profile.
- Sand and pebbles are deposited on the inside bank where the current is slower, forming a gentle slip-off slope.

A sinuous river is one with many meanders. The loops increase in size as erosion continues on the outside bank and deposition continues on the inside bank. As meanders grow, they move or migrate over the flood plain. A river today may have been in a completely different part of the valley in the past (Figure 11.23).

What are the characteristics of an oxbow lake?

An **oxbow lake** is a small, horseshoe-shaped lake that is located several metres from a fairly straight stretch of river in its middle and lower courses. These landforms may be seen near Welshpool on the River Severn.

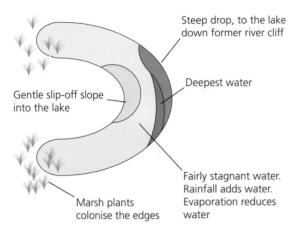

▲ Figure 11.24 Characteristics of an oxbow lake

Labels:
- Steep drop, to the lake down former river cliff
- Deepest water
- Gentle slip-off slope into the lake
- Fairly stagnant water. Rainfall adds water. Evaporation reduces water
- Marsh plants colonise the edges

→ Activities

1 Draw a cross-profile of a meander. On it, label the following: lateral erosion, vertical erosion, greatest velocity, slow-flowing water, shallow water, deposition, slip-off slope, river cliff, deepest water.

2 a) Suggest how the cross-profile you have drawn may differ from that of a straight section of a river.

 b) Explain why the cross-profiles would differ.

3 a) Sketch or trace the course of the river shown in Figure 11.23. In a different colour, sketch how you think this river would look in 200 years' time.

 b) Explain why you have given the river this new route.

4 Use Figure 11.25 to write down at least ten phrases, in sequence, to explain the formation of an oxbow lake. Make sort cards using these phrases. Shuffle these, then test yourself to re-order them.

5 Without using a diagram, describe the characteristics of an oxbow lake.

How is an oxbow lake formed?

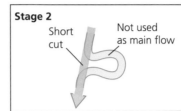

Stage 1 — Narrow neck being eroded
- Meander loop becomes very large.
- Only a narrow strip of land separates the river channel (the meander neck).
- Continued lateral erosion.
- Neck becomes increasingly narrow.

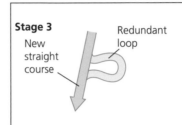

Stage 2 — Short cut / Not used as main flow
- River floods, so main flow of water cuts straight across the neck.
- This 'shortcut' begins to break down the banks and carve a new channel.

Stage 3 — New straight course / Redundant loop
- Floods recede, so the river reverts to its normal meandering channel.
- Process is repeated over and over again with every flood event.
- This new channel becomes so established by the continued lateral and vertical erosion that it becomes the main channel.

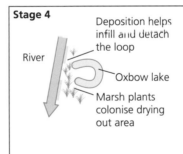

Stage 4 — River / Deposition helps infill and detach the loop / Oxbow lake / Marsh plants colonise drying out area
- Loop of the old river channel is increasingly detached as it is no longer receiving river water.
- Subsequent flooding causes deposition on the new river banks. This aids the detachment of the old loop.
- Marsh plants colonise the area, which further widens the gap.
- In time, only the far end of the meander loop is left, sometimes several metres from the main channel. This is the oxbow lake.

▲ Figure 11.25 Stages in the formation of an oxbow lake

Fieldwork: Get out there!

Imagine you have been given a tape measure, ranging pole, chain or rope and a metre rule (see Chapter 27). Describe, in detail, how you would use this equipment to gather data from which to draw the cross-profile of a meander in a river's middle course.

⭐ KEY LEARNING

➤ The characteristics of levées
➤ How levées are formed
➤ The characteristics of a flood plain
➤ How a flood plain is formed

Levées and flood plains

A period of prolonged heavy rain will cause an increase in a river's discharge, so water rises over its banks and floods over the surrounding land. Repeated annual flooding eventually builds up **levées** and a flood plain. These are landforms of fluvial deposition, found in the middle and lower courses of a river.

What are the characteristics of levées?

Levées are naturally raised river banks found on either or both sides of a river channel that is prone to flooding. The lower course of the River Severn has many levées such, as those at Minsterworth near Gloucester (Figure 11.26).

Levées typically:

- have raised river banks (about 2–8 metres high in the UK)
- are composed of gravel, stones and alluvium
- have grading of sediments with the coarsest closest to the river channel
- are steep-sided, but steeper on the channel side than on the land side
- have a fairly flat top, naturally covered by grass so often used as a footpath, for example the Severn Way.

▲ Figure 11.26 Levée on The River Severn at Minsterworth, near Gloucester

How are levées formed?

When a river bursts its banks, friction with the land reduces velocity and causes deposition. Heavy sediment is deposited closest to the river. The size of sediment then becomes progressively smaller with increased distance from the river. With each successive flood, the banks are built up higher (Figure 11.27, stages 1 and 2).

Although it may seem that levées may make it more difficult for the river to flood next time, this is not the case. This is because over time the bed of the river develops a thicker layer of sediment, which raises the river in its channel (stage 3).

Stage 1 Before levée

Silt deposits on flood plain

River

Bedrock

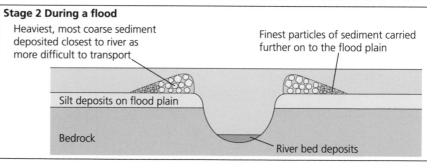

Stage 2 During a flood

Heaviest, most coarse sediment deposited closest to river as more difficult to transport

Finest particles of sediment carried further on to the flood plain

Silt deposits on flood plain

Bedrock

River bed deposits

Stage 3 After many floods

With every flood, the river banks are built a little higher

The raised banks either side of the river are natural levées

Silt deposits on flood plain

River

Bedrock

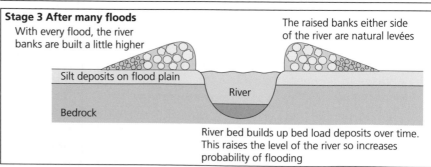

River bed builds up bed load deposits over time. This raises the level of the river so increases probability of flooding

➤ Figure 11.27 The formation of levées

What are the characteristics of a flood plain?

A flood plain is a large area of flat land either side of a river that is prone to flooding. Figure 11.28 shows the River Severn in flood over part of its flood plain at Tewkesbury, Gloucestershire. Another settlement on the River Severn that is prone to flooding is Gloucester. The characteristics of a flood plain are shown in Figure 11.29.

▲ Figure 11.28 Flooding at Tewkesbury, January 2014

→ Activities

1. Draw a sketch of the Minsterworth levée in Figure 11.26. Label it to show its characteristics.
2. Draw an annotated diagram or sequence of diagrams to show how the height of a flood plain is built up.

How is a flood plain formed?

The width of the flood plain is due to meander migration (see page 162), where the outside bends erode laterally into the edges of the valley. Their position is also gradually moving downstream. Eventually, this cuts a wider valley. When floods have receded, the flood plain is slightly higher and more fertile due to the deposits of silt and alluvium caused by the river flooding. Alluvial deposits also infill old meander scars. A flood plain is built up over hundreds of years. Each flood makes the flood plain a little higher.

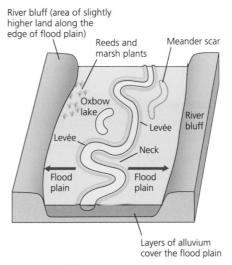

▲ Figure 11.29 The characteristics of a floodplain

Geographical skills

1. Study Figure 11.30, the OS map.
   a) Place a piece of tracing paper over the map. Trace its frame, the river and the first contour line either side of the river. Colour the river blue, the flood plain green (within the first contour) and the area beyond the flood plain brown. Mark and label the levées.
   b) Explain why levées make good footpaths.
   c) Give two pieces of map evidence to show that Minsterworth Ham has poor natural drainage.
   d) What is the width of the flood plain measured along northing 17? (On this map 2cm represent 1km.)

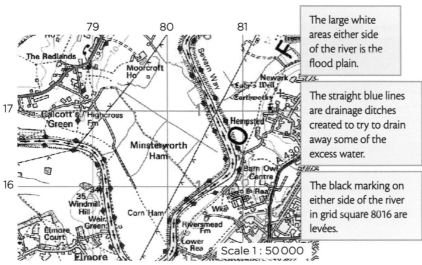

The large white areas either side of the river is the flood plain.

The straight blue lines are drainage ditches created to try to drain away some of the excess water.

The black marking on either side of the river in grid square 8016 are levées.

▲ Figure 11.30 OS map of part of the west side of Gloucester.
© Crown copyright and database rights 2020 Hodder Education under licence to OS

Example

⭐ **KEY LEARNING**

➤ The characteristics of an estuary
➤ How an estuary is formed
➤ How estuary mud flats are formed

The River Severn and its estuary

The River Severn completes its 354-kilometre journey where its estuary enters the Bristol Channel. An estuary is the tidal part of a river where the channel broadens out as it reaches the sea.

What are the characteristics of an estuary?

- An estuary is the tidal part of a river where freshwater from the river merges with salt water from the sea. It is therefore affected by both fluvial and marine processes.
- It may have a high **tidal range**. The River Severn has a tidal range of 15 metres, which is one of the highest in the world.
- It may be very wide. The Severn Estuary is 3.2 kilometres wide at the old Severn Bridge Crossing.
- It will have mudflats (Figure 11.34) that are visible at low tide and some of the mud will be covered by salt marshes.
- It may have tidal bores, which are huge waves that funnel up the river. The Severn Estuary has a tidal bore which travels as far as Gloucester on very high spring tides. Large bores occur about 25 days a year. Bores travel at 8–21 kilometres per hour, getting faster upstream. They cause great damage to the river banks and vegetation.

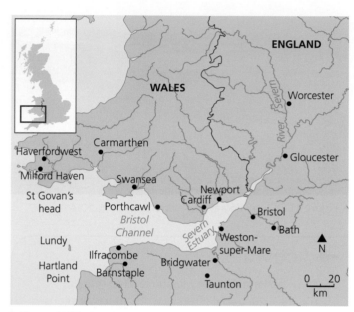

▲ Figure 11.31 The Severn Estuary

How is an estuary formed?

Figure 11.32 shows how a valley was flooded by the post glacial rise in sea level to create an estuary.

Figure 11.33 shows how an estuary is influenced by both fluvial and marine processes. The river is flowing from east to west. Notice how:

- the salinity (saltiness) increases towards the sea
- there are two sources of sediment (from the river and from the sea)
- the estuary is tidal, so fluvial and marine processes operate.

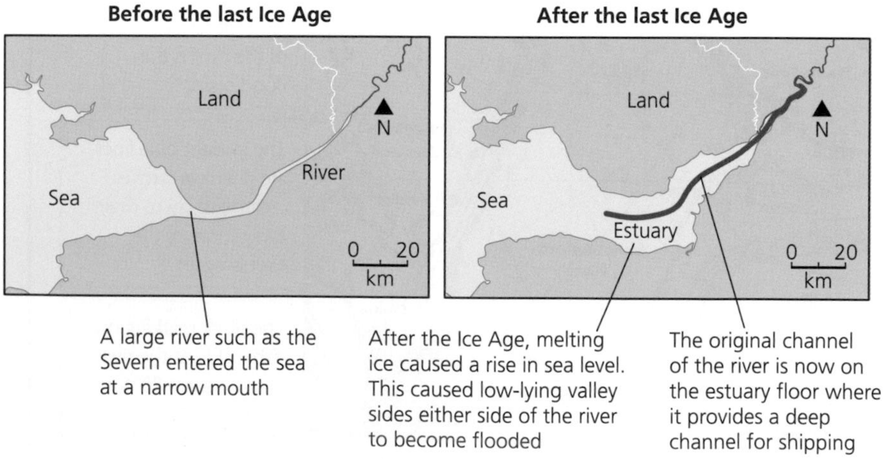

Before the last Ice Age

Land

Sea

River

N

0   20
km

After the last Ice Age

Land

Sea

Estuary

N

0   20
km

A large river such as the Severn entered the sea at a narrow mouth

After the Ice Age, melting ice caused a rise in sea level. This caused low-lying valley sides either side of the river to become flooded

The original channel of the river is now on the estuary floor where it provides a deep channel for shipping

▲ Figure 11.32 Estuary formation

How are estuary mudflats formed?

Mudflats form in sheltered areas where tidal water flows slowly. As a river transports alluvium down to the sea, an incoming tide transports sand and marine silt up the estuary. Figure 11.33 shows that just downstream of the tidal limit, fresh river water begins to mix with salty sea water. Where the waters meet, velocity is reduced, which causes deposition. This builds up layers of mud called mud flats (Figure 11.34). They will be covered at high tide, but exposed at low tide.

Within the mud flats there are many small streams (creeks). After a while, the mudflats may become colonised by salt-marsh vegetation such as cordgrass.

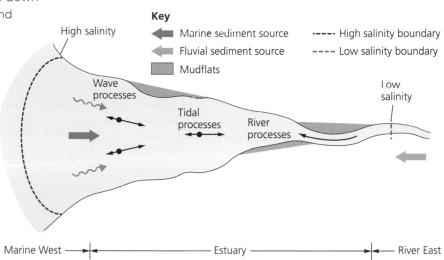

Key

- ← Marine sediment source
- ← Fluvial sediment source
- ▨ Mudflats
- ---- High salinity boundary
- --- Low salinity boundary

High salinity

Wave processes

Tidal processes

River processes

Low salinity

Marine West ⟶ ⟵ Estuary ⟶ ⟵ River East

▲ **Figure 11.33** Processes operating in an estuary

▲ **Figure 11.34** Mudflats off Chepstow, Wales, with view to M4 Severn Bridge crossing

→ Activities

1   Define an estuary in your own words.

2   Study Figures 11.31 and 11.32.
    a) Describe the location of the Severn Estuary.
    b) Explain how the estuary was formed with the help of Figure 11.32.

3   Study Figure 11.34.
    a) Describe the mudflats near Chepstow.
    b) Explain why they are not always visible.

4   Create your own fact file about the Severn Estuary. Give at least six facts.

➤ How water gets into a river

➤ How precipitation increases flood risk

➤ How geology and relief can increase flood risk

Physical causes of flooding

How does water get into and out of a river?

The drainage basin **hydrological cycle** explains how precipitation falling in a catchment area gets into a river (Figure 11.35).

Precipitation is any form of moisture reaching the ground. It includes snow, rain, sleet and hail. The risk of flooding depends on how quickly precipitation gets into a river channel. Surface runoff is the fastest route. If there is a lot of runoff then the discharge of a river will increase quickly. Other flows into a river are throughflow and groundwater flow (see Figure 11.35).

Precipitation: any source of moisture reaching the ground, e.g. rain, snow, frost

Interception: water being prevented from reaching the surface by trees or grass

Surface storage: water held on the ground surface, e.g. puddles

Infiltration: water sinking into soil/rock from the ground surface

Soil moisture: water held in the soil layer

Percolation: water seeping deeper below the surface

Groundwater: water stored in the rock

Transpiration: water lost through pores in vegetation

Evaporation: water lost from ground/vegetation surface

Surface runoff (overland flow): water flowing on top of the ground

Throughflow: water flowing through the soil layer parallel to the surface

Groundwater flow: water flowing through the rock layer parallel to the surface

Water table: current upper level of saturated rock/soil where no more water can be absorbed

Flood risk is increased by:

■ Continuous heavy rain caused by bands of depressions passing over the UK at frequent intervals, especially in winter, which may saturate the soil. The soil can no longer store water so surface runoff is increased. Rainwater will therefore enter the river quicker, resulting in higher discharge and floods.

■ Sudden bursts of heavy rain often result in the infiltration rate being too slow to cope. This may occur after a period of drought that has baked the soil hard. Surface runoff occurs, discharge increases quickly and **flash floods** occur.

■ Prolonged light rainfall may cause floods if there has been a lot of previous (antecedent) rainfall that has saturated the soil.

■ Sudden snow melt causes a release of stored water that flows over the ground as surface runoff.

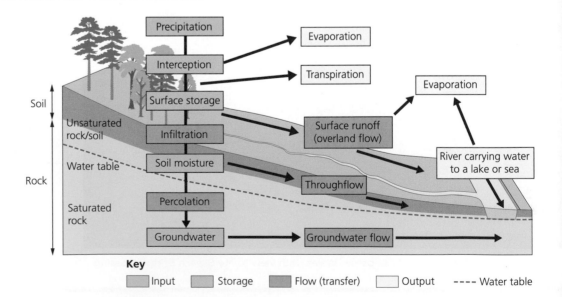

➤ Figure 11.35 The drainage basin hydrological cycle

How can geology and relief increase flood risk?

Geology can increase flood risk because:

- The type of rock found in mountains is usually impermeable rock, such as slate, which does not allow water to pass through it. The rock is often bare, with thin soils and little vegetation to intercept the rain.
- Low-lying areas often contain an impermeable clay soil. It is usually vegetated, but the soil is so compacted that it is difficult for infiltration to occur.
- Flooding is much less likely in areas of permeable rock, such as chalk and limestone, as water passes through these rocks.

Relief is the height and slope of the land. Steep slopes mean that surface runoff occurs on mountainsides before rain has had time to infiltrate the soils. A valley floor with steep sides such as the Llanberis Pass in Snowdonia (Figure 11.36) therefore has a high flood risk.

Low-lying, flat flood plains also have a high flood risk as there is not enough gradient to remove the water. The flood risk is increased further by the impermeable clay soils. Notice how the flat, low-lying relief in Surrey is prone to flooding (Figure 11.37).

Finally, it must be noted that precipitation, geology and relief are interconnected. They combine to increase the risk of flooding. The high relief of Snowdonia causes air from onshore winds to rise to cross the mountains. As air rises, it cools and condenses to form rain. This is called relief rainfall. In Snowdonia, the rainfall runs over steep-sided, impermeable slate to flood the valleys below.

▲ Figure 11.36 Llanberis Pass, Snowdonia, North Wales, an area of impermeable rock

▲ Figure 11.37 Flooding in the low-lying, flat relief of Surrey

→ Activities

1. The definitions below have been mixed up. Can you sort them out?

Precipitation	Water flowing on top of the ground
Infiltration	Moisture reaching the ground, e.g. rain and snow
Interception	Water lost through pores in vegetation
Percolation	Water flowing through the soil layer parallel to the surface
Transpiration	Water seeping deeper below the surface in the rock
Groundwater flow	Water being prevented from reaching the surface by trees and grass
Surface runoff	Water sinking into the soil from the ground surface
Throughflow	Water flowing through the rock layer parallel to the surface

2. Write these phrases in the correct order to show how rain falling on an impermeable surface may cause flooding: full river channel, heavy rain, increased discharge, flooded land, water gets into the channel quickly, much surface runoff, river bursts its banks.

3. a) Draw a labelled diagram to show how high relief increases precipitation.

   b) Which of these types of precipitation is most likely to cause flooding: a short, heavy rain storm; several days of drizzle; several days of snow? Explain your answer.

4. Explain how geology and relief have caused the flooding in Figure 11.37.

Geographical skills

Draw a field sketch from Figure 11.36. Annotate it to show how geology and relief may cause flooding.

✪ KEY LEARNING
➤ How urban land use increases flood risk
➤ How rural land use increases flood risk

Human causes of flooding

How can urban land use increase flood risk?

In the UK, changing urban land use is the main cause of increased flood risk. This is due to **urban sprawl** associated with **urbanisation** (see Section 15.9).

New infrastructure

Urbanisation leads to the growth of towns and cities. As the UK's population increases, new homes, roads, shopping centres, schools and leisure centres are built. The greater the area covered by buildings and roads (with impermeable surfaces), the greater the potential flood risk.

New houses

With the increased demand for homes in the UK, thousands of new houses are built each year (see Chapter 15), many on **greenfield sites**. Between 2001 and 2011 there was an increase of 72 per cent in the average density of new dwellings in England. Property developers are squeezing several houses into a plot formerly occupied by only one house.

There are strict planning controls regarding building on flood plains, but even so, seven per cent of new dwellings in England in 2011 were built in areas of high flood risk. Large areas of flood plains are now covered with impermeable tarmac roads and concrete pavements. Cities therefore have few natural areas in which to store excess water. Water runs off quickly through gutters, drains and culverts, and this leads to a speedy rise in a river's discharge – hence the increased flood risk.

Disappearing gardens

The increase of impermeable surfaces is greatest in our large cities, where people pave over back gardens to save mowing the lawn; this is often seen in rented accommodation. Parking is also a problem: around 47 per cent of UK households have two cars. With an absence of garages and little 'on road' parking in cities, many households have resurfaced their front gardens to accommodate their cars (Figure 11.38).

How can rural land use increase flood risk?

The increased risk of flooding in rural areas is more localised and, in the UK, does not have much effect on areas downstream. However, changing land use and farming practices can increase flood risk.

Forestry

Felling (chopping down) trees reduces interception and roots no longer take water from the soil (Figure 11.39). The impact of felling could be considerable, as a dense forest uses up 40 per cent of any precipitation. After felling, the soil soon gets saturated, runoff occurs, river discharge increases quickly and so the risk of flooding increases. Felling trees also causes exposed soil to wash into rivers, building up their beds. This reduces the capacity of channels, so rivers are more likely to flood.

▲ Figure 11.38 Concreted front gardens create more impermeable surfaces

▲ Figure 11.39 Forestry reduces interception

Farming

Since the First World War, hedges have been ripped out to make way for huge fields that are more efficient for highly mechanised arable farming. Loss of hedges means less interception.

Farming has become more intensive and there has been a further increase in arable farming (crops) at the expense of pastoral farming (animals). Once crops have been harvested, they leave the soil bare in winter. This means there is no vegetation to intercept the rainfall.

Additionally, when fields are ploughed up and downhill, the furrows create channels for water to flow down easily (Figure 11.40). More soil is transported into rivers, raising their beds and so increasing the flood risk.

Disappearing fields

Fields intercept rainfall and soak up excess water through infiltration. Just as UK gardens are disappearing, so too are fields. Some fields near towns may be sold off to property developers, while others may be converted to riding stables. As large-scale factory farming increases, fields have also been replaced by huge sheds and concrete yards. Pastures have been over-grazed. This has compacted soil and degraded pastures, resulting in muddy runoff into rivers and an increased risk of flooding.

Additionally, to extend the growing season of fruit and salad crops, vast areas of polythene like those near Hereford in Figure 11.41 cover the fields of crops. While there may be some interception and evaporation from the polytunnels, the ability of the area to soak up water is considerably reduced.

▲ Figure 11.40 An example of downhill ploughing

▲ Figure 11.41 Fields of polytunnels reduce grassy areas

→ Activities

1 Explain how building a new housing estate on a flood plain increases the risk of flooding.

2 Explain why a change from pastoral to arable farming increases the risk of flooding.

3 Create a flow diagram to show how forestry can increase the risk of flooding.

4 'Urban land use causes a more significant risk of flooding than rural land use.' Do you agree with this statement? Justify your answer.

Fieldwork: Get out there!

For your house or the house of someone that you know:

1 Calculate the total area of the front garden.

2 Calculate the percentage of the front garden taken up by impermeable surfaces.

3 Present this information as a pie chart or bar chart.

4 Considering both the front and back garden, write a report for the house owner suggesting how they might reduce the risk of flooding.

From precipitation to discharge: hydrographs

How does precipitation link to discharge?

Rivers remove excess water from the land. The speed at which precipitation reaches a river is determined by physical and human factors. A river's discharge can vary depending on many factors, including the amount, type and intensity of precipitation. The flows to the river are shown in Figure 11.42. For detail, refer back to page 168.

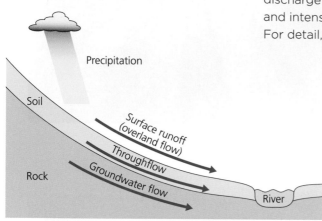

▲ Figure 11.42 How water gets into a river

What is a hydrograph?

A **hydrograph** shows how a river's discharge changes in response to a precipitation event. The vertical axis measures precipitation (usually rainfall) in millimetres and discharge in cubic metres per second (cumecs). The horizontal axis measures time, usually in hours or days. The bars represent rainfall and the line graph shows discharge. To understand how hydrographs work, it is worth taking time to understand the terminology.

Peak rainfall: the highest amount of rainfall per time unit (the highest bar)

Rising limb: shows how quickly the discharge rises after a rain storm (the first part of the line graph)

Peak discharge: the highest recorded discharge following a rainfall event (the top of the line graph)

Lag time: the time difference between peak rainfall and peak discharge (measure the horizontal distance between the top of the highest rainfall bar to the top of the discharge line and note the difference in hours)

Falling limb: shows the reduced discharge once the main effect of runoff has passed (the last part of the line graph which is going down)

Base flow: the normal flow of a river when its water level is being sustained by groundwater flow (usually shown on the hydrograph as a separate line)

Bankfull discharge: (does not always appear on hydrographs) will be drawn as a horizontal line marking the level of discharge above which flooding will occur as the river will burst its banks

On any hydrograph, the rising limb will be steeper than its falling limb. The rising limb is fed by surface runoff, which reaches the river quickly over impermeable surfaces. The gentler slope of the falling limb reflects how discharge is steadily falling once surface runoff has stopped. Water is now reaching the river mostly through the soil as throughflow, which is slower than surface runoff. Eventually, this flow stops and the river returns to normal conditions, receiving water slowly through the rocks from groundwater (base flow).

How hydrographs differ

Hydrographs may be classified as having either a flashy response or a slow response.

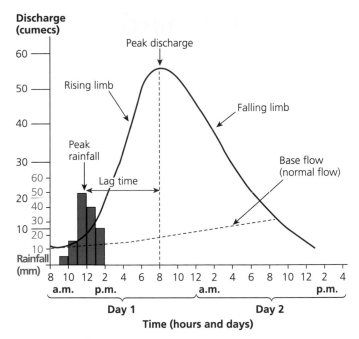

▲ Figure 11.43 A typical flashy response hydrograph

Slow response hydrograph

On a slow response hydrograph, an identical rainfall event will result in a less steep rising limb. The peak discharge is lower and the lag time longer. On this type of hydrograph, the flood risk is low.

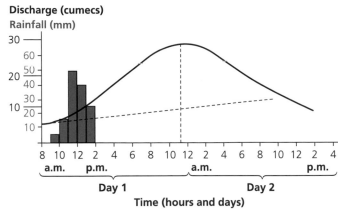

▲ Figure 11.44 Slow response hydrograph

Flashy response (storm) hydrograph

The flashy response hydrograph is associated with sudden flooding called **flash floods**.

In Figure 11.43, the rising limb is steep because rainfall has occurred in conditions that have caused a lot of surface runoff. This means water gets into the channel quickly so there is a short lag time, giving the river a high peak discharge which puts it in danger of flooding.

Several conditions may result in a flashy hydrograph. For example:

- There may have been prolonged rainfall so the soil is saturated, or a long drought so soil is baked hard and cannot absorb the water.
- It may be a clay soil, which means water is unable to infiltrate.
- It may be on a steep-sided valley floor where water runs down the hillside, or on a flat flood plain where water cannot drain easily.
- It may have a small river basin so tributaries soon link with the main river to swell its discharge.
- It may have little vegetation, because of **deforestation**, to intercept precipitation.
- It may be in an urban area, with large areas of impermeable tarmac and concrete.
- It may be in a rural area that has poor farming practices, like ploughing downhill.

Geographical skills

1 Study Figure 11.43, a flashy response (storm) hydrograph.
   a) What is the peak discharge?
   b) What is the peak rainfall?
   c) What was the total amount of rain that fell during this rainfall event?
   d) How many hours was the lag time?
2 Study Figure 11.44, a slow response hydrograph.
   a) What is the peak discharge?
   b) How much lower is this peak discharge than that in Figure 11.43?
   c) What is the lag time?
   d) How much longer is the lag time than that in Figure 11.43?

→ Activities

1 Define these terms: peak discharge, peak rainfall and lag time.
2 a) How do the rising and falling limbs differ between a flashy hydrograph and a slow response hydrograph?
   b) List the conditions that might result in a slow response hydrograph.

River management: hard engineering

Hard engineering uses heavy machinery to build artificial structures which work against nature to reduce the risk of flooding. **Dams and reservoirs** and **channel straightening** are two methods of hard engineering.

What are dams and reservoirs?

A dam is a large concrete barrier built across a river to impede its flow. This causes the valley behind the dam to flood, forming an artificial lake called a reservoir. This restricts the supply of water downstream. Water is released in a controlled manner through sluice gates in the dam. Carefully controlling and monitoring releases means there should be no risk of flooding downstream.

What are the costs and benefits of dams and reservoirs?

The benefits of the Kielder Dam and Reservoir in Northumberland, completed in 1981, can be seen in Figure 11.45 and the costs in Figure 11.46.

Boosts tourism – reservoirs are attractive. Kielder Dam attracts 300,000 tourists a year, which boosts the local economy by £6 million.

Forestry – areas around reservoirs may be planted with forests. Over 150 million trees were planted at Kielder, providing a valuable source of employment.

Highly effective against floods – release of water is highly controlled so there is virtually no risk of flooding.

Promotes new habitats – these develop in and around a reservoir. At Kielder, there are **conservation** areas. The area has rare red squirrels.

Provides **hydroelectric power** – turbines may be placed in a dam. Kielder Dam generates 6MW of electricity, enough to serve a town of 10,000 people.

Source of drinking water – the Kielder Dam created a ten kilometre long reservoir, holding almost 200,000 million litres of water, to supply drinking water to industrial North East England.

◄ Figure 11.45 Benefits of a dam and reservoir

Social costs	Economic costs	Environmental costs
• The flooding of a valley displaces people, usually farmers from their homes. At Kielder, 58 families were displaced. This causes distress and breaks up communities.	• Dams are expensive. Kielder Dam cost £167 million and may have been a waste of money. Loss of industry in North East England meant the demand for water and HEP was less than expected. • Soils downstream can become less fertile through lack of sediment from floods, which reduces crop yields.	• A concrete dam interferes with the path of migrating fish. Sediment is trapped behind the dam and this interferes with fish spawning grounds. • Algae often collects behind a dam which deoxygenates the water. • If there should be a sudden release of water through the sluice gates, this can cause river bank erosion downstream. • The building of a dam may trigger an **earthquake**. • **Landslides** often occur on the sides of a reservoir; this increases sediment and creates shock waves which damage buildings. • Reservoirs often flood areas of outstanding natural beauty. At Kielder, 1.5 million trees were lost along with 2,700 acres of farmland. This had a negative effect on habitats. New plantings are confined to Sitka Spruce.

▲ Figure 11.46 Costs of a dam and reservoir

What is channel straightening?

Channel straightening is when a meandering section of a river is engineered to create a widened, straightened and deepened course. This more efficient course improves boat navigation and reduces flood risk. In the nineteenth century, a new course was cut across a large meander loop on the River Tees to improve navigation (see Figure 11.47). Centuries of straightening have also taken place on the River Parrett, to reduce flood risk in the low-lying Somerset Levels.

What are the benefits and costs of channel straightening?

Benefits

Social:

- A straightened river reduces flood risk by moving water out of the area more quickly, as there is less friction with the bed and banks. The faster-flowing water also removes sediment that would otherwise build up the height of the river bed.

Economic:

- The historic cuts on the River Tees reduced the length of the river by 4.4 kilometres. This straightened course improved navigation considerably and increased **trade** at Stockton's port.
- Home owners gain confidence to invest in their property as they no longer expect to be flooded. Insurance costs also go down due to the lower flood risk.

Costs

Social:

- Water flows through a straightened section quickly, but when it meets a meandering section downstream, such as at Burrowbridge in the Somerset Levels (2014), velocity is reduced. This causes sedimentation of the channel, so the river is more likely to flood, causing problems in another area.

Economic:

- River straightening is expensive. Dredging a river to remove silt accumulation downstream is also expensive. After the 2014 flood damage in Somerset, the EU authorised the £5.8 million dredging of a five-mile section of the Rivers Parrett and Tome near Burrowbridge.

  In some cases, the impact of a straightened section downstream has been so severe that the river is restored to its original course. At Lewisham, London, £1.1 million was spent putting meanders back in the River Quaggy.

Environmental:

- The changes in hydrology and flooding downstream that can occur endanger animals and destroy habitats. The river's **ecosystem** is changed.
- A straightened river may have a concrete lining. This is visually unattractive and it deprives burrowing river bank animals of their habitat.
- In straightened sections, there is some evidence of increased **pollution** on the land from agro-chemicals, as runoff cannot drain into the river so easily.

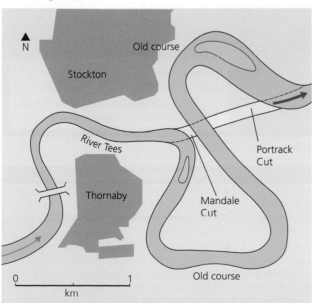

▲ Figure 11.47 Straightening the River Tees

→ Activities

1. Use information from Figure 11.46 to create your own table assessing the social, economic and environmental benefits of dams and reservoirs.

2. Imagine you live in the valley that is to be flooded. Write a diary entry describing your thoughts on hearing of the proposal.

3. Roleplay in a group of four. Take on the roles of an environmentalist, a local forestry worker, a farmer living in the area to be flooded and a member of the local tourist board. Engage in a discussion about the proposal to dam the valley to make a reservoir.

4. Draw an annotated diagram to explain how straightening a river reduces flood risk in an area.

River management: more hard engineering

Two other methods of hard engineering are **embankments** and **flood relief channels**.

What are embankments?

An embankment is an artificially raised river bank. In raising the banks, more water is contained in the channel. This reduces the flood risk. In Figure 11.48 the embankment at Bridge of Allan in Stirling, Scotland, is protecting the houses on the right that have a high flood risk. An embankment is made by bulldozers moving huge mounds of impermeable soil on to the river banks to build up their height. Some embankments are reinforced by gabions (wire cages filled with stones) or lined with concrete.

What are the benefits and costs of embankments?

Figure 11.48 shows the benefits of embankments while Figure 11.49 shows the costs.

Safer from flooding – the channel now has an increased carrying capacity. It is less likely to burst its raised banks, so the risk of flooding to settlements behind the embankment is reduced.

Cheap – compared to other methods of hard engineering, the cost of building embankments is quite low.

Habitats – earthen embankments provide habitats for riverbank animals such as water voles, kingfishers and otters.

Walking routes – embankments are often used for riverside footpaths. Some follow long-distance walks. This embankment makes an attractive walkway for local people.

▲ Figure 11.48 The embankment protects houses in Bridge of Allan, Scotland

Social costs	Economic costs	Environmental costs
• Embankments deprive people of easy access to the river for fishing and boating. • Although they reduce the risk of flooding, embankments are not as reliable as other types of hard engineering. Their presence gives people a false sense of security, which means they may not be prepared for floods.	• Embankments have higher maintenance costs than other hard engineering methods, as they need constant monitoring and repair. • Earthen embankments are prone to erosion and this increases sedimentation downstream, which will incur a dredging cost if flooding is to be avoided.	• If the embankment is breached, water lies on the land for a long time, as it has a restricted overland route back to the river. • Gabions and concrete linings displace riverbank animals from their habitats. These reinforced sections are unattractive and, if they break, wire mesh or huge slabs of concrete litter the river bed.

▲ Figure 11.49 The costs of embankments

What is a flood-relief channel?

A flood-relief channel is an artificially made channel that is designed as a backup channel for a river that frequently floods. It works like a bypass. The newly engineered channel runs roughly parallel to the main river. The River Exe at Exeter has three relief channels, which were constructed at a cost of £8 million following devastating floods in 1960. The largest of these channels is the Exwick spillway. A gate has been built across the River Exe which automatically closes off the river in times of high discharge and diverts water along the Exwick spillway, thus reducing the risk of the River Exe flooding.

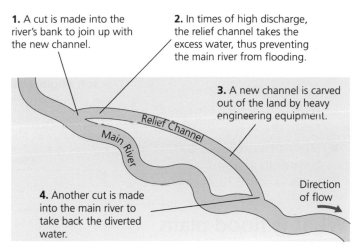

1. A cut is made into the river's bank to join up with the new channel.

2. In times of high discharge, the relief channel takes the excess water, thus preventing the main river from flooding.

3. A new channel is carved out of the land by heavy engineering equipment.

4. Another cut is made into the main river to take back the diverted water.

Direction of flow

▲ Figure 11.50 How a flood-relief channel works

	Benefits	Costs
Social	<ul><li>A relief channel removes the risk of flooding from a designated area. Exeter's relief channels protect around 3,000 properties.</li><li>Footpaths and cycle tracks are often built along a new channel.</li><li>Calm water provides areas for model boating and canoeing.</li><li>Where reed beds have been included, birdwatching and nature reserves may be set up.</li></ul>	<ul><li>People living in the path of a relief channel have to be moved, causing disruption.</li><li>Settlements downstream of a relief channel suffer from increased flooding, as the merging of water from the relief channel swells this part of the river. This raises the question of the ethics of protecting some settlements to the detriment of others.</li></ul>
Economic	<ul><li>Insurance costs are lower in the vicinity.</li><li>The value of homes increases and houses are easier to sell.</li><li>There is a more secure environment for setting up business ventures.</li></ul>	<ul><li>Flood-relief channels are expensive.</li><li>Sometimes, as in the case of the Jubilee River (see Section 11.15), they run out of funds. They also need to be maintained and repaired.</li><li>The schemes take a long time to come into effect; Exeter's relief channels took 12 years to build.</li></ul>
Environmental	<ul><li>Some relief channels include artificial reed beds and grass-covered concrete sides. These provide new habitats.</li><li>When full of water, they produce a tranquil setting.</li></ul>	<ul><li>In the construction of relief channels, habitats are disturbed.</li><li>The level of water in a relief channel varies considerably. This provides an unreliable habitat.</li><li>Relief channels look unattractive in times of low flow, when vast expanses of concrete and gabions are exposed.</li></ul>

▲ Figure 11.51 Benefits and costs of flood-relief channels

→ Activities

1 Study Figure 11.48.

   a) List the social, economic and environmental benefits of embankments.

   b) Consider the social, economic and environmental benefits and costs of embankments. In each case, do you think the cost or the benefit is greater?

2 Imagine you live in an area prone to flooding and the council is considering two proposals. One is to embank the river. The other is to build a relief channel.

   a) What four questions would you raise at the council meeting?

   b) Which proposal are you more likely to favour, and why?

➤ What flood plain zoning is

➤ What flood warnings, planting trees and river restoration involve

➤ Benefits and costs of these strategies

River management: soft engineering

A **soft engineering** strategy involves adapting to a river and learning to live with it. It is cheaper, but often less effective than hard engineering. Soft engineering includes **flood plain zoning**, **flood warnings** and preparation, planting trees and river restoration.

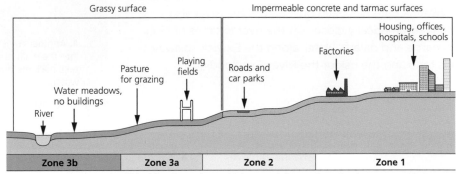

▲ **Figure 11.52** Flood plain zoning

The land use zones shown in Figure 11.52 show how land can be used sustainably by having different land uses parallel to the river. Land use which involves no buildings, such as playing fields or parks, are usually closer to the river where flooding is more likely to occur. More permanent structures can be installed further away from the river without any substantial economic or social costs, should flooding occur.

What is flood plain zoning?

Flood plain zoning is where land in a river valley is used in such a way as to minimise the impact of flooding. In England and Wales, the **Environment Agency (EA)** categorises land into four flood-risk zones and issues flood risk maps.

Local authorities are required to use these maps to produce flood-risk assessments and to guide decisions regarding new building applications.

How do flood warnings and preparation work?

The EA and other agencies, such as district councils, and the Water and Highways Authorities co-ordinate efforts to devise and carry out action plans for areas at risk. Distinct roles are identified for the emergency services, the armed forces and voluntary groups such as the Royal National Lifeboat Institution.

The meteorological office analyses data from its 200 automated weather stations and passes this to the EA, who uses it, along with river level data, to provide updated flood alert information. The media, and occasionally sirens or loudspeakers, publicise this information. The EA provides a flood map website, a three-day flood forecast, and personalised warnings. The EA also provides information on how to prepare oneself for a flood.

How does planting trees help?

Planting shelter belts of trees across slopes and woodland in floodplains (rewilding) reduces the risk of flooding, as trees intercept water by taking it up through their roots. Wales plans to plant 10 million trees over the next five years.

How does river restoration help?

River restoration is when a river that has previously been hard engineered is restored to a natural channel. For example, near Sutcliffe Park in Greenwich, River Quaggy had previously been re-routed through underground drains, but by 2007 it was brought back to the surface and restored close to its original meandering course. Part of the flood plain was lowered to create a floodwater storage area, and wildflower meadows and avenues of trees were planted.

What are the benefits and costs of these strategies?

Strategy	Benefits	Costs
Flood plain zoning	<ul><li>By restricting building on the active flood plain, local authorities do not increase impermeable surfaces, so the risk of flooding is reduced.</li><li>It is low-cost: only administration costs are involved.</li><li>Traditional water meadows by a river (Figure 11.52, zone 3b) are protected from development. In some meadows, cows may graze there when it is not flooded.</li><li>By conserving the flood plain, planners provide a welcome green space in UK towns.</li></ul>	<ul><li>This approach has limited impact as many UK cities have already sprawled over the active flood plain.</li><li>It is very difficult to get planning permission to extend or rebuild homes in the flood plain.</li><li>There is a housing shortage in the UK. Restricting building makes the problem worse. Restricted supply will inflate house prices.</li><li>Habitats are destroyed due to increased building on other greenfield sites.</li></ul>
Flood warnings and preparation	<ul><li>This is a very cheap way of protecting people and their property, as it is largely dependent on communication via the internet. The EA's personalised flood warning option makes people feel more secure and more in control.</li><li>If people are warned in advance of a flood, then they protect their valuables earlier.</li><li>It is a way of ensuring people's safety without having to invest in high-cost hard engineering.</li></ul>	<ul><li>Flood warnings are only effective if people listen and take action. Education is needed: not everyone listens to, or has access to, the media or the internet.</li><li>It does not prevent flooding.</li><li>The clear-up operation is distressing, people may have to move to temporary accommodation, their insurance costs will increase and their houses will be difficult to sell.</li></ul>
Planting trees	<ul><li>Reduces water flowing downstream as shelter belts of broad-leaf trees can reduce surface runoff.</li><li>More carbon dioxide is absorbed.</li><li>Adds variety to the landscape and new habitats. Increases **biodiversity**.</li><li>Relatively inexpensive.</li></ul>	<ul><li>Changed appearance: countryside wooded rather than open grass, arguably artificial looking and less aesthetically pleasing.</li><li>Loss of potential grazing land.</li></ul>
River restoration	<ul><li>Creates new wetland habitats and increases biodiversity, e.g. damselflies.</li><li>Increased water storage areas reduce risk of flooding downstream – the River Quaggy scheme has protected 600 homes and businesses from flooding.</li><li>Aesthetically pleasing – visitor numbers to Sutcliffe Park have increased.</li></ul>	<ul><li>Possible loss of agricultural land and flooding of crops near the river.</li><li>Can be expensive: the initial cost of restoring River Quaggy was estimated at £1.1 million.</li><li>Not always the most effective or practical strategy.</li></ul>

▲ Figure 11.53 The benefits and costs of different types of soft engineering

▼ Figure 11.54 Environment Agency flood warnings

FLOOD ALERT

Flooding is possible.
Be prepared

FLOOD WARNING

Flooding is expected.
Immediate action required

SEVERE FLOOD WARNING

Warning
Severe flooding. Danger to life

→ Activities

1. Explain the land use zoning shown in Figure 11.52.
2. Draw a flow chart to show how information is gathered and communicated to the public to warn them about a risk of flooding.
3. Go to the website www.environment-agency.gov.uk/flood. Read the 'Personal Flood Plan' leaflet. Assume you live by a river. Make a flood action plan for your family.
4. Evaluate the UK government's policy of not building on flood plains. Consider the costs and benefits of flood plain zoning. Then write a short report (300 words) to give your conclusions.

Fieldwork: Get out there!

Select a 50-metre stretch of river that runs through a settlement and which has a path alongside it.

1. With an adult, walk along the path and gather information to demonstrate how land use is zoned parallel to the river (how it changes on both sides as you look further away from the river). Take a photo.
2. Draw an annotated sketch map and annotate your photo to present your findings.

Example

🟢 KEY LEARNING

➤ The characteristics of the scheme

➤ Why the scheme was needed

➤ The issues that arose from the scheme

Jubilee River flood-relief channel

The Jubilee River is a relief channel for the River Thames in South East England. The relief channel runs through Berkshire and Buckinghamshire, flowing roughly parallel to the River Thames. It starts to the southeast of Maidenhead and flows in a south-easterly direction, passing just to the north of Eton. Once it has passed Eton, it re-joins the River Thames (Figure 11.55).

Jubilee River flood-relief channel

Flooding in Wraysbury in 20

Windsor Castle

▲ **Figure 11.55** Location of the Jubilee River

What are the characteristics of the scheme?

It was funded by the Environment Agency (EA), and cost £110 million. It opened in 2002, and at 11.7 kilometres long and 50 metres wide, it is the UK's largest artificial channel. The channel was designed to look like a natural river, so it has meanders and shallow reed beds, and a nature reserve with bird hides has been created in the area. It has five weirs, or large dams, along its course. Only two of the weirs are navigable by paddle craft. Under normal conditions, the level of water in the river is low but when discharge is high, the Jubilee River effectively diverts water from the River Thames, thus preventing the River Thames from overflowing its banks. This reduces the flood risk in southeast Maidenhead, Eton and Windsor.

Why was the scheme required?

This area of the Thames flood plain is low lying and prone to flooding. It contains the royal settlement of Windsor, which attracts many international visitors, as well as Eton, home of a prestigious public school. The impermeable surfaces of the built-up areas have historically resulted in flooding following high rainfall events. Given the high-value property in this area, the EA decided to increase the level of flood protection.

What measures were taken?

The Jubilee River was created to take overflow water from the River Thames in times of high discharge following heavy rainfall.

What issues arose from the scheme?

Social

- Is it ethical to protect some properties at the expense of others? Three thousand properties were protected in affluent Eton and Windsor, but to the detriment of the less wealthy settlements of Old Windsor and Wraysbury downstream. The Thames at Old Windsor now suffers from a much higher discharge due to the merging of the two channels just upstream. The scale of the problem came to light in 2014, when the area suffered its worst floods since 1947.
- Paddle boaters had been promised a navigable river. However, on two weirs they have to carry their boats around them, and Taplow Weir is considered too dangerous to cross.

Economic

- The Jubilee River scheme is the most expensive flood-relief scheme in the UK. Yet, one year after its completion, the weirs were damaged by floods. The initial repair bill for Slough Weir alone was £680,000. Maintaining the channel is a huge economic burden.
- At a projected cost of £330 million, the Jubilee River was one of four flood-relief channels planned for the lower course of the Thames. However, the EA ran short of funds. If further engineering is to alleviate flooding downstream, local councils and businesses will have to pay. Is this fair, when Windsor and Eton residents did not have to pay?

- Until a solution is found, small businesses such as shops stand to lose money, as they cannot open when their premises are flooded. Insurance costs are high. Business repair costs for Wraysbury alone were around £500 million in 2014. This will cause future insurance premiums to increase.

Environmental

- In 2014, there was extensive flooding immediately downstream from where the flood-relief channel rejoined the Thames. The built environment suffered from flooded roads and buildings. Fields were inundated and habitats were disturbed.
- The concrete weirs are rather ugly, especially under normal flow conditions, when more concrete is exposed. Ongoing repair work has made the matter worse, such as at Manor Farm Weir (Figure 11.56).
- There is also the problem of algae collecting behind the weirs. This disrupts the natural ecosystem.

▲ **Figure 11.56** Unsightly repairs on Manor Farm Weir

→ Activities

1   Assess the social, economic and environmental costs and benefits of the Jubilee River scheme. Try to limit each cost and benefit to six words.

2   a) Group work: debate the success of the Jubilee River scheme. Possible roles: Windsor resident, Wraysbury resident, insurance company manager, Wraysbury shopkeeper, environmentalist and a canoeist.

    b) Write a report summarising the views expressed.

    c) In your view, was the scheme worthwhile? Justify your answer.

12 Glacial landscapes

The power of ice in shaping the UK

Ice was a powerful force in shaping the physical **landscape** of the UK. Today, there is no permanent ice cover in the UK. It was a very different story 20,000 years ago (see Chapter 4), when ice covered most of the UK, as part of a vastly expanded Arctic ice sheet. In places it was three kilometres thick (Figure 12.1). Gravity caused large bodies of ice called glaciers to flow slowly from highland into lowland areas. Around 10,000 years ago, Earth's climate warmed again. As the Ice Age ended, the ice melted and retreated, revealing a transformed upland landscape, chiselled into steep peaks and sharp ridges.

What are the main processes of ice erosion?

Chapters 10 and 11 explain **erosion** in relation to the moving forces of water. In this chapter, the moving, eroding force is ice. When gravity causes ice to move down a mountainside, there are two main ways in which the rock below becomes eroded (Figure 12.2):

■ **plucking** – as the ice moves over the rock surface below, meltwater freezes around loose sections, pulling them away. Plucking is especially effective when the rock contains many joints (cracks) which the water can seep into. One reason why meltwater is present under a glacier is the sheer weight of ice above. Ice at the base of the glacier melts because of the great pressure it is under (this is called pressure

melting). Also, meltwater has travelled from the surface of the glacier to its base through crevasses (giant cracks) in the ice.

■ **abrasion** – erosion is caused by rocks and boulders embedded in the base of the glacier. These act like sandpaper, scratching and scraping the rocks below. Very large boulders can do enormous damage this way, scarring the landscape with features called striations. These are still visible in the UK today.

When a lot of plucking has taken place, large numbers of rocks and boulders become embedded in the ice. This increases the rate of abrasion.

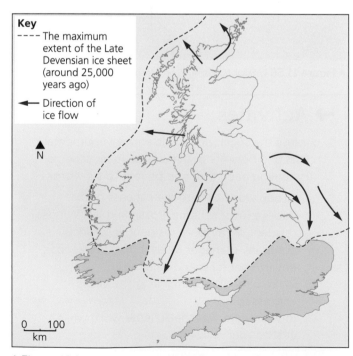

Key
---- The maximum extent of the Late Devensian ice sheet (around 25,000 years ago)
⟵ Direction of ice flow

N

0   100
km

▲ Figure 12.1 Map showing extent of main ice sheet coverage

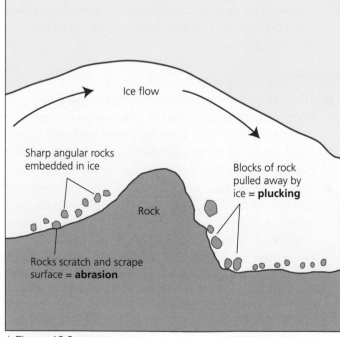

Ice flow

Sharp angular rocks embedded in ice

Blocks of rock pulled away by ice = **plucking**

Rock

Rocks scratch and scrape surface = **abrasion**

▲ Figure 12.2 The ice erosion processes of plucking and abrasion

How does freeze-thaw weathering take place?

Erosion requires a moving force like ice to break apart rock and carry it away. In contrast, weathering describes the destruction of rock that occurs in a particular place. The remains of the rock do not move; they remain *in situ*. Weathering is caused by temperature and moisture changes, along with any chemical processes caused by mild acids in rainwater. You can see the effects of weathering all around you on a daily basis on roads and buildings.

In glaciated areas, **freeze-thaw weathering** (or frost shattering) takes place on rock surfaces above the surface of the ice and at its margins. This is a physical weathering process and does not involve chemical changes:

- Water seeps into cracks in a rock face. (This may be water from summer rainfall or snowmelt.)
- The temperature falls at night, causing the water to freeze.

- Water expands by about ten per cent when it turns to ice. This expansion puts pressure on the rock either side of the crack, prising it apart and causing the crack to tear wider open.
- During the next 24-hour cycle, the ice melts, sinks deeper into the crack, and then freezes again.
- Over time, large blocks of rock can be shattered apart by repeated cycles of this weathering process.

The evidence for freeze-thaw weathering is seen in landscape features called scree slopes and blockfields. These are piles of rock debris that blanket large upland areas in the UK. Some debris dates from the last ice age, but some is more recent. Freeze-thaw weathering is still an important process in areas where many repeated freeze-thaw cycles take place during the winter months. It can also be classified as a weather hazard because of the damage that can be done to houses.

▼ Figure 12.3 How freeze-thaw weathering operates in cycles

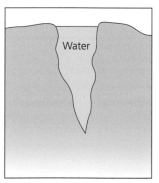

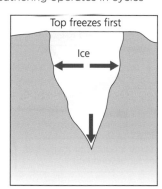

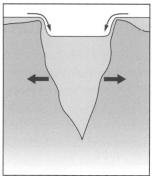

If the ice thaws the next day, the resulting water will not fill the crack, which is now both wider and deeper because of its 10 per cent expansion. Dew or rainfall on the rock surface can refill the crack.

→ Activities

1. Explain the difference between weathering and erosion.
2. a) State what is meant by plucking.
   b) Explain how the rate of ice erosion can be affected by the type and angle of the rock a glacier is moving over.
3. a) Explain how freeze-thaw weathering is affected by changing air temperatures.
   b) Using information from these pages and your own understanding, explain two challenges which freeze-thaw weathering may create for people. In your answer, you could discuss any of the following themes:
      - problems caused by freezing water pipes in homes if the weather turns cold
      - rocks falling from a cliff near a road
      - damage to buildings caused by freeze-thaw weathering.

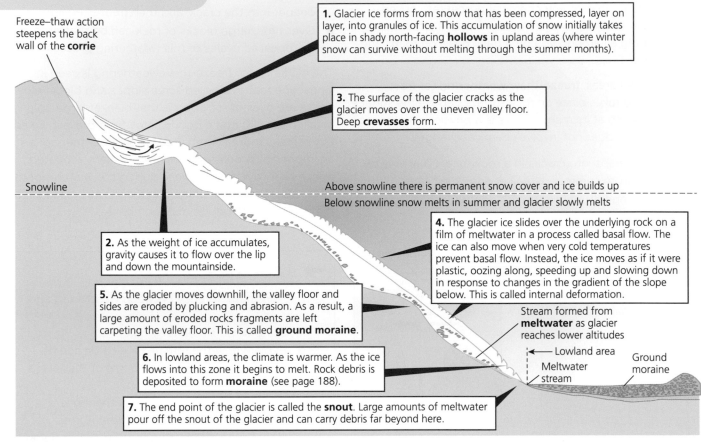

➤ How glaciers move
➤ Ways in which a glacier transports material
➤ Why glacial deposition takes place

Glacial movement and sediments

How do glaciers move?

Freeze–thaw action steepens the back wall of the **corrie**

1. Glacier ice forms from snow that has been compressed, layer on layer, into granules of ice. This accumulation of snow initially takes place in shady north-facing **hollows** in upland areas (where winter snow can survive without melting through the summer months).

3. The surface of the glacier cracks as the glacier moves over the uneven valley floor. Deep **crevasses** form.

Snowline

Above snowline there is permanent snow cover and ice builds up
Below snowline snow melts in summer and glacier slowly melts

4. The glacier ice slides over the underlying rock on a film of meltwater in a process called basal flow. The ice can also move when very cold temperatures prevent basal flow. Instead, the ice moves as if it were plastic, oozing along, speeding up and slowing down in response to changes in the gradient of the slope below. This is called internal deformation.

2. As the weight of ice accumulates, gravity causes it to flow over the lip and down the mountainside.

5. As the glacier moves downhill, the valley floor and sides are eroded by plucking and abrasion. As a result, a large amount of eroded rocks fragments are left carpeting the valley floor. This is called **ground moraine**.

Stream formed from **meltwater** as glacier reaches lower altitudes

⬅ Lowland area
Meltwater stream

Ground moraine

6. In lowland areas, the climate is warmer. As the ice flows into this zone it begins to melt. Rock debris is deposited to form **moraine** (see page 188).

7. The end point of the glacier is called the **snout**. Large amounts of meltwater pour off the snout of the glacier and can carry debris far beyond here.

▲ Figure 12.4 Characteristics and processes of glacial movement with the resulting landforms (in bold)

▲ Figure 12.5 The snout of a glacier bulldozes material when ice advances

How does a glacier transport material?

The front of a glacier is called its snout (Figure 12.5). As ice from upland areas descends into lowland areas, the snout **bulldozes** material. Soil, rocks and boulders are shoved forwards by the sheer force of the moving ice. Material is also carried on the surface of the glacier (see Figure 12.6, page 185). Freeze-thaw weathering takes place on mountainsides above the glacier, causing rock to become detached and fall onto the ice below. Some material is also carried inside the glacier for two main reasons:

- Plucking has torn away rock at the bed of the glacier that is now embedded in the base of the moving ice.
- Some rocks fall into crevasses at the surface of the glacier. Some crevasses reach deep into the glacier, resulting in a build-up of material inside the moving ice.

Why does glacial deposition take place?

Glaciers can carry ice far from the regions where the snow falls. As glaciers move from upland to lowland areas they enter a warmer climatic zone. During summer months, meltwater pours off the snout of some glaciers. Meltwater rivers can transport vast quantities of water from glaciers and ice sheets into the oceans. These rivers carry large amounts of **sediment** called glacial **outwash**. Because it has been carried by water, outwash material has been rounded and reduced in size by attrition. It has also been deposited sequentially and sorted, with fine material carried furthest from the glacier.

Under normal conditions, the snout of a glacier does not actually retreat, even though constant melting is happening. This is because new ice continually flows down from upland areas to replace and balance the meltwater loss. The following examples illustrate this point:

- A glacier moves forwards at four metres a day. Old photographs show its snout has neither advanced nor retreated. This means that a four-metre length of the glacier is being melted away each day.
- Another glacier moves forwards at ten metres a day. Old photographs show its snout has retreated (Figure 12.6). This means that more than a ten-metre length of the glacier is being melted every day.

The behaviour of a glacier is similar to that of an escalator. Rocks and boulders are constantly moved down the mountain side. Eventually, the section of ice they are carried on reaches the warmer lowland areas. The same section of ice temporarily becomes the snout of the glacier – and is promptly melted away too! Deposition occurs then.

Constant transport of new, debris-laden ice into lowland areas results in the widespread deposition of all of the eroded and weathered material from the uplands that the glacier has carried. The dumped material is called glacial **till** (Figure 12.6).

▲ Figure 12.6 A retreating glacier and the till (deposition) it leaves behind, Greenland

→ Activities

1. a) Study Figure 12.4 Describe where the snowline is found.
   b) Outline one way in which environmental conditions are different above and below the snowline.
2. Outline how snow is converted into glacier ice.
3. Explain how material is transported by the top and the underside of a glacier.
4. For revision, draw a table with two columns, labelled 'Landforms and features above the snowline' and 'Landforms and features below the snowline'. Add as many landforms and features as you can from Figure 12.4, together with a brief description of each one.
5. Explain how glacial transport and deposition processes are affected when a glacier retreats. This is quite a tricky question to answer. A glacier which is retreating is still transporting downhill. Can you explain why?

Landforms resulting from ice erosion

How are upland areas affected by ice erosion?

Glaciated places in the world are home to unique landforms that give them a special character that is not found elsewhere. These special features were revealed in the UK at the end of the last ice age in the regions shown in Figure 12.18 (see page 193).

One such landform is a **corrie**. This is a deep armchair-shaped hollow found on the flank of a mountainside, where glaciers begin. A corrie is formed by the growth of a large snow patch in a hollow. Over time, the snow deepens and becomes compressed, eventually forming a dense mass of ice. As its mass increases, the ice pulls away from the walls of the hollow, due to gravity and a process called **rotational slip**. This leads to the plucking of blocks of rock from the back wall. Once embedded in the base of the ice, these blocks abrade the hollow, causing it to get wider, deeper and steeper (Figure 12.7).

Freeze-thaw weathering takes place on the back wall of the hollow, above the surface of the ice, which soon becomes covered with fallen, weathered rock. Over time, the back wall retreats backwards, cutting deep into the side of the mountain.

When two corries develop side by side or back to back, an **arête** (a steep, sharp ridge) develops between them. When three or more corries grow in hollows on all sides of a mountain, a **pyramidal peak** is produced. As the corries erode the mountain behind them, the remaining rock is weathered into a sharp point (Figure 12.8).

The deepest corries are often found on the northeast side of mountains where least sunlight is received. When the ice disappeared at the end of the last ice age, some deep corries filled with water to create corrie lakes.

How are upland river valleys modified by ice erosion?

At the start of the ice age, small corrie glaciers developed in the UK's mountainous regions. As the climate grew colder, ice began to spill out of the corries and flowed into the numerous river valleys that are a feature of upland areas. Because ice flows far less quickly than water, the river valleys were soon filled entirely with slow-moving but powerful ice.

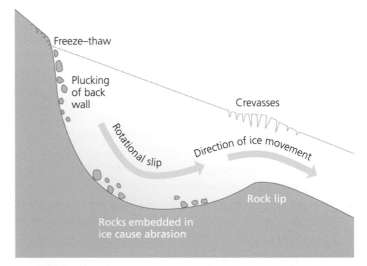

▲ Figure 12.7 The formation of a corrie

▲ Figure 12.8 Pyramidal peak, arête and corrie

The result was that the shape and appearance of these river valleys changed completely.

- Before glaciation, a river valley would have been V-shaped in profile. River tributaries flowed down the gentle valley sides to reach a meandering valley floor.
- During glaciation, the rock in the valley sides is torn away by a combination of plucking and abrasion. The result is a U-shaped valley or **glacial trough**. Its very steep, almost vertical sides lead down to a straight and wide valley floor.
- The tributaries that used to flow down the river valley sides now exit abruptly through a gap in the new cliff-shaped valley wall. The water cascades down from a high altitude and creates a waterfall. The portion of the original tributary valley that remains is now called a **hanging valley**.

- While rivers meander around spurs of land, ice has the erosive power to remove any obstacles in its path. **Truncated spurs** can usually be identified along the sides of a glacial trough.

These changes to the river valley are clearest to see in an area that has only recently been de-glaciated. The UK's glaciated landscapes have continued to change in the 10,000 years since the ice retreated. Figure 12.9 shows a typical U-shaped valley today. Parts of the valley floor that were over-deepened by plucking have filled with water to create a **ribbon lake**.

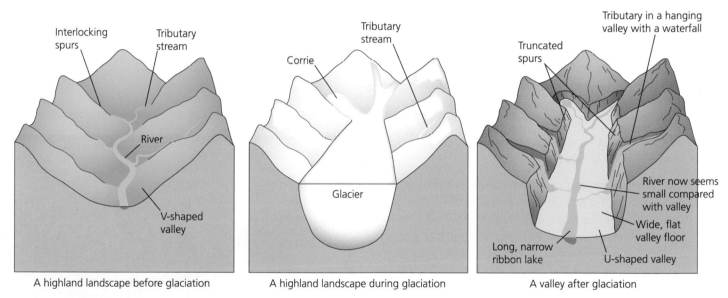

A highland landscape before glaciation · A highland landscape during glaciation · A valley after glaciation

▲ Figure 12.9 How an upland valley and its landforms are modified by glaciation

→ Activities

1. Sketch Figure 12.8 and label these features: corrie, arête and pyramidal peak.

2. Explain how a hanging valley is formed.

3. a) Give one difference between the appearance of a V-shaped valley and a glacial trough.
   b) Explain how the movement of ice is responsible for changing a V-shaped valley into a glacial trough.

4. Draw a picture of a corrie filled with ice. Add labels to show the characteristics and formation of this feature. Use two different coloured pens to make these annotations (one for the characteristics and a different one for the formation). You can base this task on Figure 12.7 but aim to add more annotations that use detailed information drawn from the text.

5. For revision, draw a table with two columns, labelled 'Before glaciation' and 'After glaciation'. Fill it with as many matched pairs of landforms as you can, for example a V-shaped valley and a glacial trough.

➤ Landforms formed by moving or melting ice
➤ What 'glacial erratics' tell us

Landforms resulting from ice transport and deposition

Which landforms result from moving or melting ice?

Depositional landforms are produced when a glacier loses the ability to carry material.

■ When a melting ice mass reduces in size, the material it is carrying drops to the ground.

■ Material is sometimes deposited underneath a moving glacier. Plucking can sometimes result in very large amounts of rock fragments being carried along in the base of the ice. Later, some of this debris gets dropped back onto the valley floor. This could be because it gets lodged, or stuck, behind obstacles the ice is flowing over, such as bands of hard rock.

Much of the material that gets deposited is till (see page 185). Where ice has retreated, we can see moraines and **drumlins**.

The different types of moraine

Moraines are accumulations of rock debris (Figure 12.10) and have distinct shapes:

■ *Lateral moraine* – a ridge of material that runs along the edges of a glacial trough close to the valley side. The source of the material is freeze-thaw weathering, high on the valley sides, causing shattered blocks of rock to fall onto the glacier below. As ice melts and the glacier gets smaller, this material is slowly lowered to and deposited on the valley floor.

■ *Medial moraine* – when glaciers meet, something very interesting happens. Two lateral moraines merge together to form a very large ridge of rock debris: the medial moraine. In Figure 12.11, you can see this by the thick dark stripe running straight down the middle of the main glacier below the point where the two tributary glaciers have met.

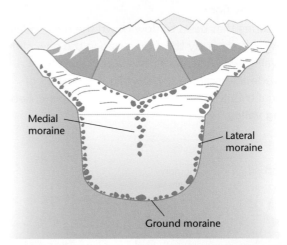

▲ Figure 12.10 The formation of different types of glacial moraine

▲ Figure 12.11 An aerial view of a glacier showing how the medial moraine is produced when two tributary glaciers meet

■ *Ground moraine* – the material that gets lodged and deposited underneath the glacier is simply called ground moraine. Vast amounts of ground moraine can be produced when glaciers disappear entirely in response to **climate change**. Ground moraine covers large areas of the UK as a legacy of the last ice age. Your home might even be built on ground moraine!

■ *Terminal moraine* – this is the enormous ridge of material that gets bulldozed by the snout of the glacier (see page 184). These features can still be identified in the landscape today. They allow us to work out how far the ice advanced during the last ice age.

Drumlins

Drumlins are egg-shaped hills composed of mounds of till. Two processes are involved in their formation:

- First, material is deposited underneath a glacier as ground moraine.
- Second, this ground moraine is sculpted to form drumlin shapes by further ice movements (think of the ridges that fingers leave when they move through sand).

Drumlins show the direction of ice movement in the past. For landscape detectives, they are an important clue! They can be 100 metres or more in length. A large group is said to resemble a 'basket of eggs'. Some drumlins may have a very large rock fragment at their core, which could help explain why they formed in some places and not others.

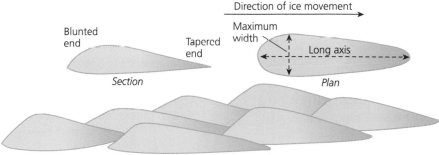

Drumlin swarm – 'basket of eggs' topography

▲ Figure 12.12 The characteristics of drumlins

What do 'glacial erratics' tell us about past ice movements?

Glacial **erratics** are another brilliant clue for landscape detectives. We have learned a great deal about ice movements in the past from the study of erratics in the UK. An erratic is a large boulder that stands out like a sore thumb in the landscape. This is because it is composed of a rock type that is nowhere else to be seen.

Figure 12.13 shows granite erratics resting on a sandstone platform in Arran. This site is far from the nearest outcrop of granite. The boulders provide clear evidence that a glacier flowed here in the past because only ice has the power to move them so far.

▲ Figure 12.13 Glacial erratics on the Isle of Arran, Scotland

Fieldwork: Get out there!

'A glacier used to be here.' How could this hypothesis be investigated for a lowland area that was once glaciated?

- Make a list of landforms that you might be able to identify and any measurements you could take.
- Why might it be hard to identify landscape features such as moraine?
- See Figure 12.16 on page 191 for an example of an OS map. How could maps provide additional evidence to help with your investigation?

→ Activities

1. a) State what is meant by a drumlin.
   b) Using Figure 12.12, describe the characteristics of a 'swarm' of drumlins.
2. Explain how medial moraines are formed. In your answer, you could draw an annotated diagram based on Figure 12.11.
3. Explain how erratics can be identified in a glacial landscape (look for 'landscape clues' in Figure 12.13).
4. Explain how different ice processes contribute to the formation of different types of moraine.

   In your answer try to use all the key words below. (You can use them more than once.)

   - plucking
   - abrasion
   - freeze-thaw
   - bulldozing
   - lateral moraine
   - medial moraine
   - terminal moraine
   - ground moraine

   If it helps, present your answers in a table with the column headings: Ice process; Types of moraine; How the process contributes to the formation.

Example

⊕ KEY LEARNING

➤ Major landforms of ice erosion

➤ How ice deposition has affected the Lake District

Glacial landforms in England's Lake District

The Lake District is one of numerous upland areas that were previously glaciated. Reaching 1,000 metres at Scafell Pike, it is England's highest mountainous region. Tough volcanic rocks were scoured and re-shaped by glaciers to produce the landforms you now see.

Which major landforms of ice erosion can be seen in the Lake District?

The Lake District is well-known for its mountains and ribbon lakes.

Its many mountains and hills were thoroughly researched by the walker and writer Alfred Wainwright in the 1950s. Wainwright counted 214 significant peaks. The Lake District lacks a truly good example of a pyramidal peak, but nonetheless has some spectacular arêtes and corries. Figure 12.15 shows a corrie lake and arête called Red Tarn and Striding Edge respectively. They are located just to the east of Lake Thirlmere. There is a hanging valley at Grisedale.

Red Tarn provides us with landscape evidence of how rotational slip in a corrie eroded deep into the mountain side.

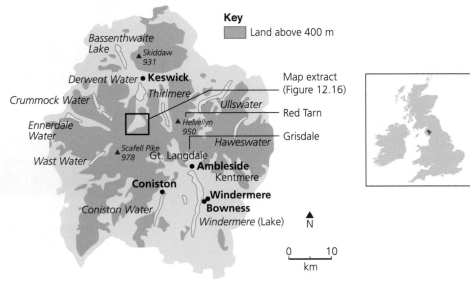

▲ Figure 12.14 Glaciated upland areas and ribbon lakes in the Lake District

You can see the steep back wall that forms part of the arête. The lake is a clue that erosion once took place. Glacial erosion has left behind a deep, wide hollow that filled with water in the post-glacial period. Immediately after the ice first retreated, the corrie's sides would have been even steeper than they are today. Over time, the Lake District's features have softened. This is because rain and running water, rather than ice, are now the main influences on landscape development.

There are many ribbon lakes in the Lake District. They mark the position of over-deepened glacial troughs. Like Scotland's lochs, these lakes are much deeper than you might expect from looking at photographs. In the 1930s, Lake Coniston gained notoriety when Donald Campbell was killed there.

▲ Figure 12.15 Striding Edge with Red Tarn to the left, Helvellyn, Cumbria

He was attempting to break a world water speed record in his high-powered boat, the Bluebird. Sadly, the boat overturned. Campbell had chosen Coniston because it was long, straight and deep.

In other parts of the Lake District, settlements have developed on dry, flat sections of the area's wide and U-shaped valley bottoms. Keswick is situated on the floor of a glacial trough that the River Derwent now flows into (see Figure 12.14), as is Rosthwaite (see Figure 12.16).

How did ice deposition affect the Lake District in the past?

Ice deposition features are still visible in the landscape even after 10,000 years. Farming in lowland areas takes place on ground moraine, though you cannot see it now that vegetation is present. Some relief features are visible, however.

- Fields in Borrowdale use terminal moraines as boundaries. You may think you are looking at an artificially created earth embankment, but it is entirely natural.
- Swarms of drumlins can be seen in some places, such as Swindale in the northeast Lake District.
- Glacial erratics are strewn across low-lying areas of the Lake District. Deposited by melting ice between 10,000 and 30,000 years ago, some glacial erratics may have travelled all the way from Scotland!

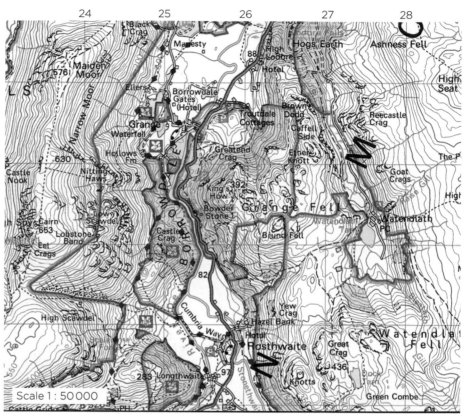

▲ Figure 12.16 OS map extract showing part of the Lake District. © Crown copyright and database rights 2020 Hodder Education under licence to OS

Geographical skills

Map work

The OS map (Figure 12.16) has a scale of 1:50,000. This means that two centimetres on the map represents one kilometre.

1 Approximately how large is the map area in square kilometres?

2 Identify (a) a high area shown on the map, (b) a low area and (c) a flat area where slope angle is very low.

3 Give the four-figure grid reference of a corrie lake.

4 What features are found at squares 2415 and 2416?

→ Activities

1 Make a table with two columns. In one column, list all the erosion and deposition features found in the Lake District. In the second column, write the name of where the feature can be seen.

2 a) Give three uses that humans have found for glacial landforms in the Lake District.

 b) Explain what the characteristics of these landforms are that make them useful.

➤ Important land uses in glaciated upland areas

➤ The importance of tourism for glaciated upland areas

➤ Why quarrying takes place in some glaciated upland areas

Economic activities in glaciated upland areas of the UK

Why are farming and forestry important land uses in glaciated upland areas?

Glaciated upland areas can be extreme environments. For rural landowners, it is difficult to farm crops because of the:

■ steep slopes, due to past ice erosion, which makes using machinery difficult

■ thin soils with limited fertility, due to resistant underlying rocks and steep slope gradients

■ low temperatures at high altitudes, resulting in a short growing season

■ heavy relief rainfall, especially in western areas bringing waterlogging to flat sites and **soil erosion** on slopes.

Extensive agriculture such as animal grazing is well suited to glaciated upland areas. In the Scottish Highlands, sheep were introduced to many upland estates during the 1800s, when there was a growing demand for wool from textile factories in the UK's growing industrial cities. More recently, some landowners have introduced deer for venison meat, and Highland cattle for speciality beef. Some farmers have even raised more exotic species, such as ostriches.

Another competing land use is commercial forestry. Coniferous woodland occupies 2 million hectares of land in the UK, much of it in upland areas. Around half is managed by the Forestry Commission, which was established in 1919 to ensure that the UK would never run short of timber supplies. In the 1980s, forest cover increased further, thanks to private investors. Many upland areas of Scotland, such as the Isle of Arran, were soon carpeted with fast-growing pine and spruce trees. Wood is used as timber for furniture and building construction, and to make wood chips for gardens and **biofuel**.

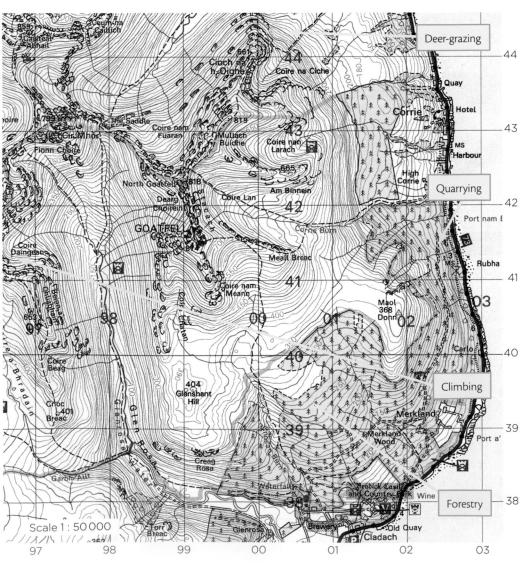

▲ Figure 12.17 OS map extract showing upland land uses in Scotland's Isle of Arran. © Crown copyright and database rights 2020 Hodder Education under licence to OS

How important is tourism for glaciated upland areas?

Tourism is a major draw in many upland areas (Figure 12.18). It employs more people than any other industry and is often the most important source of an upland area's income. Glaciated landscapes attract mountain climbers and walkers. Some glaciated upland areas fare especially well from tourism because of their location. The Lake District is highly accessible, for instance. It also has a milder climate than more northerly upland areas. The Lake District's glacial features were explored on pages 190–91. The benefits of tourism for Scotland's Isle of Arran are covered in detail on pages 196–97.

Why does quarrying take place in some glaciated upland areas?

Why are rocks quarried in upland areas, far from the towns and cities where they are used for construction? It is because the **geology** of upland areas tends to be different from lowland areas. By their very nature, upland areas are composed of tough, resistant rocks that are not found in lowland areas. A lack of population means that there are fewer dangers (and objections!) when explosives are used to shatter rock into blocks that can be transported easily. The rock types shown in Figure 12.19 all have a high economic value.

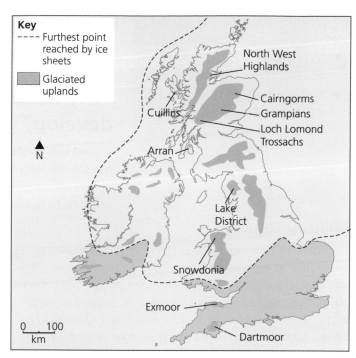

▲ Figure 12.18 Glaciated upland areas in the British Isles (UK and Eire)

▼ Figure 12.19 Quarrying in glaciated upland areas

Lake District slate	This distinctive blue-grey rock is used around the world as a roofing and decorative material. The Lake District has 13 active quarries.
Pennines limestone	Limestone is a widely used building material. Limestone fragments and gravel are a popular landscaping material for gardens.
Highlands granite	This tough, resistant rock has a range of uses, from pavement edges to kitchen work surfaces. Granite from the glaciated Scottish island of Ailsa Craig is used in the sport of curling due to its unusually uniform hardness.

→ Activities

1. a) State what is meant by 'extensive' agriculture.
   b) Explain three physical factors that mean only extensive agriculture can be carried out in glaciated upland areas.

2. Outline one reason why the amount of forest cover in upland areas has increased during the last 100 years.

3. Using Figure 12.18 and your own understanding, suggest why some glaciated upland areas may receive more tourists than others.

4. Explain the costs and benefits of carrying out quarrying in upland areas.

   Think about different types of costs and benefits, such as the economic or environmental impacts of quarrying.

Geographical skills

Identifying land uses

Figure 12.17 shows part of the Isle of Arran in Scotland (also see pages 196–97).

1. Identify possible land uses shown in grid squares 0038 and 9843.

2. Give the grid reference for the quarry.

3. Identify a grid square likely to be popular with mountain climbers. Use map evidence to explain your choice.

4. Suggest two reasons why forest has not been planted in some places.

✪ KEY LEARNING

- ➤ Why land use conflicts sometimes develop
- ➤ How development and conservation needs can be balanced

Land use conflicts in glaciated upland areas in the UK

Why do land use conflicts sometimes develop?

A **land use conflict** develops when the activities of two different groups of people are incompatible. Figure 12.20 shows a conflict matrix for glaciated upland areas. Conflict can develop in relation to all the following land uses, in addition to tourism (page 196).

Military training

During their training in the Second World War, young pilots flew at high speeds around pyramidal peaks and into glacial troughs. The exercises are dangerous though: six military aircraft crashed into the mountains of Arran in the 1940s, with the loss of 51 lives (see Figure 12.22). Today, some walkers object to the jet engine noise spoiling the upland's tranquillity.

Reservoirs

Glacial troughs can be dammed to help create deep and wide reservoirs (see Chapter 11). The downside is that local people may be forced to relocate.

Wind turbines

Increasingly, upland areas are seen as suitable for turbines as there are few residents. Tourists and walkers may object, however, that turbines ruin the landscape.

Hunting

Many Scottish upland areas are privately owned land where hunting is allowed. On the Isle of Jura, visitors pay £500 to shoot a stag. Some hunters visit upland areas to achieve a 'Macnab': they stalk a red deer, shoot a grouse and catch a salmon on the same day. Many nature lovers are opposed to hunting.

Forestry

Conifer plantations block out the view and acidify the soil where they grow. Often they are planted very close together and light cannot reach the forest floor. As a result, very few animal or bird species live there.

After the forest has been cut down, there is no vegetation left to intercept rainfall. Worse still, machinery used by the loggers compacts the soil, which can result in localised flooding. The wood has many vital uses, however.

Key
- Little conflict (or activities can co-exist)
- Strong conflict
- Very strong conflict

		Development/exploitation						Conservation/recreation			
		Sheep and deer farming	Forestry	Quarrying	Reservoirs	Military training	Wind turbines	Riding	Walking and climbing	Hunting and shooting	Photography and filming
Conservation/recreation	Wildlife conservation	Little conflict	Little conflict	Strong conflict	Very strong conflict	Strong conflict	Strong conflict	Little conflict	Strong conflict	Little conflict	Little conflict
	Photography and filming	Little conflict	Little conflict	Little conflict	Little conflict	Very strong conflict	Little conflict	Little conflict	Little conflict	Little conflict	
	Hunting and shooting	Strong conflict	Very strong conflict	Very strong conflict	Very strong conflict	Strong conflict	Strong conflict	Little conflict	Strong conflict		
	Walking and climbing	Little conflict	Strong conflict	Strong conflict	Strong conflict	Strong conflict	Little conflict	Little conflict			
	Riding	Little conflict	Strong conflict	Strong conflict	Very strong conflict	Strong conflict	Little conflict				
Development/exploitation	Wind turbines	Little conflict	Strong conflict	Little conflict	Very strong conflict	Strong conflict					
	Military training	Strong conflict	Strong conflict	Little conflict	Little conflict						
	Reservoirs	Very strong conflict	Very strong conflict	Very strong conflict							
	Quarrying	Strong conflict	Strong conflict								
	Forestry	Very strong conflict									

▲ Figure 12.20 A matrix of possible land use conflicts in glaciated upland areas

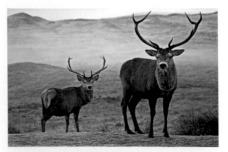

▲ Figure 12.21 Stag on the Highlands

▲ Figure 12.22 A present-day clue that RAF pilots used to train for war here –the remains of a crashed aeroplane

▲ Figure 12.23 The Hogwarts Express is filmed passing through a glaciated upland area

How can development and conservation needs be balanced?

Many people who live in upland areas would like more types of employment, yet others do not want exploitation of the environment. Compromise is possible by allowing some development to take place, but adopting strict management to make sure **ecosystems** do not become permanently damaged, and landscapes are not spoilt by too much activity and noise. Common management measures include:

- Maximum visitor numbers – it is important that visitor numbers do not exceed the **carrying capacity**.
- Signing – signs can be used to show people areas that are permanently or temporarily out of bounds.
- Seasonal closure – some visitor attractions are closed in winter months to give each site a chance to recover.
- Restricted activities – landowners are allowed to ban the use of motorbikes or horses. Camping is restricted to designated campsites.

An excellent strategy that balances development and **conservation** needs is film-making. Glaciated upland areas, both in the UK and elsewhere, are very popular locations for television and movie making. The Harry Potter film series used upland glaciated areas in Yorkshire, Northumberland and Scotland (Figure 12.23).

Places that feature in films benefit in several ways. Actors and production staff may stay in the area for a long time, generating revenues for local hotels and businesses. Landowners may charge a fee for the use of the location. Once the film is released, it may capture the interest of the public, causing more people to visit: this can be evaluated as an important long-term benefit.

→ Activities

1 a) Give two ways in which upland glaciated landscapes can be used to help generate electricity.
  b) Suggest two reasons why upland areas are well-suited to these ways of energy production.

2 Explain two advantages and two disadvantages of bringing forestry to upland glaciated areas.

3 Using Figure 12.23 and your own understanding, explain why glaciated upland areas are often used for film-making.

4 'Land use conflicts in glaciated upland areas cannot be avoided.' Do you agree? Include at least one conflict that you think can be resolved easily and at least one which cannot. Consider:
  - a wildlife conservation area heavily damaged by erosion from walkers' boots
  - plans to build a reservoir in an area popular with walkers and riders
  - any other possible conflicts based on Figure 12.20.

Geographical skills

Conflict matrix

Study the conflict matrix in Figure 12.20.

1 Identify five pairs of strongly conflicting activities and five pairs of activities with little conflict.

2 Suggest reasons for the conflicts you have identified. How might the levels of conflicts be resolved?

Example

Tourism in the upland Isle of Arran, Scotland

What are the tourist attractions in Arran's glaciated upland areas?

The glaciated Isle of Arran is reached easily from the Scottish coast, making it a popular destination for day-trippers (Figure 12.25). The following impressive glacial features provide striking views and pose an active challenge for those who like to spend their leisure time outdoors:

■ Goatfell – Arran's highest mountain and most popular natural visitor attraction. From the 874 metre summit of its pyramidal peak, visitors can see all of the way to Ireland on a clear day. Goatfell is flanked by many other dramatically shaped mountains.

■ A'Chir ridge – a knife-edged glacial arête that divides two corries. It is Arran's most exciting and challenging ridge. Climbers must be extremely careful as there is a long, vertical drop on either side.

■ Glacial troughs – the past action of the ice has left Arran with several deep and wide, U-shaped valleys. They include Glen Rosa, Glen Sannox (Figure 12.24) and Glen Catacol. Glacial striations (see page 182) and polished rock surfaces can still be seen in these valleys.

Walkers of all ages take in the views at their own pace. Younger adults may also be interested in sporting activities. Among the island's many visitors are people who have come with the intention of climbing, running, biking, paragliding and abseiling. Helicopter tours of the mountains are also available, for those who can afford it.

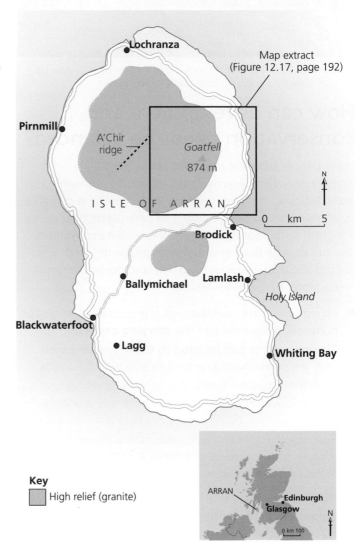

Key

High relief (granite)

▲ Figure 12.25 Map of Scotland and the Isle of Arran

▲ Figure 12.24 A view of Arran's mountains from a glacial trough, North Glen Sannox

What are the impacts of tourism in Arran?

The carrying capacity of popular visitor sites has been exceeded, resulting in footpath erosion and other problems (Figure 12.26). Additionally, Arran's upland areas experience heavy frontal and relief rainfall for much of the year, resulting in a markedly seasonal pattern of tourism. Many tourist workers are left unemployed in winter.

▲ Figure 12.26 Footpaths such as this can suffer erosion over time due to hiking boots and exposure to wind and rain

▼ Figure 12.27 Costs and benefits of tourism in Arran's upland areas

Negative impacts (costs)	Positive impacts (benefits)
● Footpath erosion occurs on popular paths at 'honeypot' sites. One of the worst affected routes is the North Goatfell Ridge. The problem is made worse by steep slope angles in the valleys and corries. High altitude brings relief rain and more soil erosion.	● The island has grown to rely on tourism as a large part of its income. Visitor numbers reach 400,000 annually. Many used to cater for themselves or camp outdoors, but more are now using the island's hotels and restaurants.
● There is a lot of congestion, especially at popular times of the year. Arran's upland roads are narrow single-track roads due to the difficulties of construction in upland areas.	● A range of new visitor attractions have helped to create more jobs, generating around £160 million revenue annually. Whisky visitor centres were opened in the village of Lochranza in 1999 and in Lagg in 2019.
● Injuries and fatalities are frequent. Many people have died walking and climbing Arran's mountains after under-estimating the challenge. This is upsetting for the islanders and costly too. Arran's Mountain Rescue Service is staffed by volunteers but sometimes requires the help of expensive RAF helicopters.	● Tourism is helping to tackle the island's ageing population problem. Visitors often fall in love with the island's dramatic upland landscape and decide to move there permanently. This brings young families and children to the island. In turn, this ensures that the island's schools have a sustainable future.

→ Activities

1 For revision or exam preparation, draw a table with three columns, labelled 'Social impacts', 'Economic impacts' and 'Environmental impacts'. Rearrange the information given in Figure 12.27 to fill the three columns.

2 Suggest why a glaciated upland area might be an attractive holiday destination for different groups of people.

In your answer, write about at least two different 'types' of person. Here are some examples to get you thinking.

■ A day-tripper from Glasgow who is keen on outdoor adventures, but does not have a lot of money to spend.
■ Two married geography teachers and their teenage children, who are spending one week at a hotel on the island.
■ A very wealthy 75-year-old who used to enjoy walking in the mountains when she was younger.

Explain the activities they might participate in while visiting a glaciated upland area. Make reference to specific landforms or ideas from physical geography wherever you can.

⭐ KEY LEARNING

➤ How the impacts of tourism in Arran are managed

➤ Judging the success of the management strategies

Evaluating tourism management in the Isle of Arran

How are the impacts of tourism in Arran being managed?

There are three main concerns in Arran: footpath erosion in tourist 'hot spots', climbing accidents and the seasonal character of the island's tourist industry.

Footpath erosion

The National Trust for Scotland owns some of the worst-affected parts of the island. They have established a mountain path team to restore mountain paths, and have raised money through the Footpath Fund Appeal. They carry out maintenance and small-scale restoration work on the mountain paths. Some paths are stabilised using paving stones to create steps. This stops soil erosion and mud flows, which allows natural regeneration of vegetation to occur.

The National Trust's aim is to maintain mountain paths for future generations, using techniques with a low **environmental impact**. Wherever possible, they use locally sourced natural materials so that the area still looks good. However, the work costs up to £140 per metre.

Climbing and walking accidents

The Arran Mountain Rescue Team was formed in 1964. It is funded by a combination of grant aid and public donations. The team provides search and rescue assistance to walkers and climbers on the Isle of Arran. All members are experienced hill-walkers or mountaineers who know Arran's glaciated upland areas like the backs of their hands. They are on call 365 days a year. All are unpaid volunteers who give their time freely to help people in need. The team have rescued many people, although fatalities still occur, most recently in 2015 and 2019. The Arran Outdoor Education Centre gives safety talks to visiting school students and staff.

Seasonal tourism and visitor numbers

Purpose-built visitor attractions, such as the Balmichael Centre (Figure 12.29) and Auchrannie Resort, encourage people to visit the island in winter, when walking in the upland areas becomes difficult. This means more local people have permanent full-time tourism jobs. In addition, local businesses collaborate to create and maintain a new web site called *VisitArran*. It promotes Arran as an ideal short break or one-day holiday destination.

Each year, around 400,000 visitors generate around £160 million for the island economy. Almost four out of every five tourists visit again, which shows that people often fall in love with Arran's upland scenery and want to return. The growth in recent years of online bookings has helped, too. Many Americans and Canadians are descended from Scottish migrants and they are keen to visit Arran and see the landscape of their ancestors.

Have the management strategies in Arran been a success?

Although the costs of footpath repairs are high, there have been tremendous improvements overall. Special attention has been paid to making sure that water drains from the paths quickly, meaning that the repairs are

▲ Figure 12.28 Mountain rescue over Goatfell

intended to be lasting and sustainable. The popular coastal path, the land owned by the National Trust (including Goat Fell), and the Forestry Commission estates have all benefited the most. In contrast, some privately owned upland land still suffers from eroded footpaths.

Additionally, visitor numbers and spending are still rising. Some local people are unimpressed, however. From their perspective, it was better when there were fewer visitors and cars, and there was more peace and quiet!

Yet Arran's three tourist issues – footpaths, accidents and visitor numbers – are interrelated with one another. Precisely because more visitors than ever are using improved footpaths, more walkers may be getting into danger than in the past. In 2015, there were a record number of call-outs for the mountain rescue. Therefore, one unexpected result of successful management of the tourist industry seems to be that more people are becoming exposed to risk in upland areas.

▲ Figure 12.29 Tourist activities in an upland area of Arran (the Balmichael Centre)

Geographical skills

Evaluating success

The focus of an evaluation exam question will often be on the level of success for a **management strategy** or action. There are three steps to take when carrying out an evaluation:

- Work out what information is needed.
- Decide which facts can be categorised negatively as 'costs' ('failures' or 'challenges'), and which facts are better described positively as 'benefits' ('successes' or 'opportunities').
- Weigh up the positives and the negatives in order to arrive at an overall judgement.

On balance, do you feel the people responsible for the management have done a good job or not?

Try answering: 'Evaluate the success of tourism management in a glaciated upland area you have studied.' Consider:

1 What were the management strategies?
2 In each case, what worked and what didn't work? Have some strategies worked better than others?
3 On balance, is it your judgement that the area as a whole has been successfully managed? What, if anything, might you advise is done differently?

→ Activities

1 Outline two physical causes of increased footpath erosion in upland areas.
2 a) State what is meant by seasonal tourism.
   b) Explain why seasonal tourism creates challenges for communities in glaciated upland areas.
3 Using Figure 12.28 and your own understanding, suggest how the physical environment creates risks for tourists in glaciated upland areas.

In your answer, think about:
- landscape features which the tourists might hope to see, explore or climb
- challenges which the rescue team and helicopter might encounter while trying to save tourists who have run into trouble (think about the weather and the slope angles).

Coastal landscapes

3.1 Study Figure 1, a 1:50,000 Ordnance Survey map of the Isle of Purbeck, Dorset.

Give the four-figure grid reference for a headland. Choose **one** answer from:

A 0383

B 0176

C 0582

D 0181   [1 mark]

3.2 Which of the following statements best describes the features of square 0379 in Figure 1?

A a sandy beach with groynes

B a headland with cliffs

C a shingle beach with a sea wall

D a sandy beach with sand dunes   [1 mark]

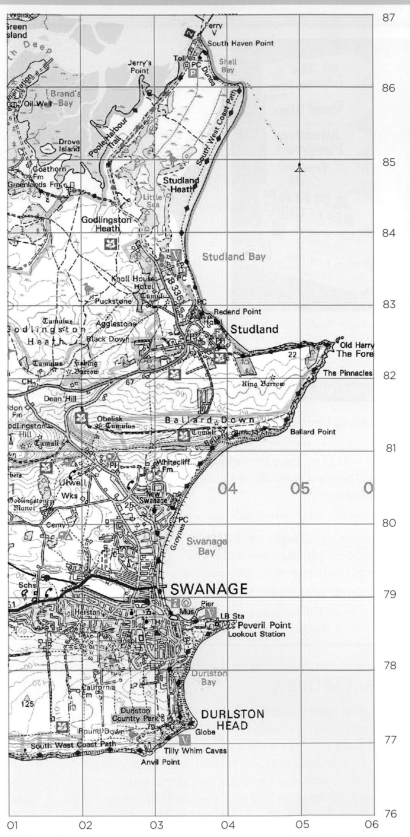

▲ Figure 1 1:50,000 OS map of the coastline between Durlston Head and Studland Bay on the Isle of Purbeck, Dorset. © Crown copyright and database rights 2020 Hodder Education under licence to OS

3.3 Study Figure 10.25 on page 135. Compare it with Figure 1 on page 200.

    a) What type of landforms are Old Harry Rocks? [1 mark]

    b) In which direction was the camera pointing in the photo? [1 mark]

    c) Give the six-figure grid reference for Old Harry. [1 mark]

> The easiest way to do this is to rotate the map to match the photo.

3.4 Study Figure 10.11 on page 128. Compare it with Figure 1 on page 200.

    Which type of coastline is found between Durlston Head and Studland Bay – concordant or discordant? [1 mark]

> Outline the arguments for and against using groynes and come to your own conclusion.

3.5 Looking at Figure 1, study the groynes along the beach in Swanage. Which physical process do groynes help to prevent, or slow down? [1 mark]

3.6 Explain how groynes work, to prevent or slow down this process. [4 marks]

3.7 'Groynes are an effective way to manage the coastline in a seaside resort like Swanage.' To what extent do you agree with this statement? [6 marks]

River landscapes

4.1 Study Figure 2, an Ordnance Survey map of a section of the River Severn in Shropshire.

Give the four-figure grid reference for a meander. Choose **one** answer from:

A 2510

B 2411

C 2712

D 2613 [1 mark]

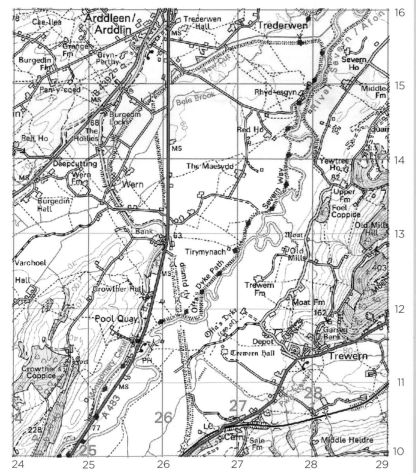

▲ Figure 2 1:50,000 OS map of the River Severn in Shropshire.
© Crown copyright and database rights 2020 Hodder Education under licence to OS

4.2 Which of the following statements best describes the features of grid square 2812?

    A a steep, west-facing valley side, rising to over 400m

    B a flat valley with meandering river

    C a steep, east-facing valley side, rising to over 200m

    D a flat, narrow valley with steep slopes on each side   [1 mark]

4.3 Figure 3 shows meanders on a river.

On which stage of a river would you be most likely to find this landform? Choose **one** answer from:

    A the source   **B** the upper course   **C** the middle course

    D the lower course   [1 mark]

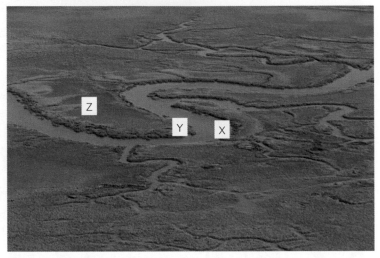

▲ **Figure 3** An aerial view of meanders on a river

4.4 Match the letters X and Y in Figure 3 with two features of a meander – a river cliff and a slip-off slope.   [1 mark]

4.5 What change would you expect to happen at Z in Figure 3 in the future?   [1 mark]

4.6 Explain the processes which would cause the change at Z.   [3 marks]

4.7 Name the landform which may be formed as a result of the changes.   [1 mark]

4.8 'The UK is short of housing, so building on river flood plains would help to solve this problem.' To what extent do you agree with this statement?   [6 marks]

> Outline the arguments for and against building on flood plains and come to your own conclusion.

Glacial landscapes

5.1 Study Figure 4. Name **one** glacial landform shown.

◄ **Figure 4** Striding Edge with Red Tarn to the left, Helvellyn, Cumbria

[1 mark]

5.2 Study Figure 12.16 (page 191). Name the glacial feature found in grid square 2518.

[1 mark]

5.3 Study Figure 12.17 (page 192). Identify the glacial landform at grid reference 999426. Select **one** letter only.

    A Corrie   **B** Arête   **C** Drumlin   **D** Hanging valley

[1 mark]

5.4 Name one glacial feature found in grid square 9938.

[1 mark]

5.5 Identify the grid square that shows the highest point of the land.

[1 mark]

5.6 Using Figure 12.26 (page 197) and your own understanding, explain how tourist activities may affect the environment.

[4 marks]

5.7 Using Figures 5 and 6 and your own understanding, explain how depositional landforms are created in glacial environments.

[6 marks]

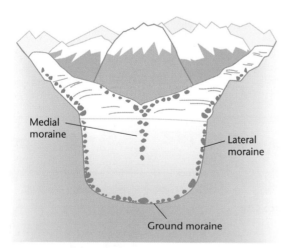

▲ **Figure 5** The formation of different types of glacial moraine

▲ **Figure 6** An aerial view of a glacier

13 The global pattern of urban change

⭐ KEY LEARNING

➤ How the world's urban population is growing
➤ Where the world's largest cities are
➤ How rates of urbanisation differ between continents

Urban trends

How is the world's urban population changing?

Over half the world's population now lives in cities. By 2030, it is expected that 60 per cent of the world's population will live in urban areas and, by 2050, it will be 70 per cent. This is **urbanisation**: when an increasing proportion of people live in towns and cities.

Where are the world's largest cities?

The world's largest cities, with populations over 10 million, are known as **megacities**. In 1975, there were only four megacities – Tokyo, New York, Mexico City and São Paulo. Today, there are 29 (Figure 13.1) and the number is rising year by year. You may notice that London, the UK's largest city, is not among the world's megacities. Its population is not predicted to reach 10 million until 2030. That might give you an idea of how big megacities really are.

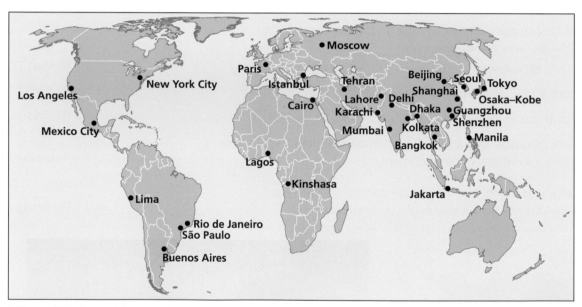

▲ Figure 13.1 The world's megacities

The world's largest city is Tokyo, with a population of around 38 million. Unlike many other cities, its population is no longer growing. Other cities are expected to overtake it eventually. Globally, megacities are home to about 12 per cent of the world's urban population – over 4 billion people. Most urban dwellers live in cities with less than 5 million people. Many of those cities are growing rapidly.

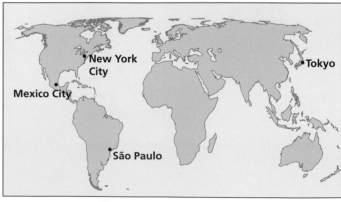

▲ Figure 13.2 The world's megacities in 1975

How do rates of urbanisation vary around the world?

Rates of urbanisation differ between continents (see Figure 13.3). The highest rates of urbanisation (the steepest lines on the graph) are in countries with the lowest income in Asia and Africa. In most of these countries, a majority of the population still live in rural areas and the rate of **rural–urban migration** is high. The population of cities is younger, so the rate of **natural increase** is also high.

There are lower rates of urbanisation in countries with the highest income in Europe, North America and Oceania. In these countries, urbanisation has slowed down as the majority of the population already live in cities. The urban population is ageing, and this may even head to **natural decrease**.

One exception to this pattern is South America, with many **newly emerging economies** like Brazil. Here, urbanisation happened earlier and has slowed down, even though these countries are not yet among the countries with the highest income.

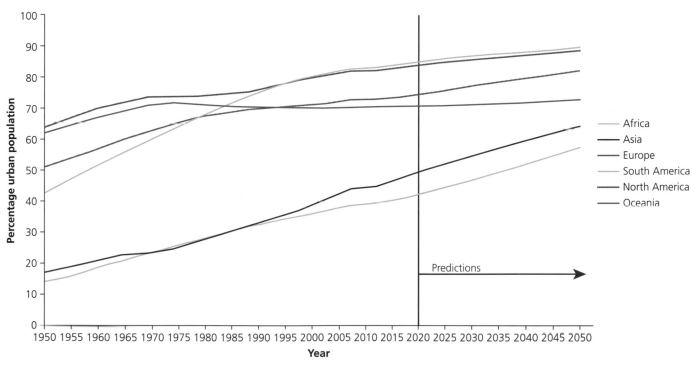

▲ Figure 13.3 The urbanisation of the world's continents, 1950–2050

→ Activities

1 Look at Figure 13.1. Describe the distribution of the world's megacities by continent. Which continent has most megacities? How many megacities are in each continent?
2 Compare Figures 13.1 and 13.2.
   a) How has the number of megacities changed since 1975?
   b) How has the distribution changed?
3 Make a list of the world's megacities in Figure 13.1 and name the countries in which each one is found. Try to do this without an atlas. Then check the atlas to see if you named the correct countries.

Geographical skills

1 Look at Figure 13.3.
   a) Which continents had the highest rates of urbanisation in 1950?
   b) Which continents had the highest rates of urbanisation by 2015?
   c) Explain why your answers to (a) and (b) are different.

➤ Factors affecting the rate of urbanisation

➤ How the world's megacities have grown

➤ Where the world's megacities will be in future

How urbanisation happens

Urbanisation is the result of rural-urban migration, as people move from the countryside into the cities, and natural increase, as **birth rates** exceed **death rates** in cities. These two processes are connected as most migrants are young, which leads to higher birth rates in cities.

Rural-urban migration is a result of people thinking about push and pull factors. **Push factors** are the disadvantages of living in the countryside, such as lack of jobs and services. **Pull factors** are the attractions of living in cities, such as better jobs and services (see Section 14.2).

What factors affect the rate of urbanisation?

Recent urbanisation has led to the large numbers of megacities in Asia:

- Asia is where over half the world's population lives. China and India both have more than a billion people.
- The majority of Asia's population is still rural, although this is changing as people move to cities. Over 50 per cent of China's population now live in cities compared to just 20 per cent in 1980.

▼ Figure 13.4 Factors affecting the growth of Shanghai, a megacity in China

Migration

Rural–urban migration is the main driver of urbanisation. Most of these migrants are young. They migrate from the countryside to cities because of pull factors, like jobs and the chance of a better education.

Natural increase

The young population in many cities leads to high rates of natural increase as they have babies. Cities also tend to have better health care than rural areas, so death rates are lower and **life expectancy** is higher.

Location

Historically, cities have grown on rivers, coasts and other busy transport routes where **trade** can thrive. Even today, many of the world's megacities are ports, which are a good location for trade.

Economic development

Cities that trade are also a good place for business, so they grow economically. It is economic growth that creates jobs, which attract people, and it is people who bring the ideas and enterprise on which cities thrive.

How have the world's megacities grown?

Cities do not grow at a constant rate. Some cities that grew rapidly in the twentieth century, such as Tokyo in Japan, have now slowed down. Meanwhile, other cities that grew slowly in the twentieth century, such as Lagos in Nigeria, are now urbanising rapidly (Figure 13.5).

Where will the world's megacities be in future?

By 2050, the world's largest megacity is likely to be one that does not even exist at the moment! At least, it does not have a name. China has plans to merge cities in the Pearl River Delta to create one large megacity with a population of 120 million. The existing cities of Hong Kong, Shenzhen and Guangzhou would merge to form an urban area 20 times the size of London, with a population 12 times bigger. Most of the new megacities in future are likely to be in Asia, particularly in China and India.

Many of the world's fastest growing cities, such as Lagos, are now in Africa, where population growth and rural-urban migration rates are still high. Most cities with low growth rates are in Europe, North America and Japan.

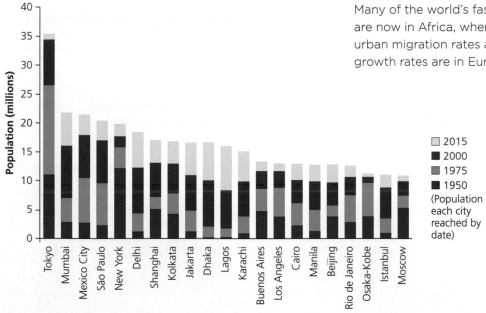

▲ Figure 13.5 Growth of the world's megacities

→ Activities

1. Define these terms in your own words:
   - rural-urban migration
   - natural increase.

2. Look at Figure 13.4. Explain how each of these factors affects the rate of urbanisation. Write a sentence for each one.
   a) migration   c) location
   b) natural increase   d) economic development

3. Look at Figure 13.5. By which date – 1950, 1975, 2000 or 2015 – did each of these cities reach a population of ten million?
   a) Tokyo   c) Shanghai
   b) Mexico City   d) Lagos

4. Suggest why some megacities are growing faster than others in the twenty-first century. For example, why is Lagos growing faster than Tokyo? (You can find out more about Lagos in Chapter 14.)

5. Find out more about a city you know. It could be a city in the UK, for example London or Birmingham. Answer these questions for your city to explain how urbanisation happened.
   - Why did the city first grow?
   - When did its most rapid growth occur? What caused this?
   - What is the present day population?
   - How is the population changing now, and why?

14 Urban growth in Nigeria

⊛ KEY LEARNING

➤ How Lagos compares with your image of Africa
➤ Where Lagos is located
➤ The importance of Lagos to Nigeria and Africa

Welcome to Lagos

What is Lagos like?

Lagos is Africa's biggest city and one of the fastest-growing cities in the world. In this chapter, you will get to know it well. You will discover an exciting, vibrant, crowded, enterprising megacity – sometimes overwhelming but never boring!

From above, Lagos could be any modern city. The city centre is dominated by modern, high-rise offices, surrounded by miles of sprawling suburbs, linked by busy roads (Figure 14.1). But, get down to street level and the noise around you will be unlike anything you have heard before.

In the background is the constant drone of generators that power the city (Lagos does not have a reliable electricity supply). Even louder is the roar of traffic. Groups of motorcycles, fleets of yellow minibus taxis and old trucks belching fumes from their exhausts are gridlocked with cars, all honking their horns in a cacophony of noise. Above it all rises the chorus of street vendors selling their wares on every corner, loudspeakers blaring a mixture of traditional Nigerian music and modern afrobeat, and the call to prayer from the city's mosques. Welcome to Lagos!

▲ Figure 14.1 Lagos, Nigeria

Where is Lagos located?

Lagos is the largest city in Nigeria, itself the most populous country in Africa. It lies in the southwest of the country, on the coast of the Gulf of Guinea, close to the border with Benin (Figure 14.2).

Until the fifteenth century, it was a small fishing village on an island. In 1472, Portuguese settlers gave it the name 'Lagos' (or 'lakes' in Portuguese) after the water that surrounds it. In the early twentieth century, by then under British control, Lagos was made the capital of Nigeria. It remained the capital after independence from Britain in 1960.

What is the importance of Lagos?

In 1991, the Nigerian government moved to Abuja, which became the new capital of Nigeria, though Lagos retained its importance as the country's centre of trade and commerce. About 80 per cent of Nigeria's industry is based in and around Lagos, and it is now the main financial centre in West Africa. The city also has a major international airport and a busy seaport. The population of Lagos continues to grow (Figure 14.3). It is estimated that, by 2040, Lagos will be the world's third largest city, after Tokyo and Delhi, with a population of 30 million.

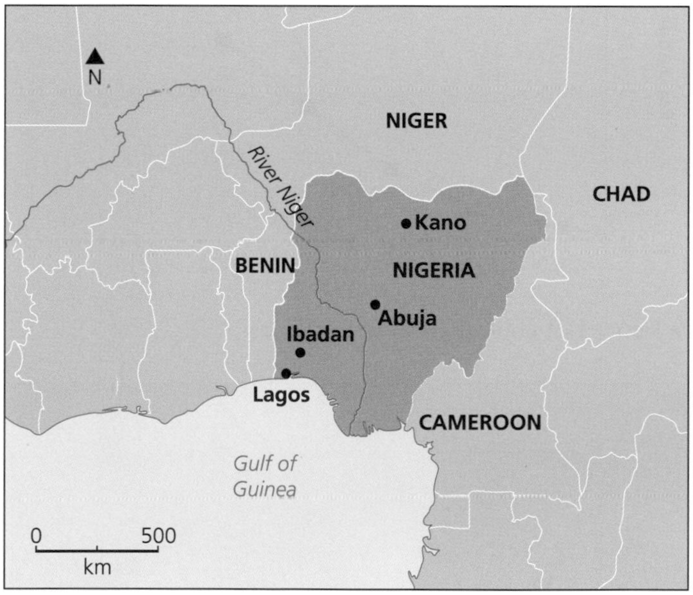

▲ Figure 14.2 Lagos' location in Nigeria

Key facts

- Largest city in Africa, with an estimated 17 million people
- Many schools, hospitals and universities
- Generates 25 per cent of Nigeria's wealth
- Many large companies have their headquarters in Lagos
- The main financial centre in West Africa
- Nollywood, Nigeria's film industry, is based in Lagos
- A transport hub, with an international airport and a port
- Has a thriving arts and culture scene

▲ Figure 14.3 Lagos fact file

→ Activities

1 Study Figure 14.1.
   a) Describe Lagos from above. Mention its size, the buildings (their age and height), the roads and its surroundings.
   b) Compare the photo with a UK city. What similarities and differences do you see?

2 Look at the maps in Figures 14.1 and 14.2. Describe the location of Lagos (a) within Africa and (b) Nigeria. Mention the Equator, the Atlantic Ocean, the Gulf of Guinea and Benin.

3 Look carefully at Figure 14.2.
   a) Why do you think Abuja was chosen as the new capital of Nigeria?
   b) Was this a good decision for Lagos? Give reasons for your answer.

4 Read the facts about Lagos. Try to classify the facts into three groups – regional, national and international – to show the importance of Lagos. List each fact under a heading. You may decide that some facts come under more than one heading.

⭐ KEY LEARNING

➤ How fast Lagos is growing
➤ Causes of population growth in Lagos
➤ How push and pull factors lead to rural–urban migration

Growing Lagos

How fast is Lagos growing?

Lagos is growing so fast that no one can agree what its population really is! In 1960, the city still had a population of less than a million people, but this grew to about 4 million by 1990 and about 17 million by 2019 (Figure 14.4). Some estimates are higher than this and, if you include the surrounding area, the population is over 20 million.

As Lagos' population has grown, so has the area of the city (Figure 14.5). The original site was on Lagos Island, surrounded by Lagos Lagoon. By 1960 the city had expanded northwards onto the mainland, following the line of the main railway.

Lagos' expansion really took off during the oil boom in Nigeria in the 1970s, which drew thousands of people to the city for work. The city continued to grow despite a fall in living standards during the 1980s and 1990s. It has expanded around the Lagoon to the north and west, and eastwards on the Lekki Peninsula.

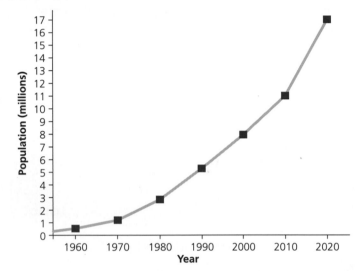

▲ Figure 14.4 Population growth in Lagos

▲ Figure 14.5 The growth of Lagos

What causes population growth in Lagos?

The main driver of growth in Lagos over the past 50 years has been rural–urban migration. People are encouraged to leave the countryside by push factors such as the lack of job opportunities and low wages (Figure 14.6). They are brought to the city by pull factors such as the prospect of well-paid work and the attraction of an urban lifestyle.

Another reason for Lagos' population growth is the high rate of natural increase in the city's population. This is due to the city's youthful population, since most migrants to the city are young.

Nigeria is becoming an increasingly urbanised country. Just over half the population are still living in rural areas, but as rural–urban migration continues, the majority will be urban within the next few years.

How do push and pull factors lead to rural–urban migration?

Education and health services are poor in rural areas.

There is a land shortage due to population growth. Despite urbanisation, the rural population continues to grow.

Changing climate is making the weather less predictable. Droughts and floods occur more often now.

Farming pays low wages but requires a lot of hard work.

Land is degraded due to farming and other activities. Land in the Niger Delta region is polluted by the oil industry.

Few job opportunities exist other than farming.

Political unrest creates insecurity in rural areas. The terrorist group, Boko Haram, is active in the north of Nigeria.

▲ Figure 14.6 Push factors that lead to rural–urban migration in Nigeria

→ Activities

1 Study Figure 14.4. Describe how Lagos' population has changed since 1960. Use data from the graph in your description.
2 Study Figure 14.6. Read the list of push factors for rural–urban migration in Nigeria. Write an equivalent list of pull factors that would attract people to Lagos, such as more job opportunities (you might get more ideas in Section 14.3). Draw a table like this to list the push and pull factors for Lagos.

Push factors for rural areas	Pull factors for Lagos
Few job opportunities	More job opportunities

3 Imagine you are one of the people in the photo in Figure 14.6. Send a text message to your cousin in Lagos to explain why you want to move there. (Keep a copy of your message until Section 14.8, so you can see how your expectations of Lagos might have changed after you have lived there for a year.)

Geographical skills

1 Look at Figure 14.5.
   a) Describe how Lagos has grown in size since 1960. Use the scale on the map to estimate its size in 1960, 1994 and 2012.
   b) Suggest how the growth of the city has been affected by its coastal location and physical geography.

✪ KEY LEARNING

➤ How urban growth has created opportunities and challenges in Lagos
➤ What the opportunities are
➤ What the challenges are

Lagos – opportunities and challenges

What opportunities has urban growth brought to Lagos?

At first sight, Lagos might not look like the sort of place where you would choose to live. The city is straining under the pressure of a growing population. Congested roads, electricity in short supply, a sewage system that hardly works – these are problems that Lagos residents regularly have to contend with. Yet, despite the problems, there are plenty of social and **economic opportunities**.

Social opportunities

Social opportunities are the chances for people to improve their quality of life through services, like education and healthcare. In cities, there is often better access to education and healthcare, as well as resources like water and energy, even though these are far from perfect in Lagos (see page 219).

Education

There are more schools and universities in Lagos than you find outside the city. Lagos State Education offers all children nine years of basic education, so the city has a higher **literacy rate** than the rest of Nigeria (Figure 14.7). The number of universities in Lagos is growing to meet the needs of the economy, including the top-ranked University of Lagos with 57,000 students.

Healthcare

Although it is not always free, at least healthcare is available in Lagos. The nearest clinic or hospital is a lot closer than if you live in a village, though you might have to queue in a public hospital if you can't afford a private one. Average life expectancy in Nigeria is 55, but there is no evidence that it is higher in Lagos.

Economic opportunities

Economic opportunities are the chances for people to improve their living standards through employment. More jobs are available in Lagos than anywhere else in Nigeria. The city generates a quarter of Nigeria's **gross domestic product.** It is the centre of most of the country's manufacturing and also for the finance industry, not just for Nigeria but for much of Africa. In addition, there is a thriving music, film and fashion scene and, more recently, Lagos is the hub for a growing tech industry.

Even if migrants can't find work in the **formal economy,** paying tax, it is possible to work in the **informal economy,** for example as a street vendor or recycling waste (see Section 14.4), paying no tax.

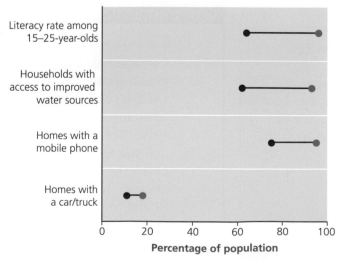

Key
● Nigeria
● Lagos

▲ Figure 14.7 Comparing economic and social indicators for Lagos and Nigeria

What are the challenges in Lagos?

Managing urban growth

Two-thirds of the population live in **squatter settlements**, sometimes referred to as slums, like Makoko (Figure 14.8). People are forced to build their homes on land they do not own, using whatever materials they can find (read more in Section 14.5).

Meanwhile, Lagos' growing economy has led to fabulous wealth for some and a widening gap between the rich and poor. Inequality brings its own challenges. Victoria Island, close to the city centre, is a wealthy neighbourhood where luxurious mansions are gated to provide security from intruders (Figure 14.9).

▲ Figure 14.8 Housing for the poor in Makoko

▲ Figure 14.9 Housing for the rich on Victoria Island

Providing water, sanitation and energy

Water and sanitation

Despite improvements, Lagos still lacks adequate clean water for drinking and household use. Only the wealthiest homes have a piped water supply. Others use public taps or boreholes, or buy their water from street vendors, with the risk of contamination by sewage (read more in Section 14.6).

Energy

Lagos suffers from regular power cuts. Many households and businesses rely on back-up generators when the network fails. There are plans to build mini-power plants around the city to provide power, 24/7.

Reducing unemployment and crime

Unemployment

With no unemployment benefit, people are forced to find whatever work they can. About 40 per cent of the workforce is in the informal economy, often in poorly paid, unregulated, dangerous work, earning less than $1.25 (about £1) a day (read more in Section 14.4).

Crime

Despite its reputation for violence and corruption, the crime rate is falling. Violent clashes occasionally break out between street gangs known as 'Area Boys'. Cyber-crime and scams are a problem for the financial industry in Lagos.

→ Activities

1 What are the main economic and social opportunities and challenges in a city? Think about Lagos, but also think about any city you know well. Draw a large table like this and list at least three ideas in each box.

	Opportunities	Challenges
Economic		
Social		

2 Study Figure 14.7.
   a) Compare the indicators for Lagos and Nigeria. Write four sentences, one about each indicator, using data from the chart.
   b) For each indicator, is it an economic or social indicator? You may decide it is both. In each case, give your reasons.
3 Look at Figures 14.8 and 14.9. In which area of Lagos – Victoria Island or Makoko – would you be most likely to,

   a) drive a car?
   b) use a kerosene lamp?
   c) buy water from a street vendor?
   d) pay for healthcare when you are ill?

   In each case, explain your answer.

Case study

✪ KEY LEARNING

➤ The opportunities Lagos offers for industry
➤ The new economic opportunities in Lagos
➤ The opportunities and challenges of the informal economy

Economic opportunities and challenges

What opportunities does Lagos offer for industry?

Like many other megacities around the world, Lagos has a coastal location. Over the centuries, it has transformed from a small fishing village into a busy seaport. Lagos Lagoon provides a good, sheltered harbour for shipping. In addition to its port, Lagos now has a major international airport, which is the main arrival point for 80 per cent of flights to West Africa (Figure 14.10).

Good transport connections have helped Lagos to develop into a major industrial centre. Its growing population also provides a large market for goods and services, encouraging more industry to locate there. With many schools and universities, Lagos also has a well-educated and skilled workforce, attracting more companies.

What are the new economic opportunities in Lagos?

Now, Lagos is building a new city on the coast called Eko Atlantic, destined to be the new financial hub of West Africa. Inspired by Hong Kong, or perhaps Canary Wharf in London (see Section 15.5), it is a joint project between the city government of Lagos and international private investors. It will be home to a quarter of a million people and employ 150,000 more.

A new **sea wall**, over 6 km long, has been built to reclaim land and protect it from the sea. Eko Atlantic will have independent, reliable electricity, advanced fibre optic telecoms and clean water (see Figure 14.11). Further east, in rapidly expanding Lekki, a new industrial project is taking shape. A US$12 billion oil refinery is being built, the largest of its type in the world. It will produce enough oil and kerosene to meet demand for the entire Nigerian economy (see Section 19.3).

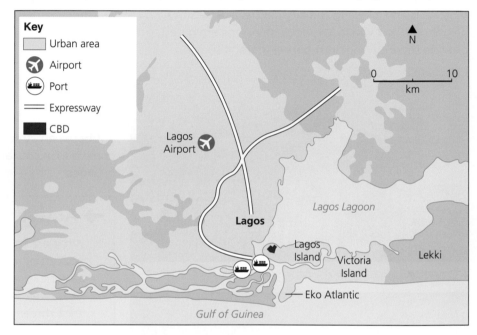

▲ Figure 14.10 Lagos' transport connections

▲ Figure 14.11 The Eko Atlantic land reclamation and development project

What are the opportunities and challenges of the informal economy?

Unemployment in Lagos is much lower than in the rest of Nigeria at a rate of about ten per cent. However, unlike the UK, there is no unemployment benefit for those without work. Most people need to find work in the informal economy in order to survive. They do jobs like street vending, car washing, shoe shining or **waste recycling**; 90 per cent of new jobs created in Lagos are in the informal sector.

Although the formal economy in Lagos is growing, there is a limit to the number of jobs it can create. About 40 per cent of the workforce work in the informal economy. It plays a vital role in both providing employment and helping the city to function. Lagos has earned the reputation of being an enterprising city where people are prepared to do anything to make a living such as sorting rubbish. The Olusosun dump in Lagos is the largest **landfill site** in Africa and one of the largest in the world (Figure 14.12).

Olususon is a huge landfill site near the heart of Lagos. The city has grown up around the site.

Around 500 people work at the dump. Each day they sort 3,000 tonnes of waste by hand, picking out valuable items to sell.

Workers live at the landfill site, building their homes out of discarded materials.

There are also shops, restaurants, bars, cinemas and a mosque at the landfill site.

Municipal governments collect just around 40 per cent of the 10,000 tonnes of waste produced in Lagos every day. It is taken to landfill sites like Olusosun and separated for recycling.

Rubbish can be turned into energy by harnessing methane gas, emitted from rotten waste. A new project by the Lagos State Waste Management Authority is planned to produce 25MW of electricity, enough to power a town.

Electric waste, imported to Nigeria, is also brought to the site. It is treated with chemicals to extract the reusable materials, but toxic fumes are released.

Without the dump, a lot of reusable items would go to waste. People in Lagos can save money by buying recycled goods.

Natural gases build up under the decomposing waste, especially when it is dry. This often leads to fires that are hard to extinguish.

Olusosun is an example of the way people in Lagos find solutions to problems – seeing an opportunity where others might just see junk.

▲ Figure 14.12 The Olusosun rubbish dump in Lagos

→ Activities

1 Look at Figure 14.10.
   a) Draw a sketch map to show Lagos' location.
   b) Annotate your map to list the opportunities for industry. Mention at least four factors.
2 Study Figures 14.10 and 14.11.
   a) Describe the location of Eko Atlantic in relation to the rest of Lagos.
   b) Outline the opportunities and challenges of Eko Atlantic as the site for a new financial hub.
3 Classify each of these jobs into formal or informal employment. In each case, give a reason for your classification:

- Recycling at Olusosun
- Selling mobile phones
- Working in a car factory
- Fishing at Makoko
- Driving a minibus taxi
- Working for a bank

4 Look at Figure 14.12.
   a) Draw a table to list the opportunities and challenges of working at Olusosun dump.
   b) Do you think that Olusosun is an effective way of managing waste disposal in Lagos? Give reasons for your answer.

⭐ KEY LEARNING

➤ The challenges of urban growth in Lagos
➤ The problems of living in squatter settlements
➤ Whether squatter settlements should be demolished or improved

The challenges of urban growth in Lagos

Where are squatter settlements found in Lagos?

An estimated 2,000 new people arrive in Lagos every day. The lack of properly built homes in Lagos has forced many of them to build their own homes on land (or even water!) they do not own (Figure 14.15). These so-called squatter settlements are found all over the city, particularly on marshy, poorly drained land where no one else wants to build (Figure 14.13).

Squatter settlements are not unique to Lagos. They are found in cities in low-income countries and newly emerging countries all around the world. They can also be called informal settlements, shanty towns or slums.

▼ Figure 14.14 Housing conditions in squatter settlements in Lagos

Condition	%
Housing density	
Households living in more than one room	25
Households living in one room	75
Household facilities	
Kitchen, bath and toilet	10
Lacking either kitchen, bath or toilet	52
No kitchen, bath or toilet	38
Water supply	
Piped water	11
Public tap	14
Well or borehole	55
River	4
Water vendor	16
Toilets	
Septic tank (underground tank in which sewage collects)	10
Pit latrine (sewage soaks straight into the ground)	55
Pail latrine (sewage is poured into a drain or river)	33
Bush (i.e. no toilet!)	2

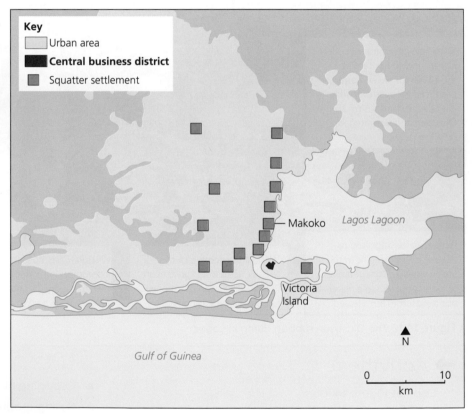

▲ Figure 14.13 Large squatter settlements in Lagos

What are the problems of living in a squatter settlement?

Squatter settlements in Lagos are densely populated due to the shortage of available land. In the case of Makoko, on the edge of Lagos Lagoon, homes extend into the water, built on stilts (Figure 14.15).

The homes are usually makeshift shelters built from materials like tin sheets and wooden planks. They lack basic facilities and good **sanitation** (Figure 14.14). The population of Makoko is estimated at up to a quarter of a million people. Most of them make a living in the informal economy and by fishing. This goes back to Makoko's origins as a fishing village outside Lagos. As the city grew, it was swallowed up in the urban area.

Should Makoko be demolished or improved?

For all its problems, people in Makoko are fiercely protective of their homes. In 2012, the authorities wanted to demolish the area, but residents had nowhere else to go, and resisted the demolition (Figure 14.16). People have also taken their own initiatives to solve some of the challenges of living in Makoko.

▲ Figure 14.15 Makoko

The demolition of Makoko

In the shadow of the Third Mainland Bridge in Lagos, fragile wooden huts have stood for decades on stilts above the water like long-legged birds. Beneath steaming traffic jams, the people of Makoko drift across the muddy water in canoes, casting nets for fish.

But after giving residents 72 hours to leave their homes, state authorities began demolishing the shanty town a week ago.

This is not the first attempt to wipe out Makoko. Lagos authorities call the shanty town 'unwholesome' and out of keeping with Lagos' 'megacity status'. Lagos governor said there were plans to build something much grander. 'We have a plan to turn that place into the Venice of Africa', he told protestors from Makoko.

They have built pipes to bring in clean drinking water from boreholes (see Section 14.6). Without any official clinics or hospitals in Makoko, a network of unregistered clinics has been opened to deal with health problems. As a result, for such a densely populated area, Makoko has remarkably few incidences of infectious diseases (such as cholera).

▲ Figure 14.16 A news article about the demolition of Makoko, July 2012

→ Activities

1 Look at Figure 14.13.
  a) Describe the distribution of squatter settlements in Lagos. Mention their location in relation to the Lagos Lagoon.
  b) Try to explain the distribution. (Hint: think about what the land near the lagoon would be like.)

2 Study Figure 14.14. Identify the main problems of living in a squatter settlement in Lagos. Mention at least four.

3 Look at Figure 14.15. Draw an annotated sketch of the photo. In your labels, describe some of the problems faced by people in Makoko, including:
  ■ its location
  ■ housing density
  ■ building construction
  ■ sanitation.

4 Read the article in Figure 14.16.
  a) With a partner, discuss whether Makoko should be demolished or improved. One of you should play the role of a Makoko resident and the other should play the role of the Lagos governor.
  b) Together, decide how the growth of squatter settlements should be managed. Should Makoko be demolished or improved? Write a short report giving reasons for your decision.

The challenges of clean water, sanitation and energy

How does Lagos obtain its water supply?

Case study

★ **KEY LEARNING**

➤ How Lagos obtains its water supply
➤ The problem with sanitation in Lagos
➤ The problem with energy supply in Lagos

Among the common sights in Lagos are water vendors on the street selling water in containers (see Figure 14.17). It is often difficult to obtain drinking water from any other source, even though the city is surrounded by water!

Only 11 per cent of the population in Lagos have a piped water supply that has been treated and purified. The rest of the population either rely on water vendors or dig their own wells or sink boreholes to reach **groundwater** supplies that lie below the water table (see Figure 14.18). Water in the lagoon or tidal creeks around the city is not suitable for drinking because it is salty (not to mention polluted!).

In 2012, the newly formed Lagos State Water Regulatory Commission began the job of regulating the water supply and water vendors, and issuing licences for boreholes. It is responsible for ensuring a safe water supply at a reasonable price for consumers. Given the rapidly growing population of Lagos, and increasing demands for water, this is a challenging responsibility.

▲ Figure 14.17 A water vendor in Lagos

▼ Figure 14.18 Water supply around Lagos

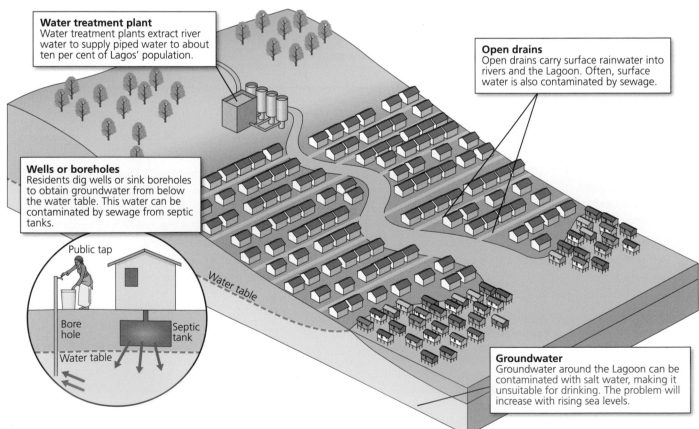

Water treatment plant
Water treatment plants extract river water to supply piped water to about ten per cent of Lagos' population.

Open drains
Open drains carry surface rainwater into rivers and the Lagoon. Often, surface water is also contaminated by sewage.

Wells or boreholes
Residents dig wells or sink boreholes to obtain groundwater from below the water table. This water can be contaminated by sewage from septic tanks.

Public tap

Bore hole

Septic tank

Water table

Water table

Groundwater
Groundwater around the Lagoon can be contaminated with salt water, making it unsuitable for drinking. The problem will increase with rising sea levels.

What are the problems with sanitation in Lagos?

Drinking water in Lagos often contains bacterial or chemical **pollution** that can lead to diarrhoea. The number of cases of diseases like cholera and dysentery is high. One of the main causes of pollution is the lack of a proper sanitation system in the city. Sewage is sometimes disposed of with rainwater through open drains. It is carried into rivers and the Lagoon, which also become polluted.

Sewage may also soak into the ground from pit latrines or leaking septic tanks. Here it can find its way into the water supply through wells and boreholes. Even water from vendors can be contaminated because they also obtain water from the same sources.

One of the Lagos State Government's solutions to the problems is to hold a 'Sanitation Day' once a month where residents are compelled to take part in a clean-up, unblocking drains and clearing rubbish. People complain about doing this because it stops them earning money, thereby damaging the economy.

▲ **Figure 14.19** A resident unblocks a drain to improve sanitation in Lagos

What are the problems with energy supply in Lagos?

There is no reliable energy supply in Lagos. Forty per cent of the population have no access to the electricity grid. Even for those who do, power cuts happen regularly. Unsurprisingly, 80 per cent of households rely on diesel generators as their main source of energy, or for backup. This is a source of air pollution and adds to carbon emissions. There are plans to future-proof Lagos by developing independent 'power projects' to provide more reliable energy and to lower air pollution. **Renewable energy sources** such as sun or wind will provide at least 20 per cent of this energy. The aim, by 2030, is for 100 per cent of households to have access to energy and for there to be street lighting for the whole city.

→ Activities

1   Explain why Lagos has a shortage of drinking water, despite being surrounded by water.

2   Study Figure 14.18. Describe how water is supplied from: (a) rivers, (b) ground water.

3   Look at Figures 14.17 and 14.18. Compare three sources of water in Lagos – water vendors, piped water and wells or boreholes.

   a)   Draw a large table like this. Complete the boxes in the table.

Water source	Availability	Purity	Sustainability
Water vendors			
Piped water			
Wells or boreholes			

   b)   Which source of water would you prefer to use? Give your reasons.

4   Do you think having a 'Sanitation Day' once a month is a good idea? Give arguments both for and against the idea before coming to your own final judgement.

5   Explain how rising sea levels in Lagos could lead to:

   a)   more flooding

   b)   less drinking water.

⭐ KEY LEARNING
➤ The impact of traffic congestion and air pollution on Lagos
➤ The efforts made to reduce congestion
➤ How a transport master plan could help Lagos

Traffic congestion and air pollution

What impact does traffic congestion and air pollution have on people in Lagos?

The average Lagosian commuter spends over three hours in traffic every day. It makes Lagos one of the most congested cities in the world. Perhaps that is not surprising when you consider that 40 per cent of new cars in Nigeria are registered in Lagos, which occupies just one per cent of the country's total area.

Traffic congestion causes many problems for people in Lagos. It is not just a matter of the inconvenience, though that is considerable (see Figure 14.20). The fatal accident rate in Lagos is 28 per 100,000 people – three times higher than the rate in European cities. Air pollution rates in Lagos are five times higher than the internationally recommended limit, mainly due to traffic congestion and the use of so many diesel generators.

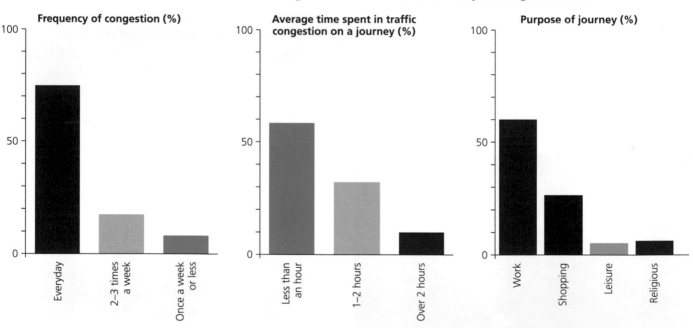

▲ Figure 14.20 The impact of traffic congestion on people in Lagos

What efforts have been made to reduce traffic congestion?

In 2003, the Lagos state government set up the Lagos Metropolitan Area Transport Authority (LAMATA) to improve transport in the city. One of its first achievements was to introduce a bus rapid transit (BRT) system on a north–south route from the suburbs to the CBD on Lagos Island (see Figure 14.20). It provides a separate lane for buses to reduce travel times. 200,000 people use the service each day – a quarter of all commuters in Lagos.

However, a single BRT route is inadequate in a city the size of Lagos. The public transport system has to be supplemented by a large fleet of minibus taxis, known as 'danfos'. They are designed to carry 10–15 passengers, but demand is so high that they often carry 20–30 people. It is estimated that 2 million vehicles get stuck in traffic every day in Lagos, with drivers and pedestrians inhaling the polluted air. Rates of respiratory disease are high.

How could a transport master plan help Lagos?

A scheme that opened in 2016 is a new light railway on a west–east route into the CBD, designed to carry seven times as many passengers as the BRT. Eventually, there are plans for a network of seven new rail lines, known as Lagos Rail Mass Transit (LRMT).

This is part of a wider Strategic Transport Master Plan for Lagos which includes:

- an **integrated transport system** where road, rail and waterway networks link together to make journeys easier
- a new waterway network of ferries to make better transport use of the water areas around Lagos
- a more efficient road network with separate bus lanes, and without obstacles like markets and street vendors, to speed traffic flow
- better urban planning with mixed-use developments, (for example, residential and commercial), to reduce the number of journeys people need to make
- a new airport on the Lekki Peninsula, further from the congested urban area
- better walking and cycling facilities (like pavements for pedestrians).

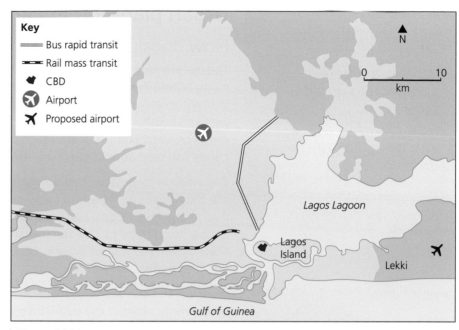
▲ Figure 14.21 New and planned transport developments in Lagos

▲ Figure 14.22 Minibus taxis, or danfos, in Lagos

→ Activities

1 Why might each of these people be concerned about traffic congestion in Lagos?
   a) A Makoko resident
   b) A danfo driver
   c) A business owner in the CBD
   d) The governor of Lagos

2 Look at the graphs in Figure 14.20.
   a) Describe the impacts of congestion on people in Lagos.
   b) How does this compare with a city you know?

3 Read the six ideas for the Strategic Transport Master Plan in Lagos. Explain how each of these ideas would help to reduce congestion in the city.

4 Study Figure 14.21.
   a) On a copy of the map, create an integrated transport plan for Lagos. Draw at least:
      - two more BRT routes
      - two more LRMT routes
      - two waterway routes.

      Try to integrate the routes on your map.
   b) Write a short report of up to 200 words to support your plans. You can compare your plans with the real thing at www.lamata-ng.com.

Example

➤ How urban planning affects the poor in Lagos

➤ What is a different approach to urban planning

➤ How planning is improving life for the urban poor

Urban planning in Lagos

How does urban planning affect the poor?

The government's approach to urban planning in Lagos, in common with many governments around the world, is to attract **private investment** that will create a gleaming new city. Although there will be highly paid jobs and luxury homes for some, the poor are mainly excluded from such developments. This is the approach of the Eko Atlantic development in Lagos you read about in Section 14.4.

The attitude to squatter settlements is quite different. Governments often see these areas as reflecting badly on the city and that they may even discourage investment. The easiest solution is to demolish them and evict people living there to find somewhere else to live. This is what the Lagos State Government attempted to do in Makoko in 2012, but people resisted, as you read in Section 14.5.

However, this still seems to be official planning policy in Lagos. In 2017 the government ordered the demolition of another waterfront community near Makoko, called Otodo Gbame. Around 30,000 people were evicted, losing not just their homes but also their possessions and fishing livelihoods. Unable to afford rented accommodation in Lagos, they were forced to build new homes elsewhere. Many of them went to Makoko.

Is there another approach to urban planning?

After the government's attempt to demolish Makoko in 2012, a new 'Waterfront Regeneration Plan' has been produced as an alternative to demolition (Figure 14.23). The main aims of the plan are to:

- preserve the ancestral and historical character of the community, based on fishing
- involve the community itself in developing plans to improve Makoko
- encourage economic **development**, including tourism, to provide employment
- build sustainably (read more in Section 16.1) and with resilience to **climate change**.

At the centre of the plan are new Neighbourhood Hotspots spread around Makoko, which would provide meeting places for people and improved healthcare clinics. They would also be centres for waste collection which would generate power from biogas by burning the waste. So far, the government does not show much sign of accepting the plan.

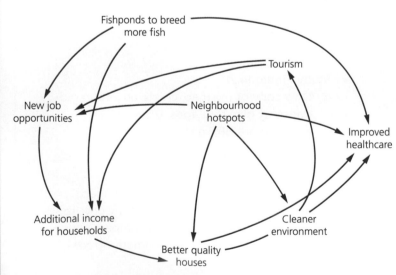

▲ Figure 14.23 Waterfront Regeneration Plan for Makoko

How is planning improving life for the urban poor?

In 2014, a new 'floating school' was built in Makoko (Figure 14.24). Built mainly from wood and other local materials, it was designed to be sustainable, generating its own energy from solar panels and able to withstand rising sea levels. The school had classrooms that could host lessons for up to 60 children at a time, much needed in Makoko where there are few other schools.

The plan was to expand the idea of a floating school into a whole floating community to help with housing Makoko's expanding population (Figure 14.25). Unfortunately, in 2017, the school collapsed but, fortunately, without any injuries. Pupils had to find other schools if they could. However, the idea lives on and there are plans to replace the school with a stronger structure.

▲ Figure 14.24 Makoko's floating school, before it collapsed

The idea of a floating community might sound far-fetched but, if coastal communities like Makoko are to withstand rising sea levels as a result of climate change, then radical new ideas may be needed. The solution at Eko Atlantic was to surround the whole development with a new sea wall costing millions of dollars. It is unlikely that anyone will suggest such an expensive scheme for Makoko or other poor communities.

▲ Figure 14.25 An artist's impression of a floating community in Lagos

→ Activities

1 Using examples from Lagos, explain why urban planning policies do not always improve life for the urban poor.

2 Study Figure 14.23. Describe at least three connections between different parts of the plan, shown by the arrows. Write a sentence for each connection you describe. For example, you could describe how fishponds might lead to new job opportunities.

3 Study Figures 14.24 and 14.25.

   a) Explain how a floating community would help Makoko to deal with

      i)   a growing population

      ii) rising sea levels.

   b) Can you think of any drawbacks to a floating community? Explain your ideas.

4 Think about what you have learned about Lagos in this chapter and how that compares with your expectations at the start.

   a) Look back at the text message you sent from a person thinking of moving to Lagos in Activity 3 on page 211. How might you feel now, having lived in Lagos for a year? Make a list of the good points and bad points.

   b) Send another text message, back to your parents in the village you came from, to describe your experiences in Lagos.

15 Urban challenges in the UK

⊛ KEY LEARNING

➤ How the population of the UK is distributed

➤ Where cities in the UK are located

➤ How UK cities are growing

Cities in the UK

How is the UK's population distributed?

The UK is one of the most urbanised countries in the world, with 82 per cent of our population living in cities. This is typical of most high-income countries (HICs) that went through the process of urbanisation during the nineteenth and twentieth centuries. It is different in low-income countries (LICs) and newly emerging economies (NEEs), such as Nigeria, which are still rapidly urbanising today (see Chapter 14).

If you live in London (Figure 15.1) or another UK city, you might get the impression that we are an overcrowded country. But when you look at a map of the UK's population distribution (Figure 15.2), you can see that people are unevenly distributed. While some areas, such as South East England, are densely populated, other areas, such as Northern Scotland, are **sparsely populated.**

Overall, the UK's population density is 260 people per square kilometre (km²), ranging from about 5,000/km² in London to less than 10/km² in northern Scotland. This makes us one of the more densely populated countries in Europe – more crowded than France, for example, but less crowded than the Netherlands.

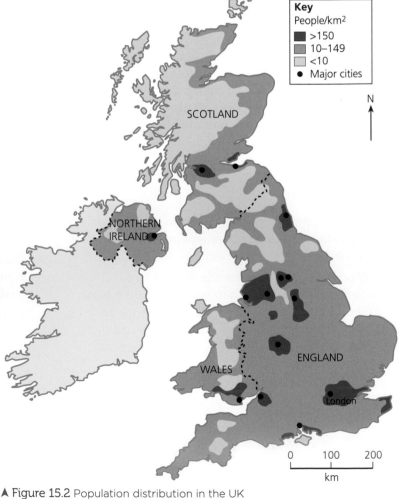

Key
People/km²
■ >150
■ 10–149
□ <10
● Major cities

N

SCOTLAND

NORTHERN IRELAND

ENGLAND

WALES

London

0   100   200
km

▲ Figure 15.2 Population distribution in the UK

▲ Figure 15.1 London is the UK's largest city. It contains the centre of the finance industry, often called the 'City of London'.

Where are cities in the UK located?

The UK's cities are found in the most densely populated areas (Figures 15.2 and 15.3). They tend to be located in flat, low-lying parts of the country, particularly on the coast or near major rivers. Historically, this is where many cities grew, supported by farming, trade and industry.

How are UK cities growing?

Today, the fastest-growing cities are in South East England, which is the region with the fastest-growing economy (read more in Chapter 20). By far the biggest growth so far in the twenty-first century has been in London, with over a million new people. At the other end of the scale, Sunderland, in the northeast of England, is the only major UK city where population has fallen. This is due to the decline of industry and loss of jobs, forcing people to move away to find work.

→ Activities

1   Study Figure 15.1.
    a) If you lived in London, why might you think the UK is overcrowded?
    b) Would you be right or wrong? Explain your answer.

2   Study Figure 15.2. Describe the distribution of population in the UK. Include each part of the UK in your description, for example South East England, northern Scotland.

3   Compare Figures 15.2 and 15.3. Explain the connection between population distribution and the location of cities in the UK.

4   Study Figure 15.3.
    a) Where are the fastest-growing cities and the slowest-growing cities, in the UK?
    b) How can you explain the pattern? (Use Figures 15.1 and 15.4 to inform your answer.)

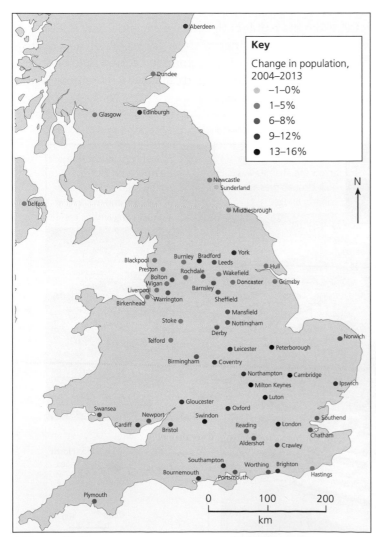

Key

Change in population, 2004–2013
- −1–0%
- 1–5%
- 6–8%
- 9–12%
- 13–16%

▲ Figure 15.3 Major towns and cities in the UK, showing how their populations are changing

▲ Figure 15.4 Sunderland, a city where ships were once built and coal was mined. The shipyards have long closed. The football stadium was built on the site of an old coal mine.

⊛ KEY LEARNING

- ➤ Where London is located
- ➤ Why London has grown
- ➤ The national and international importance of London

London on the map

Where is London located?

London is located in South East England on the River Thames. It is the site chosen by the Romans when they conquered the south of England in AD43. They built a walled settlement on the north bank of the Thames to defend themselves against the defeated Britons. They called the settlement Londinium and it became the capital of the Roman colony in Britain.

Two factors were important in London's success as a city:

- The Thames is a tidal river. At high tide, ships were able to navigate up the river to London and the city became a port.
- London was built at the lowest bridging point on the Thames – the widest point on the river where it was possible to build a bridge at that time.

Why did London grow?

Two thousand years after it was built by the Romans, London is still the capital city of the UK. From the eighteenth century onwards, new docks built along the river increased the number of ships using London as a port. London's importance as a centre of trade and commerce grew, and new manufacturing industries developed. This, in turn, attracted more people and its population increased (see Section 15.3).

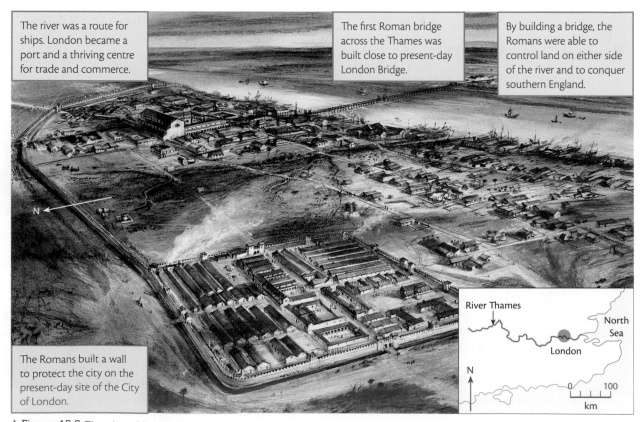

The river was a route for ships. London became a port and a thriving centre for trade and commerce.

The first Roman bridge across the Thames was built close to present-day London Bridge.

By building a bridge, the Romans were able to control land on either side of the river and to conquer southern England.

The Romans built a wall to protect the city on the present-day site of the City of London.

River Thames
North Sea
London

▲ **Figure 15.5** The site of Roman London

London's role as a port declined towards the end of the twentieth century, with the opening of new docks on the coast. However, it remains the main hub for the UK transport network. Both the UK's road and rail networks focus on London (Figure 15.6). The UK's two busiest airports – Heathrow and Gatwick – are both close to London. They help to maintain London's global connections and its importance as a tourist destination.

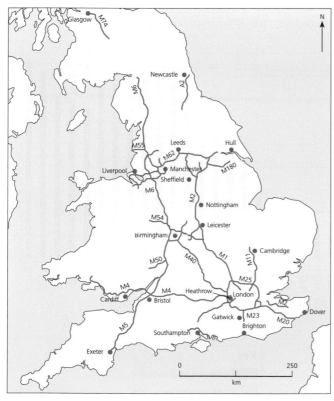

▲ Figure 15.6 The motorway network in England and Wales

What is London's national and international importance?

London is not just the UK's capital; it is also by far the UK's largest and wealthiest city. The gap between London and the rest of the UK has widened in the twenty-first century, as both earnings and house prices have risen faster in London than elsewhere (Figure 15.7).

An indication of London's modern-day importance is its status as a **world city**. A world city's influence is not just national, but also global. Along with New York, London is one of the two most important financial centres in the world. The headquarters of many large international companies, as well as most major British companies, are based there.

London is also a national and international centre for:

- media and communications networks
- education, including renowned universities and research
- legal and medical facilities
- culture, entertainment and tourism.

The city attracts investment and people from all around the world. Many of London's iconic buildings, like the Shard, (and football teams!), are owned by investors from other countries. Many migrants come to London to work in high-paid as well as low-paid jobs.

▼ Figure 15.7 London and the UK compared (2019)

Categories	London	UK
Population	8,800,000	66,000,000
Average male life expectancy	79	78
Average female life expectancy	83	82
Proportion of people from minority ethnic groups in the population	37%	14%
Average earnings	£35,677	£29,044
Average house price	£478,000	£243,000
Unemployment rate	4.2%	3.9%
Educational achievement (top GCSE grades)	25.2%	20.8%
Murder rate	1.1/100,000	1.4/100,000

Geographical skills

Study Figure 15.5. Use this to draw a sketch map to show the site of Roman London. Annotate your map to explain why this was a good site for a city.

→ Activities

1 Study Figure 15.6.

   a) Plan journeys from three other cities in the UK to London. In each case, list the motorways you would travel on.

   b) Describe the pattern of motorways in the UK in relation to London.

2 Study Figure 15.7. Compare London with the rest of the UK. Write three short paragraphs to compare (a) the population, (b) distribution of wealth and (c) quality of life.

3 What do you think are (a) the advantages and (b) the disadvantages of living in London? Give at least three examples of each.

Case study

⊗ KEY LEARNING

➤ How London's population has changed

➤ What London's population structure is now

➤ The ethnic composition of London's population

London's growing population

How has London's population changed?

London's population is higher now than it has ever been. In 2015, London's population reached 8.6 million, overtaking the peak it last reached in 1939.

For most of the past two hundred years, London's population has been growing (see Figure 15.8). In 1801, with just over a million people, it was already the largest city in the world. During the Industrial Revolution in the nineteenth century, the city grew as it attracted migrants, mainly from other parts of the UK.

London's population reached its previous peak at the start of the Second World War. The city was badly bombed during the war and its population fell after 1939. Numbers continued to decline after the war as housing was demolished and people moved out. During the twentieth century, many cities in other countries grew bigger than London.

London's population has been climbing again since 1991. National and international migration continually add to the population. Migrants mainly come to London for better paid work. London's population is predicted to reach 10 million by 2030, which will make London one of the world's megacities.

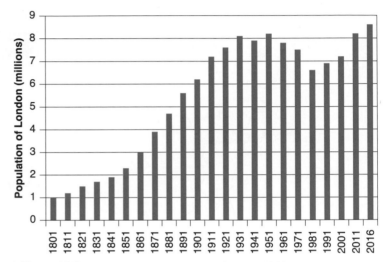

▲ Figure 15.8 London's population growth since 1801

How old is London's population?

London has a much larger population than any other UK city. It also has a younger population (see Figure 15.9). This helps to explain why its population is growing.

■ Young people in their 20s and 30s, especially university graduates, move to London for work. They are attracted by more job opportunities, higher pay and the perception of an exciting social life in London.

■ Younger people, particularly in the 20–30 age group, are more likely to have children. That leads to a higher rate of natural population increase in London.

■ Migrants from around the world add to London's population. At the same time as people arrive, others leave. The balance between the two groups is net migration.

■ Although net migration into London is quite low, most immigrants are young, while most people leaving are older. This reduces the average age of the population and leads to greater natural increase.

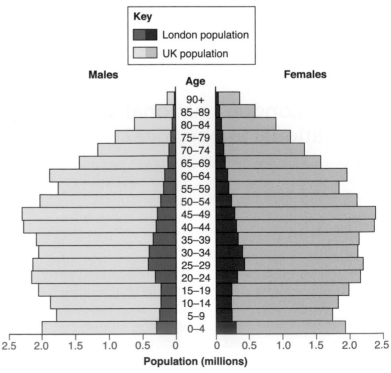

▲ Figure 15.9 London's population structure, compared with the UK (2011)

What ethnic groups make up London's population?

London is also the most diverse city in the UK. Less than half of London's population are of white British origin (see Figure 15.10), and 37 per cent were born outside the UK.

Migration to London goes back to Roman times. Later, Saxons and Normans also settled in London. In the seventeenth century, French Huguenot (Protestant) refugees arrived and settled outside the city walls to the east in Spitalfields. Later, in the nineteenth century, came Jewish refugees from Eastern Europe and more recently **economic migrants** from Bangladesh. Each group of migrants has helped to change the character of Spitalfields (see Figure 15.11).

Today, London's population comes from every part of the world. The largest numbers are from countries like India, Nigeria and Jamaica, each once part of the British Empire. After 2007, more migrants came from Eastern Europe, with the free movement of people in the **European Union** (EU). This is likely to slow down in 2020 when the UK leaves the EU.

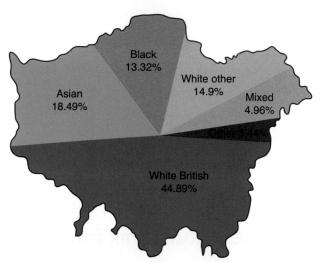

▲ Figure 15.10 The ethnic composition of London's population, 2011 census

Black 13.32%

White other 14.9%

Asian 18.49%

Mixed 4.96%

White British 44.89%

▲ Figure 15.11 Brick Lane Mosque, used by Bangladeshi Muslims in Spitalfields. It was first built as a Huguenot (Protestant) church and later used as a Jewish synagogue

→ Activities

1   Study Figure 15.8. Describe changes in London's population since 1801. Mention key figures and dates from the graph. Be as accurate as you can.

2   Study Figure 15.9. Compare the structure of London's population with the UK's. Which age groups make up a larger proportion of a) London's population and b) the UK's population?

3   a) Use the information in Figure 15.9 to predict how London's population is likely to change in the future.

    b) Explain how you made your prediction.

4   Study Figure 15.10. Turn the information into a bar chart to show the percentage of each ethnic group in London's population.

5   Study Figure 15.11. Buildings show one way in which ethnic diversity has helped to change the character of London. Think of at least five other ways in which ethnic diversity can change the character of an area.

Fieldwork: Get out there!

'Areas with a more diverse population also have more diverse shops and services.'

■   Devise at least one fieldwork method you could use to test this hypothesis.

■   Suggest two areas you know where you could carry out your fieldwork, one with a more diverse population than the other.

■   Predict what results you would expect to get from your fieldwork and why.

⊛ KEY LEARNING

➤ How an old area of London has changed
➤ The cultural mix found there now
➤ The opportunities for recreation and entertainment

The changing culture of London

How has an old area of London changed?

One old area of London close to the city centre is Shoreditch (see Figure 15.12). It typifies the sort of changes that have happened around London, and in some other UK cities.

Just 30 years ago, Shoreditch was still a run-down inner-city area, with many old factories and warehouses. Most industries had closed down and people were moving out of the area. In their place, newcomers were moving in, particularly Bangladeshi immigrants around Brick Lane (see Section 15.3).

What is the cultural mix found in Shoreditch?

Shoreditch today is almost unrecognisable from 30 years ago. Old industrial buildings have been converted into flats and offices. Pubs and bars have been brought back to life as restaurants and art galleries. Jobs have been created in new creative industries, such as web design, film-making and art.

One focus for employment is around the Old Street roundabout (Figure 15.13). So many new hi-tech companies have appeared that it is nicknamed 'Silicon Roundabout' after Silicon Valley, the centre of the hi-tech industry in California where tech companies grew. Facebook, Google and Microsoft have all invested in the area around the Silicon Roundabout.

What opportunities for recreation and entertainment are there?

The population of Shoreditch has changed too. Many older residents and Bangladeshi families are moving away, as rents and property prices go up. In their place wealthier, young professional workers, many in the finance and creative industries, are moving in (Figures 15.14–18).

This process of rising property prices and changing population is known as **gentrification**. With the new, younger population have come new forms of recreation and entertainment. Shoreditch is now one of the most vibrant parts of London. Cafés are busy during the day, while clubs and bars bring the streets alive at night. Artists have covered the walls with graffiti (or art, depending on your point of view!).

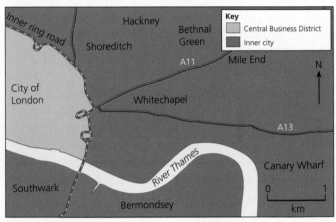

▲ Figure 15.12 The location of Shoreditch in London

▲ Figure 15.13 New, hi-tech companies around Old Street roundabout

▲ Figure 15.14 Shoreditch is now well known for its nightlife and themed cafés

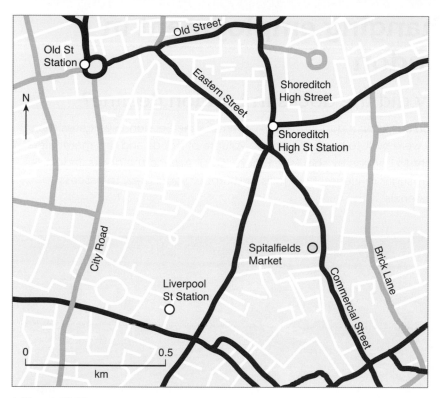

▲ Figure 15.15 Street map of Shoreditch

▲ Figure 15.16 Street art or graffiti in Shoreditch? You decide!

▲ Figure 15.17 Spitalfields Market, once a fruit and veg market, is now a fashionable place to shop

Flat to rent, Shoreditch, E2, £330/week

A unique studio flat set in a recent warehouse conversion, close to Shoreditch station. The property comprises a surprisingly spacious studio room fitted with original hardwood floors, an integrated kitchen equipped with fully functional appliances and a modern bathroom.

The property also benefits from superior travel links. You'll be a 15-minute commute to Canada Water station (serving the Jubilee Line), 10 minutes to Liverpool Street and 5 to Dalston Junction, with 24-hour bus services reaching out to almost every corner of London.

Amenities:

☑ Balcony   ☒ Garage   ☒ Patio
☑ Parking   ☒ Garden   ☒ Accessible

▲ Figure 15.18 An advert for accommodation in Shoreditch

→ **Activities**

1 Study Figure 15.12. Explain why Shoreditch is a good location in London to:
   a) start a new business
   b) live. (You could use an Underground map to help.)

2 Study the photos and map in Figures 15.14–15.17.
   a) Identify different types of recreation and entertainment in the photos and list them.
   b) Write an advert for one of the places in the photos. Think about the type of person your advert would be aimed at. Use the map to give directions.

3 Read Figure 15.18. What type of household is this advert aimed at? Consider:
   ■ their ages
   ■ their income
   ■ their occupations
   ■ if they have children.

4 How would the following be affected by the changes in Shoreditch?
   ■ A student in London
   ■ A Bangladeshi family on a low income
   ■ A young couple working in a hi-tech industry
   ■ The owner of an old warehouse.
   Explain your ideas.

Case study

⭐ KEY LEARNING

➤ Why the docks in London declined

➤ Why new industries, like finance, have grown

➤ What employment opportunities there are in London

Changing employment in London

Why did the docks in London decline?

London has been a port since Roman times (see Section 15.2). Later, the docks were built to handle the huge volume of goods and raw materials brought to London by ship (see Figure 15.19). Around the docks, industries such as sugar refineries, flour mills and timber yards grew to process the materials.

By the 1970s, the docks were in decline. New container ships were being used and the docks were no longer large enough to hold them. One by one the docks closed, until by 1980 they were lying empty, with many of the industries gone too.

➤ Figure 15.19 Docks on the Isle of Dogs, still working in the 1960s

Why have new industries, like finance, grown?

In 1981, the government set up a new body, the London Docklands Development Corporation (LDDC), to plan the **regeneration** of the docks. It was given the task of finding new ways to use the land around the docks by attracting private investment. It was hoped this would create new economic opportunities and jobs to replace those lost when the docks closed down.

What happened next became a model for other regeneration projects around the UK. At the heart of Docklands lies Canary Wharf, dominated by high-rise office blocks that are now home to many international banks. Over 100,000 people work there and, together with the City of London, Docklands has helped establish London as one of the world's leading financial centres.

▲ Figure 15.20 Canary Wharf on the Isle of Dogs today

What employment opportunities does London offer?

The number of jobs in London has been rising almost continuously since 1994 (Figure 15.21). Even the recession after 2007 did not really slow the rise. So, what are all these new jobs?

The biggest growth in jobs was in services, especially 'Professional, real estate and business services' (Figure 15.22). This includes work in company head offices, management consultancy, law and accountancy, estate agents, advertising and market research. The biggest decline in jobs over the same period was in manufacturing. London has very few factories left.

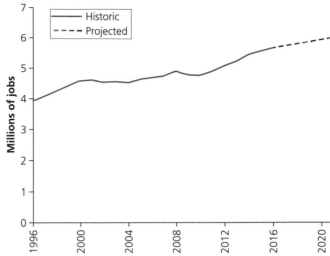

▲ Figure 15.21 London's historic and projected employment from 1992 to 2021 (calculated in 2013)

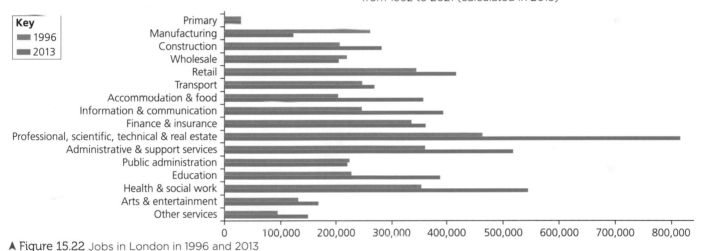

▲ Figure 15.22 Jobs in London in 1996 and 2013

→ Activities

1 Study Figures 15.19 and 15.20. Turn them into 'living photos' by adding speech bubbles to copies or sketches of the photos.

   a) Choose quotes for your speech bubbles for each photo from the ones below:

   - 'I don't think there's much of a future in this job.'
   - 'I expect a bonus this year to pay for our holiday.'
   - 'I like to go to the company gym at lunchtime.'
   - 'It's freezing doing this job in winter.'
   - 'I hope to get overtime this week to pay the rent.'
   - 'I'd like to be Chief Executive by the time I'm 40.'

   b) Summarise the way in which employment in Docklands has changed since the 1960s.

Geographical skills

1 Study Figure 15.21.

   a) Describe how London's total workforce has changed since 1996. Mention dates and figures.

   b) Calculate the percentage growth in the workforce from 1996 to 2016.

2 Study Figure 15.22, which shows the different employment sectors in London.

   a) List the sectors in which the number of jobs declined from 1996 to 2013.

   b) List four sectors with the largest increase of jobs from 1996 to 2013. Give one example of a job in each sector.

   c) Overall, say how employment patterns in London have changed.

Case study

⭐ KEY LEARNING

➤ Why there is a need for improved transport in London

➤ The planned transport improvements

➤ How Crossrail could impact on London

London's transport network

Why is there a need for improved transport in London?

London has a well-integrated transport system, but it is struggling to cope with the increase in passenger numbers (Figure 15.23). As the population grows and work opportunities increase, more people are using public transport to commute to work. Driving a car is not a sensible option for most Londoners, with limited space to park and traffic congestion causing long journeys.

In 2014, roughly 75 million passengers used underground trains (the Tube) and buses in London each week – 25 million on the Underground and 50 million on buses. The number is growing every year (see Figure 15.24).

What transport improvements are planned?

The demand for public transport in London is predicted to grow by 60 per cent by 2050, when the population will be much higher. Though 2050 may seem like a long time away, improvements to transport need long-term planning and investment.

Crossrail is a new, east–west rail route across London due to open in 2021, linking Shenfield and Abbey Wood in the east with Reading and Heathrow in the west (Figure 15.25). Once open, Crossrail will be known as the Elizabeth line. It will tunnel under the city centre, reducing journey times and increasing the total number of passenger journeys in London. Already, Crossrail 2 is being planned for 2030. It would be a similar project, this time on a north–south route across London.

▲ Figure 15.23 A crowded London Underground train

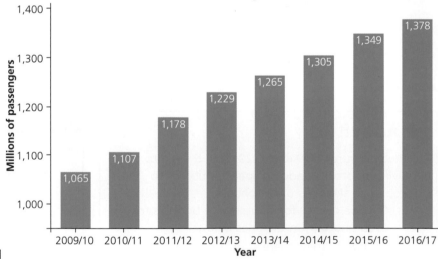

▲ Figure 15.24 Recent increase in Underground passenger journeys

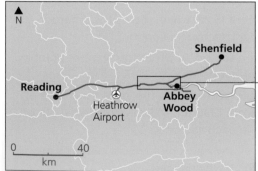

▲ Figure 15.25 The Crossrail route in London

What could the impacts of Crossrail be on London?

Crossrail is one of the largest **infrastructure** projects London has ever seen. It is expected to bring a number of benefits. It will:

- reduce journey times – for example, the journey time from Liverpool Street to Heathrow will fall from over an hour to 35 minutes
- increase the number of rail passenger journeys in London by ten per cent, that is an extra 200 million journeys a year
- bring an extra 1.5 million people within a 45-minute journey of central London, increasing the number of people who can commute to work in London
- improve the integrated transport system in London by providing more interchanges with the Underground network
- raise property values by about 25 per cent around stations along the Crossrail route
- encourage further regeneration across London, providing access to thousands more jobs
- improve accessibility for people with disabilities to new stations, with no steps from platform to street level.

▲ Figure 15.26 Canary Wharf is one of London's main financial hubs (see Section 15.5). Crossrail will make commuting quicker and easier. There are plans for a new phase of Docklands regeneration nearby at Wood Wharf, with thousands of jobs in creative industries.

▲ Figure 15.27 Custom House is one stop from Canary Wharf on Crossrail, but a world apart. It is one of the poorest parts of London, with a high proportion of social housing and people on low income. On the other side of the Crossrail track is ExCel London, London's largest exhibition centre.

→ Activities

1. Study Figure 15.24.
   a) Describe how the number of passenger journeys changed from 2010 to 2017.
   b) Explain why this change happened and predict future changes.

2. Study Figure 15.25. Explain how Crossrail will:
   a) reduce journey times in London
   b) increase the number of passenger journeys
   c) improve integrated transport in London. (You could refer to an Underground map of London to help you.)

3. a) Rank the benefits of Crossrail for London in order of importance.
   b) Explain why you ranked your first benefit as the most important.

4. Study Figures 15.26 and 15.27.
   a) Which Crossrail station do you think will bring the biggest benefits – Canary Wharf or Custom House? Consider local residents, commuters, regeneration and property values.
   b) Give reasons for your decision.

Case study

⭐ KEY LEARNING

➤ How much of London is green

➤ The benefits of green cities

➤ What strategies could make London greener

Urban greening

How much of London is green?

London is one of the world's greenest cities. Almost half the city – 47 per cent – is green space, including parks, woodlands, cemeteries and gardens (Figure 15.28). The percentage might have been higher, but in recent years many people have paved over their gardens to create patios or make space to park their cars.

London's large area of green space is a result of the way in which the city has developed.

- Central London parks – London has more big parks than many cities. They include royal parks, such as Hyde Park, which once belonged to royalty. Now everyone can enjoy them.

- Local parks – many parts of inner and outer London have municipal parks, run by the local council. They date back to the nineteenth century where there was concern about public hygiene in London and the need for people to have fresh air.

- Suburban growth – the expansion of London in the early twentieth century led to the development of suburbs. They were built on farmland, providing millions of new homes with gardens for Londoners.

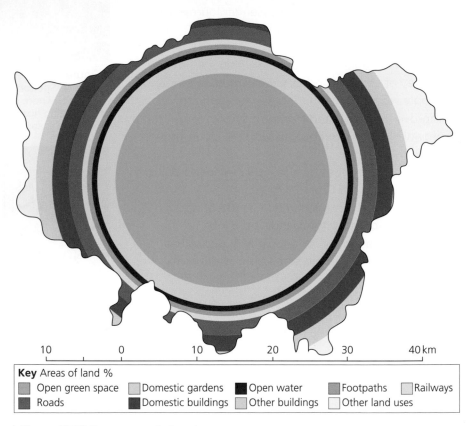

10	0	10	20	30	40 km

Key Areas of land %

- Open green space
- Roads
- Domestic gardens
- Domestic buildings
- Open water
- Other buildings
- Footpaths
- Other land uses
- Railways

▲ **Figure 15.28** Green space in London

Why is it good to have green cities?

There are many reasons for cities to be green, but not all of these were known when London was developing.

- Trees produce oxygen, clean the air and help to reduce global warming by using carbon dioxide. There are 8.3 million trees in London – almost one per person!

- Trees and green open space reduce the danger of flooding by slowing down the rate at which rainwater drains from the land.

- Parks, woodlands and even domestic gardens provide a habitat for wildlife, including birds, insects and mammals. There are 13,000 wildlife species in London.

- People enjoy green open spaces and they help us to keep healthy. We use these spaces for walking, running, cycling and for sport.

- People also use green spaces for growing food. There are 30,000 allotments in London: shared open spaces where people grow their own food.

What strategies can be used to make London greener?

Urban greening is about how we increase and protect the green spaces we have in cities. London is already a 'green city', so urban greening here is more about protection.

■ On a small scale, this is about individual actions, like encouraging people to feed birds in winter or not paving over gardens.

■ On a larger scale, it could be about connecting the green spaces we already have to help species to migrate naturally. London now has a 'green grid' to link open spaces.

In 2019, London was designated as the world's first 'National Park City'. The idea behind this is to conserve and enhance nature and to promote the parks to people. The charter for London as a National Park City has seven aims – to work for better:

■ lives, health and wellbeing
■ wildlife, trees and flowers
■ places, habitats, air, water, sea and land
■ times outdoors, culture, playing, walking, cycling and eating
■ locally grown food and responsible consumption
■ decisions, sharing, learning and working together
■ relationship with nature and with each other.

▲ Figure 15.29 Parks help to make London a green city

Fieldwork: Get out there!

How much green space is there in your area?

To answer this question, you will have to do some land-use mapping in your area.

■ Walk around the area with a large-scale outline map. Shade all the spaces on your map in different colours to show how the land is being used.

■ Calculate the percentage of the total area for each land use. You could do this by placing a grid with 100 squares over your map and counting the squares for each land use. Is there more or less green open space than you predicted?

→ Activities

1 Study Figure 15.28.
   a) Estimate the percentage area of each land use in London. Your percentage figures should add up to 100 per cent. List them in order of the amount of space they occupy.
   b) Does anything surprise you about the graph and percentages? Explain why you do or don't find it surprising.

2 a) What benefits can you think of for having green open space in a city? Make a list and include some of your own ideas.
   b) Write a letter to your council to persuade them to keep and enhance the green space that is already there. (You could use the findings from your fieldwork on the left to help you to make your case.)

3 a) Think about a National Park City. In what ways is it similar to other national parks like the Lake District? In what ways is it different?
   b) Do you think that London as a National Park City is a good idea? Give reasons for both sides and reach a final judgement.

⭐ KEY LEARNING

➤ The challenges of social and economic deprivation
➤ How deprivation varies between areas of London
➤ Why inequality is still a challenge in London

Urban deprivation and inequalities

What is social and economic deprivation?

If you live outside London, you might not think of it as a deprived city. After all, it is the wealthiest city in the UK. But economic and **social deprivation** is a major problem that over 2 million people living in poverty in London face. Deprivation is the degree to which a person or a community lacks the things that are essential for a decent life, including work, money, housing and services.

How does deprivation vary between areas of London?

London is divided into 33 boroughs. These are administrative areas that make it easier to run such a large city, and they are also a useful way to show variations within the city. Figure 15.30 shows the percentage of people on state benefits in each borough, as a measure of economic deprivation. People may rely on benefits in the UK when they face unemployment or are low paid to help them financially. The most deprived boroughs, with the highest level of benefits, are ones with more low-skilled workers with poor or fewer educational qualifications, on lower wages. This pattern may be changing with the fastest improvement in educational achievement being in inner London boroughs.

Life expectancy is a measure of social deprivation. The more deprived a person is, the lower their life expectancy is likely to be. There are many reasons for this, such as having a poorer diet, fewer opportunities for exercise and higher rates of air pollution for people living close to busy roads with less green space. The variation in life expectancy in London is clearly demonstrated by

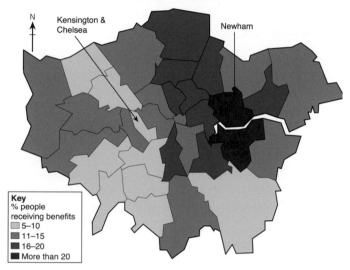

Key
% people receiving benefits
☐ 5–10
☐ 11–15
☐ 16–20
■ More than 20

▲ Figure 15.30 People on benefits in each London borough

travelling on the Underground from Green Park, in the borough of Westminster, to West Ham in the borough of Newham. Life expectancy for those living in each area falls on average by over one year for every station along the route (see Figure 15.31).

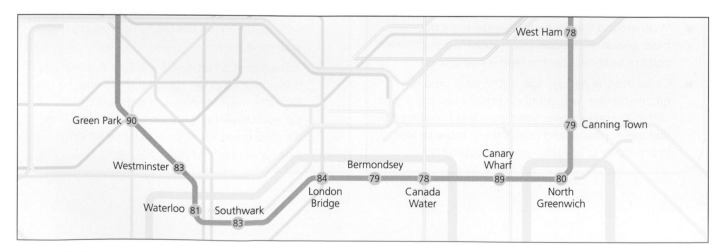

▲ Figure 15.31 Life expectancy falls on a journey from Green Park to West Ham. The numbers refer to average life expectancy in each place.

Why is inequality still a challenge in London?

Despite years of economic success, **inequalities** in housing, education, health and employment are still a challenge in London. Differences in life expectancy still exist. Low life expectancy in the most deprived parts of the city is closely linked to poor diet, housing and education, as well as lack of employment.

Kensington & Chelsea, one of London's most affluent boroughs, has better indicators for deprivation than Newham (see Figure 15.32). However, there is also inequality within each of these boroughs.

▼ Figure 15.32 Inequality between Kensington & Chelsea and Newham

Measure of deprivation	Kensington & Chelsea	Newham
Male life expectancy	83.7	75.7
Female life expectancy	87.8	79.8
Unemployment	3.9%	9.4%
Pupils achieving five + good GCSE grades	80%	62%
Households with joint income < £15,000	9%	26%
Households with joint income > £60,000	26%	7%

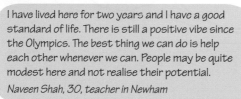

It's lovely here. Everyone is very friendly. It feels like a village. You see people running, see the mums power-walking on the school run. There's an organic food shop and good shops like Waitrose and M&S.

Jessica Kelly, 22, student in Kensington & Chelsea

I have lived here for two years and I have a good standard of life. There is still a positive vibe since the Olympics. The best thing we can do is help each other whenever we can. People may be quite modest here and not realise their potential.

Naveen Shah, 30, teacher in Newham

▲ Figure 15.33 Views from the street in Kensington & Chelsea and Newham

→ Activities

1   Study Figure 15.30. Describe the pattern of social deprivation in London. In which parts of London are the most and least deprived boroughs?

2   Study Figure 15.31.

   a)   Describe the changes in life expectancy on an Underground journey from Green Park to West Ham.

   b)   How can you explain these changes?

Geographical skills

1   Study Figures 15.32 and 15.33.

   a)   Compare life in Kensington & Chelsea and Newham using:
   - the data in Figure 15.32
   - the photos in Figure 15.33
   - people's experiences in Figure 15.33.

   b)   Which of these three sources do you think is the most reliable way to compare the two boroughs? Give reasons.

Fieldwork: Get out there!

'Levels of deprivation vary between areas of a city.'

- Devise at least one fieldwork method you could use to test this hypothesis. For example, you could take photos or conduct interviews with people.

- Suggest two areas of a city you know where you could carry out your fieldwork.

- Suggest any other data you could use to compare the two areas, apart from what you will find out by doing fieldwork.

239

Case study

⭐ KEY LEARNING

➤ Why there is a shortage of homes in London

➤ The reasons for building on greenfield or brownfield sites

➤ The impact of urban sprawl

New homes needed

Why is there a shortage of homes in London?

London's population is growing (see Section 15.3) by about 100,000 people every year, yet only about 20,000 new homes a year are being built, and of these, not enough are affordable housing. This has led to a severe housing shortage in London and the rest of South East England. The result is that house prices are rising faster in London than the rest of the country (Figure 15.34).

Another result of the housing shortage is that there is more homelessness and more overcrowded households. The visible sign of homelessness is the large number of rough sleepers on the city's streets, but there are many others sleeping on friends' sofas or in cheap hotels. There are 170,000 homeless people in London, about half of the total homeless in the UK. Each borough is responsible for housing homeless people. Sometimes people are moved to other, less crowded parts of the country instead of within London.

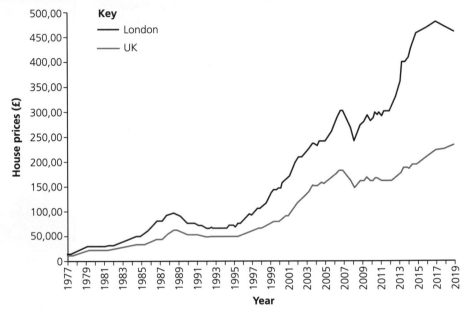

▲ Figure 15.34 House prices in the UK and London since 1977

What are the reasons for building on greenfield or brownfield sites?

One possible solution to the shortage of homes in London is to build outside London on **greenfield sites**. These are areas of land that have not previously been built on – usually farmland on the **rural–urban fringe**. However, building on these sites can lead to **urban sprawl** (the unplanned growth of urban areas) and is not very popular with those people already living in the countryside.

The alternative is to build on **brownfield sites** in the city. These are areas of previously developed land which are often **derelict** now and have potential for redevelopment. Often, this is land that was previously used for industry where the ground may be contaminated by chemicals. There are many sites like this in London due to the decline of the manufacturing industries.

▼ Figure 15.35 Issues about building on greenfield or brownfield sites

Greenfield sites	Brownfield sites
● Public transport is worse in rural areas, so more need for cars.	● Sites are available since industry declined.
● Increases urban sprawl.	● Reduces the need for urban sprawl.
● Once land is built on, it is unlikely to be turned back to countryside.	● Public transport is better in urban areas, so less need for cars.
● Land is cheaper in rural areas.	● Old buildings may need to be demolished first.
● No demolition or decontamination is needed.	● Ground may need to be decontaminated.
● Valuable farmland or land for recreation may be lost.	● New development can improve the urban environment.
● Natural habitats may be destroyed.	● Land is more expensive in urban areas.

What is the impact of urban sprawl?

Around many cities in the UK, including London, is a **green belt**. This is land on which there are strict planning controls. It was established in 1947 to prevent further urban sprawl. Since then, it has helped to preserve farmland, woodland and parkland around London (Figure 15.36).

Now, with the pressure for more housing in London, people are questioning whether we can afford to keep the green belt. They suggest that less valuable areas of green belt land could be used for building new homes on greenfield sites.

As the population of London grows and house prices rise, more people move to **commuter settlements** around London. This forces population and house prices in the rest of South East England to rise too. The problem of urban sprawl has shifted to commuter settlements outside the green belt. Cities like Reading and Chelmsford, within a half-hour train journey to London, are growing rapidly. Urban development, in the form of new housing estates and **business parks**, encroaches into the surrounding countryside.

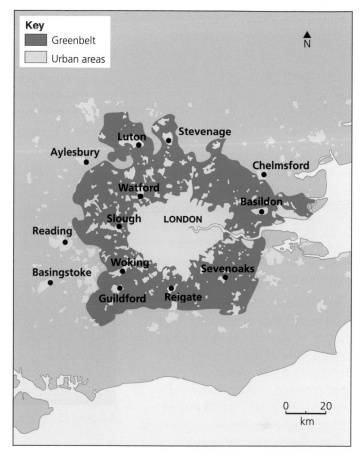

▲ Figure 15.36 The green belt around London

→ Activities

1 Study Figure 15.34.
   a) Compare the change in house prices in London and the rest of the UK.
   b) Explain why this has happened.
2 Study Figure 15.35.
   a) Classify the issues about building on brownfield and greenfield sites into advantages and disadvantages.
   b) Redraw the table like this and list the impacts in the correct boxes.

	Greenfield sites	Brownfield sites
Advantages		
Disadvantages		

3 Study Figure 15.36. Think about each of these people's interests in the green belt. Would they be for or against protecting it? In each case, explain why.
   ■ A homeless person in London
   ■ A resident in London
   ■ A resident of Woking
   ■ A farmer in the green belt
4 Prepare for a class debate about the green belt. You can base your ideas on the green belt around London or another city you know. Write a short speech for or against keeping the green belt.

⭐ KEY LEARNING

➤ How serious the air pollution in London is

➤ How new cycle superhighways will help

➤ What happens to London's waste

London's pollution problem

How serious is the problem of air pollution in London?

Compared to years gone by, pollution in London is often thought to be less of a problem than it used to be. In the mid-twentieth century, when coal was burnt to power factories and provide domestic heating, the city used to experience smog, a dense mixture of smoke and fog.

However, air pollution is still a problem (see Figure 15.37). The main problem now is emissions from road vehicles and modern heating systems. It is made worse by the dense road network in London and the tall buildings that trap air between them.

London has a worse pollution record than most other European cities, though not as bad as many cities in Asia. One of the worst modern pollutants is nitrogen dioxide (NO_2) that comes primarily from road vehicles, especially diesel engines. London regularly breaks regulations on air quality. Most of central London is above the safe limit of 40 mg/m³ for NO_2 (see Figure 15.38). There are over 4,000 premature deaths a year in London due to long-term exposure to air pollution.

▲ Figure 15.37 Air pollution over London

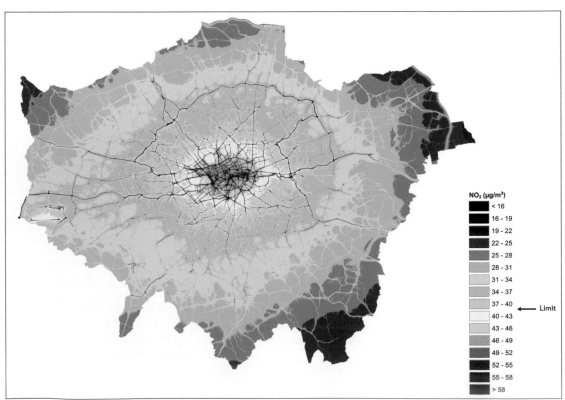

➤ Figure 15.38 NO_2 pollution levels in London, 2016

NO_2 (µg/m³)

- < 16
- 16 – 19
- 19 – 22
- 22 – 25
- 25 – 28
- 28 – 31
- 31 – 34
- 34 – 37
- 37 – 40 ← Limit
- 40 – 43
- 43 – 46
- 46 – 49
- 49 – 52
- 52 – 55
- 55 – 58
- > 58

How will new cycle superhighways help?

New cycle superhighways are fast routes for cyclists along main roads. They are planned for London and should encourage more people to cycle and reduce traffic and harmful emissions from vehicles (Figure 15.39). Cyclists have increased from 1 per cent to 15 per cent of road users in London over the past 50 years. The percentage should increase further with the new cycle superhighways.

What happens to London's waste?

Almost a quarter of London's waste still goes to landfill sites outside London. In the past this was acceptable because the waste was out of sight, out of mind. Now, we realise that landfill waste contributes to wider environmental problems, such as the production of methane that adds to the greenhouse gases in the atmosphere (see Section 4.4). And, of course, 'waste' is just that – a waste of potentially valuable resources.

More of London's waste is now recycled. The target is for zero waste to go to landfill by 2030, by focusing on waste reduction and by managing resources more efficiently. By then, the aim is for 65 per cent of waste to be recycled, an improvement on the 52 per cent in 2016 (Figure 15.40). Most of the remaining waste by 2030 would be incinerated or burned to generate electricity.

▲ Figure 15.39 Plans for one of London's new cycle superhighways. Notice the separate lane for cyclists and the reduced width of the road for traffic

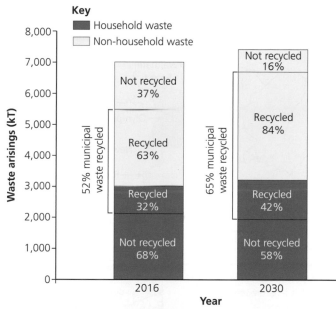

➤ Figure 15.40 London's waste management

→ Activities

1 Study Figure 15.37. Explain why air pollution in London:
    a) was a problem in the past
    b) is still a problem today.

2 Study Figure 15.38.
    a) Draw a sketch map of Greater London to show the most and least polluted areas.
    b) Identify these places and label them on your map – central London, Heathrow Airport, North Circular Road (around central London).
    c) Explain the pattern on the map.

3 Municipal waste is all the waste a city produces, including household waste.
    a) Make a list of all the types of municipal waste you can think of.
    b) How will the recycling of municipal waste in London change from now to 2030?
    c) Suggest at least three ways households could increase recycling.

Fieldwork: Get out there!

Cyclists in London represent about 15 per cent of road users. What percentage is it in your area? Would the number increase with safer cycle routes?

■ Suggest how you could find out the percentage of cyclists on the road.

■ Suggest how you could find out if more people would cycle if there were safer cycle routes.

■ Predict what results you expect to get from your fieldwork in your area.

⭐ KEY LEARNING

➤ Why the Lower Lea Valley was in need of regeneration

➤ What obstacles had to be overcome to regenerate the site

➤ Why the London 2012 Olympic bid was successful

Urban regeneration: the Olympic plan

Why was the Lower Lea Valley in need of regeneration?

The Lower Lea Valley in East London was the site for the 2012 Olympics (Figure 15.41). The River Lea is a tributary of the Thames and the Lea Valley was once one of the main industrial areas in London. You will also remember that Newham, along with the other boroughs around the site, is in one of the most deprived parts of London (Figure 15.30).

▼ Figure 15.41 The site of the Olympic Park, East London

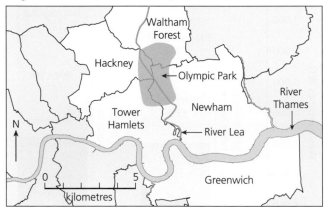

By 2007, when work began to create the Olympic Park, many of the industries had already gone and some of the site was derelict and overgrown. But the land around the River Lea was far from being empty (Figure 15.42).

Part of the site was a thriving industrial area with many businesses specialising in recycling and vehicle repair. There were also two industrial estates, a small residential area and the newly built Stratford International Station. The main reason for the lack of housing is that the Lea Valley used to flood.

Open space There were playing fields, a nature reserve and allotments.

Industry There were two industrial estates still working. Altogether there were 250 businesses on the site, employing over 5,000 workers.

Derelict land There was plenty of unused, overgrown land on previous industrial sites. The land was badly contaminated by chemicals.

Housing There was a community with 500 homes in one part of the site. Most of the site was not residential.

Transport Stratford International station was already built on land that had once been a railway freight terminal. By 2007, most of this land was a brownfield site.

Water The River Lea and Lee Navigation Canal go through the site, connected by a network of waterways to relieve flooding.

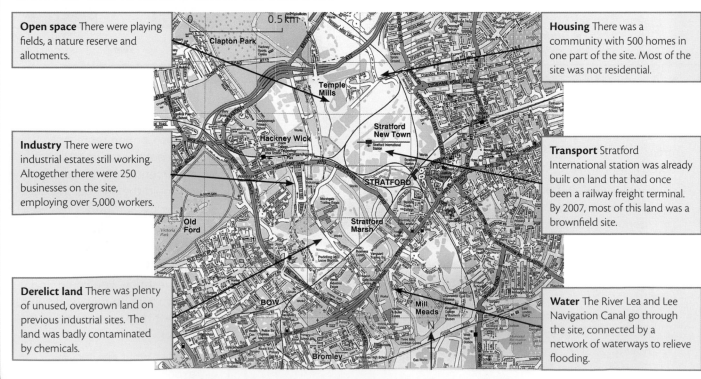

▲ Figure 15.42 The Olympic Park site before work began in 2007

What obstacles had to be overcome to regenerate the site?

The construction of the Olympic Park in just five years, from 2007 to 2012, was an impressive achievement. Eventually, it is expected to lead to the regeneration of this part of East London, though the whole process will take much longer (see Section 15.12). Before construction or regeneration could begin, obstacles had to be overcome:

■ The land had to be brought together under one new owner, the Olympic Delivery Authority (ODA), which was set up by the government.

■ Existing landowners and users had to leave the site by 2007. Some of them protested (Figure 15.44). The land was bought from them by the ODA.

■ Land that was previously polluted by industry had to be decontaminated before building could begin.

■ Electricity pylons had to be removed and overhead cables buried below ground to improve the appearance of the **landscape** (Figure 15.43).

■ Waterways and railways crisscrossed the site, so bridges were built to link the area together.

▲ **Figure 15.43** The Olympic Park site before 2007

Why was the London 2012 Olympic bid successful?

As you know, London won the bid to host the 2012 Olympics. There were a number of reasons why London beat the other rival cities around the world:

■ There was a large area of available land, even though there were also businesses and homes on the site.

■ East London has very good transport connections, particularly Stratford station, where most spectators arrived in 2012.

■ London's diverse population made it the natural city to host guests from around the world. Newham is the most diverse borough in London.

■ The Olympic bid promised to leave a lasting legacy that would help to regenerate East London.

▲ **Figure 15.44** Local protests against the Olympics

→ **Activities**

1 Study Figure 15.41. Describe the location of the Olympic Park in London. Mention the boroughs that surround the site.

2 Study Figures 15.42 and 15.43. Explain why the Lower Lea Valley was in need of regeneration.

3 a) What does the slogan on the banner in Figure 15.44 mean?

   b) How true was it? (Hint: read more in Section 15.12 to find out what happened next.)

4 East London was an area in need of regeneration. Was that an advantage or disadvantage for London's bid? Explain your answer.

5 Would each of the following have supported the Olympic bid, or not? In each case, give reasons.

   ■ A local resident living on the site
   ■ A local resident living outside the site
   ■ A business owner on the site
   ■ The Mayor of London
   ■ The British government

Example

⭐ KEY LEARNING
➤ The main features of the project
➤ What social and economic changes there have been

Urban regeneration: the Olympic legacy

What were the main features of the project?

The Queen Elizabeth Olympic Park has completely transformed the environment of the Lower Lea Valley.

Gone are the:

- old factories, industrial estates and homes
- derelict and overgrown sites
- electricity pylons and overhead cables
- contaminated soil and polluted waterways
- the Olympic Delivery Authority.

In their place have appeared:

- stunning new sports venues, including the Aquatics Centre (see Figure 15.45), stadium and velodrome
- a landscaped park with tourist attractions and natural habitats
- the Athletes' Village, now converted into a residential community
- clean soil and waterways
- the London Legacy Development Corporation (LLDC) who have taken over from ODA (see Figure 15.46).

▲ Figure 15.45 The Aquatics Centre beside the River Lea

Here East The new name for the Media Centre is now a hub for creative and media industries with 5,000 jobs.

Queen Elizabeth Olympic Park With over 100 hectares of open space, this is the largest new park in London for over a century.

Olympic stadium The new home of West Ham United FC, but still an athletics stadium in the summer.

LLDC This organisation controls an area including the park and surrounding neighbourhoods to make the Olympic legacy a reality.

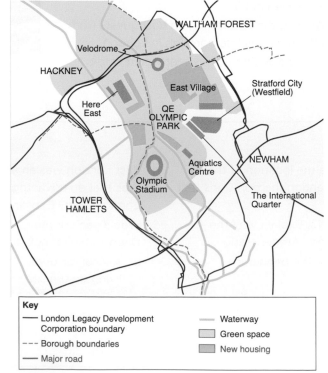

Key
— London Legacy Development Corporation boundary
--- Borough boundaries
— Major road
— Waterway
▢ Green space
▢ New housing

▲ Figure 15.46 The legacy plan for the Olympic Park

East Village The new name for the Athletes' Village, it now provides 2,800 homes for local people and newcomers.

Westfield Stratford City Not really part of the Olympic legacy, but next door to the park and employs 10,000 people.

The International Quarter A new commercial development of high-rise offices which will employ 25,000 people.

The Aquatics Centre and Velopark Two new sports venues open to the public and used by schools.

What social and economic changes have taken place?

The big promise of the 2012 Olympic Games was a lasting legacy to help regenerate one of the most deprived parts of London. The government spent £9.3 billion of public money on the Games so people expected to see long-term social and economic benefits.

While new jobs have been created and new homes have been built within the Olympic Park, benefitting some, there are others living around the Park who have seen little change to their lives. The nearby Carpenters Estate remains in urgent need of renovation ten years after the Games.

In 2012, the LLDC was set up to plan regeneration after the Games. It is likely to take until 2030 for the process to be completed, so it is still too early to judge how successful the changes will be.

However, the Athletes' Village (now East Village) has already been converted into new homes (see Figure 15.47). By 2030, another five new residential communities are planned, with a further 8,000 new homes, turning Queen Elizabeth Olympic Park into a new part of London.

- 2,800 new homes, half for private rent and half for affordable rent.
- A range of homes from one-bedroom apartments to four-bedroom town houses.
- The site occupies 27 hectares, including 10 hectares of park and public open space.
- 35 small independent shops, cafés, bars and restaurants, a supermarket and a gym.
- A new school for 1,800 students aged from 3 to 18.
- Close to bus routes, a new local station and Stratford International station.

▲ Figure 15.47 East Village – a new community in east London with a new postcode, E20

→ Activities

1 Study Figure 15.45. Compare the photo with Figure 15.43 in Section 15.11. Describe how the environment of the Lower Lea Valley has changed.

2 Study Figure 15.46.

   a) Classify the changes in the Olympic Park into social, economic and environmental changes. You can include other ideas from these two pages.

   b) Draw a table to list the three types of changes under the correct headings.

3 Study Figure 15.47.

   a) East Village was built on a brownfield site. What are the advantages of this?

   b) What are the benefits of East Village for the residents who live there?

4 Compare Olympic regeneration with Docklands regeneration in another part of East London (see Section 15.5). The Olympics is an example of partnership regeneration, involving both public and private investment. The Docklands regeneration was an example of market-led regeneration, involving mainly private investment.

   a) In your view, which project has brought, or is likely to bring, more benefits to:
      - local people
      - large companies
      - people in the rest of London and the UK
      - the environment?

      In each case, give reasons for your view.

   b) What lessons do you think could be learnt for future regeneration projects in the UK?

16 Sustainable development of urban areas

★ KEY LEARNING

➤ The impact cities have on the environment

➤ How large our urban ecological footprint is

➤ How cities could become more sustainable

Urban sustainability

What impact do cities have on the environment?

A sustainable city is one that can meet its needs without making it more difficult for future generations to meet their needs. Cities put pressure on the natural environment by using inputs, like food, water and energy and, at the same time, by producing outputs, like waste and pollution (see Figure 16.1).

Despite this, in some ways, living in a city can be more sustainable than living in the countryside. In cities:

- people need to make fewer road journeys because everything they need is closer
- careful planning of things like public transport helps to save resources
- people work together to generate ideas or produce goods and services that benefit the economy.

How large is our urban ecological footprint?

One way to think about the impact of cities on the environment is their **ecological footprint**. This is the area of land or sea that is needed to produce all the inputs a city uses and to dispose of its outputs.

A city's ecological footprint is always much larger than the city itself. In the case of London, it is estimated that each person uses six global hectares (gha). That means London's total footprint is about twice the size of the UK! London's footprint spreads globally to all the places where its inputs come from and where its outputs end up.

Food Most of it is grown outside the city on farms, or is imported.

Water It is taken from rivers or from below ground and stored in reservoirs.

Energy Most energy comes from burning fuels that are drilled or mined.

Other resources Building materials, like timber and concrete, plus other resources we consume.

Waste A lot of it ends up in landfill sites outside the city, or is burnt.

Sewage It is treated in sewage works before it is returned to a river.

Pollution It can spread beyond the city in the air or water.

▲ Figure 16.1 A city has inputs and outputs

How could cities become more sustainable?

Many cities in the UK and around the world are taking initiatives to be more sustainable. These initiatives can include:

- recycling more waste
- creating more green spaces
- water and **energy conservation** (see Section 16.2)
- improved public transport
- better cycling routes (see Section 16.3)

In 2004, the UK government devised a framework for sustainable communities (Figure 16.2). It included social, economic, political and environmental aspects of **sustainability**.

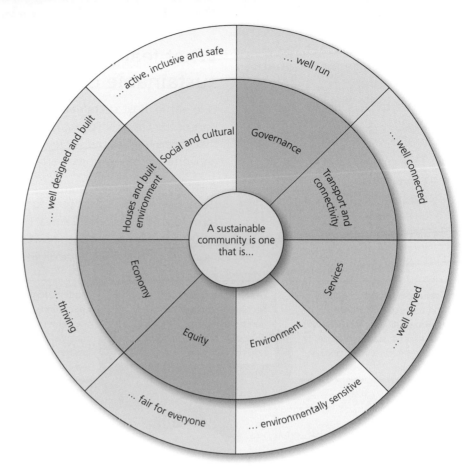

▲ Figure 16.2 A framework for sustainable communities devised by the UK government

→ Activities

1   Do you think that cities are sustainable places to live? Give at least three reasons to support your opinion.

2   Look at Figure 16.1.
    a) Draw a simple diagram of a city like Figure 16.1. List the inputs and outputs.
    b) Now, draw a model of a more sustainable city that recycles, with reduced inputs and outputs. List the things that could be recycled.

3   a) Identify at least five things that contribute to a city's ecological footprint, (for example, dumping waste in landfill sites). Suggest why London's footprint is so large.
    b) Suggest at least five ways in which cities could be more sustainable (for example, recycling more waste).

👣 Fieldwork: Get out there!

How sustainable is my community?

- Look at Figure 16.2. Make a list of the features you would expect to find in a sustainable community. List them under these headings – Transport, Services, Environment, Economy, Buildings and Social. Think of at least two features under each heading.

- Design a sheet you could use to assess sustainability in your community. List sustainable and unsustainable features on each side, with spaces to score between +2 and –2, like this:

Sustainable features	+2	+1	O	–1	–2	Unsustainable features
Transport Close to a train station						Transport Far from a train station

- Use your sheet to assess your community. Give a score for each pair of features. Work out a total score.

Sustainable urban living

What features make East Village a sustainable community?

One of the most **sustainable urban communities** in the UK is East Village, part of the Olympic legacy in London (see Section 15.12). You may remember that it was built as the Athletes Village for the 2012 Olympics and then converted into new homes after the Games.

A key aim of the 2012 Olympics was for London to be 'the most sustainable Games ever'. Structures built for the Olympics were planned to have a long-term function after the Games. In the case of East Village, it was to provide 2,800 new homes for both newcomers and local residents. It was built to high standards of sustainability (see Figure 16.3).

Transport Local bus services and trains connect to the London Underground. Stratford International station provides a fast route to central London and Europe.

Green open spaces Ten hectares of parkland, with hundreds of planted trees and ponds, encourages wildlife and helps to purify the air.

Modern, high-density apartments Built to high standards of insulation and energy efficiency. Less heat is lost from apartments than from low-rise, individual houses.

Green roofs On residential blocks, encourage more wildlife and slow down the rate at which water drains off.

Affordable housing Half of the homes are rented at lower rates so that people in East London can afford them.

Walking and cycling There are good walking and cycling routes to encourage people out of their cars. Residents pay extra for car parking spaces.

Shops and services Run by small, independent businesses, helping to keep money in the local economy and avoid the need to shop elsewhere.

Public services A school for 3– to 18-year-olds and a large health centre are essential services provided in the community.

▲ Figure 16.3 East Village: a more sustainable urban community

How are water and energy conserved in East Village?

Water

Water use is 50 per cent less than an average urban area. This is achieved by recycling water within the area (Figure 16.4). Rainwater is filtered and cleaned naturally in ponds before being recycled for toilet flushing and irrigating plants. Drinking water is part of a separate system.

Energy

Energy use is at least 30 per cent less than an average urban area because of its **combined heat and power (CHP)** system (Figure 16.5). CHP is more efficient because it generates electricity and produces heat from the same source of energy by burning **biomass**. CHP systems only work on a local scale because hot water can only be piped a few kilometres underground before it loses heat.

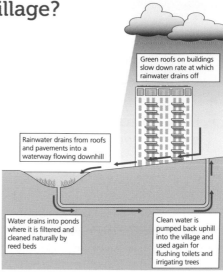

Green roofs on buildings slow down rate at which rainwater drains off

Rainwater drains from roofs and pavements into a waterway flowing downhill

Water drains into ponds where it is filtered and cleaned naturally by reed beds

Clean water is pumped back uphill into the village and used again for flushing toilets and irrigating trees

▲ Figure 16.4 Water recycling in East Village

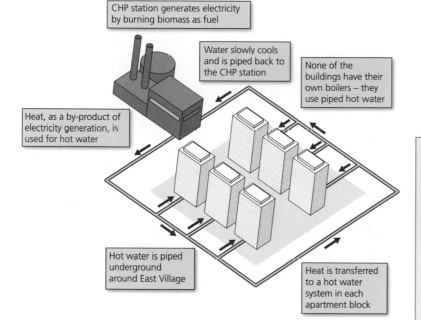

CHP station generates electricity by burning biomass as fuel

Water slowly cools and is piped back to the CHP station

None of the buildings have their own boilers – they use piped hot water

Heat, as a by-product of electricity generation, is used for hot water

Hot water is piped underground around East Village

Heat is transferred to a hot water system in each apartment block

▲ Figure 16.5 A combined heat and power system in East Village

What green spaces have been created in East Village?

East Village is a high-density urban area, yet there are 10 hectares of green open space within a total area of 27 hectares. This is equivalent to the proportion of green space in all London (see Section 15.7). It has:

- a wetland area with ponds where water is recycled, surrounded by parkland
- a large central park and an adventure play area
- green roofs on top of apartment blocks
- shared private green space for each block
- an orchard with fruit trees and a play area.

→ Activities

1 Look at Figure 16.3.
   a) Identify the sustainable features of East Village.
   b) List them under the same headings as in the framework for a sustainable community in Figure 16.2 on page 249.
   c) In what ways is East Village a more sustainable community than where you live?

2 Look at Figure 16.4.
   a) Explain how water recycling works in East Village.
   b) How does this reduce water use by 50 per cent?

3 Look at Figure 16.5.
   a) Explain how a combined heat and power system works.
   b) How does this reduce energy use by 30 per cent in East Village?

4 Draw a sketch of one section of East Village from Figure 16.3. Annotate it to show the types of green space in the village.

➤ Why Bristol needs a sustainable urban transport strategy
➤ The benefits of cycling
➤ How Bristol's cycling strategy works

Sustainable urban transport

Why does Bristol need a sustainable urban transport strategy?

Like many cities in the UK, Bristol is developing a more sustainable urban transport strategy. As a growing city with a densely populated historic centre, transport is a key issue. Thousands of daily journeys are still made by car. Reliance on cars is leading to traffic congestion, poor air quality and ill health (see Section 15.10), as well as making the streets less friendly to people.

Cycling is easy, cheap and pollution-free (Figure 16.6). The number of people cycling in Bristol has doubled over ten years, but the city still has a long way to go to achieve the levels of cycling in some European cities, like Copenhagen or Amsterdam. To help with this, there is now a Bristol Cycling Strategy.

▲ Figure 16.6 Cycling in cities is growing in popularity with more off-road routes

What are the benefits of cycling?

Cycling has many social, economic and environmental benefits, each contributing to a more sustainable city (Figure 16.7).

How does Bristol's cycling strategy work?

People cycle in cities for different reasons:

■ to get into the city centre for work or shopping
■ to go sight-seeing or shopping around the city centre
■ to travel to work, school or local shops within a zone in the city
■ to come in or out of the city from surrounding places.

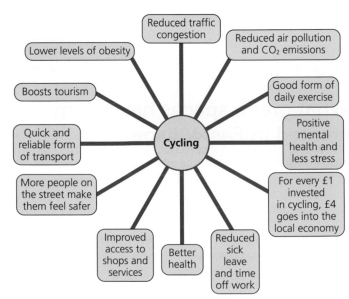

▲ Figure 16.7 The benefits of cycling

▼ Figure 16.8 Cycling in UK cities

City	Residents who cycle at least once a month (%)
Bristol	19
Newcastle	16
London	15
Manchester	15
Leeds	13
Sheffield	13
Birmingham	12
Liverpool	12
Nottingham	12

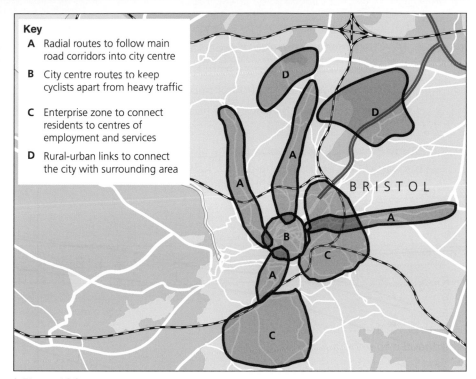

Key

A Radial routes to follow main road corridors into city centre

B City centre routes to keep cyclists apart from heavy traffic

C Enterprise zone to connect residents to centres of employment and services

D Rural-urban links to connect the city with surrounding area

▲ Figure 16.9 The Bristol Cycle Strategy

The Bristol Cycle Strategy includes all of these types of journey (Figure 16.9). It provides a network of cycle routes, both on-road and off-road, throughout the city. Wherever possible, cyclists are directed along quiet routes that avoid heavy traffic. When this is not possible, for example in the city centre, street space is split to keep traffic and cyclists apart (see Section 15.10).

Other urban transport strategies in Bristol

In addition to the cycling strategy, several other strategies exist in Bristol:

- a metro-style rail service linking Bristol with other nearby towns by reopening old railway lines (MetroWest)

- a new generation of rapid transit buses to improve journey times to Bristol (MetroBus)
- a network of charging points for electric vehicles at car parks (Source Bristol)
- 20 mph limits in neighbourhoods across Bristol, to make streets safer for cyclists and pedestrians
- and three park and rides around the city where visitors can park their cars and travel into the city centre by bus.

→ Activities

1. Look at Figure 16.6. Do you think cycling should be part of a sustainable transport strategy? Give reasons.

2. Look at Figure 16.7. Classify the benefits from cycling into social, economic and environmental benefits. List them in a table.

3. Look at Figure 16.8. Compare levels of cycling in UK cities. Can you think of any reasons they should differ? Make a list.

4. Plan a cycling strategy for your town or city. If you live in a large city, you could select one part of the city.

   ■ Think of the types of journeys people make in your town or city. These could be similar to Bristol or different.

   ■ Draw cycle zones onto a map of the city (see Figure 16.9) to show where people make these types of journeys.

   ■ Explain how you would encourage more people to cycle on routes in these zones.

1.1 What is the definition of the term 'urbanisation'? Choose **one** answer from:

    A The growth in the size of urban areas

    B The movement of people from the countryside into cities

    C The increasing proportion of people living in towns and cities

    D The way in which villages are absorbed into a growing urban area   [1 mark]

1.2 Study Figure 1, a graph showing the change in the percentage urban population for each continent from 1950–2050.

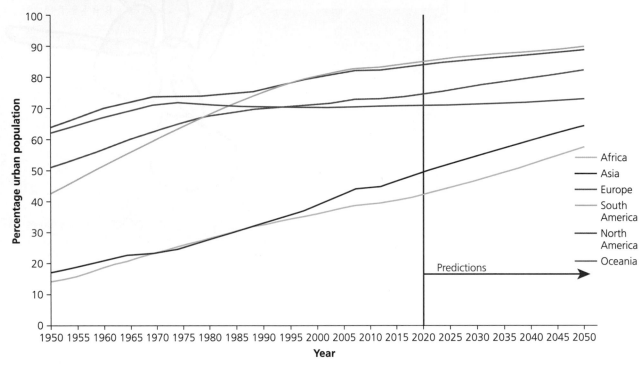

▲ **Figure 1** Change in percentage urban population for the world's continents, 1950–2050

Complete the following paragraph to describe the changes in the percentage urban population in Europe. Choose the correct answers from this list:

HICs   LICs   25%   50%   increasing   decreasing

Most of the countries in Europe are _____, which urbanised in the nineteenth and early twentieth centuries. Between 1950 and 2050 the urban population of Europe is likely to increase by over _____.
Urbanisation in Europe is now happening at a _____ rate.   [3 marks]

1.3 Using Figures 13.1 and 13.2 (page 204) and your own understanding, describe the changes in the number and distribution of megacities between 1975–2015.   [4 marks]

> Where the question asks you to use figures and your own understanding you cannot get the top level of marks without using both. In this question you need to use two figures and your own understanding.

1.4 Study Figure 2.

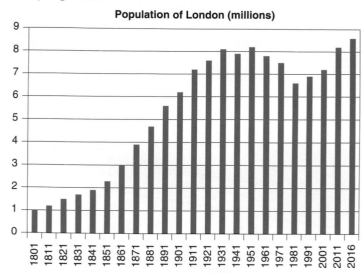

◄ Figure 2 London's population growth since 1801

Which **one** of the following statements about London's population is true?

A London's population has grown continuously since 1801

B London's population was five times greater in 2016 than it was in 1801

C London's population in 2016 was 8.6 million

D London reached its highest population in 1951 and has declined since

[1 mark]

1.5 Explain how natural increase has led to population growth in London. [2 marks]

1.6 Using Figure 3 and your own understanding, explain how **two** features shown could help regenerate East London.

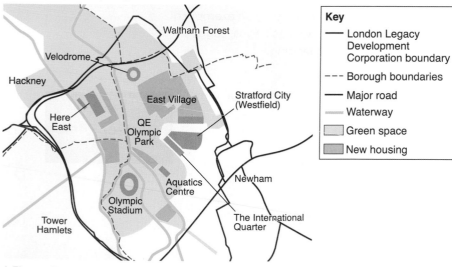

▲ Figure 3 The legacy plan for the Olympic Park

SPaG marks are extra marks given for spelling, punctuation and grammar. Credit is also given for using specialist vocabulary, so try to use geographical terms where possible.

[4 marks]

1.7 Using an example you have studied, such as the Olympic Park regeneration project, evaluate the social, economic and environmental impacts on the area.

[9 marks]
[+ 3 SPaG marks]

When evaluating, you need to think about the successes and failures of the project.

1.8 Using Figure 4, a photo of Lagos in Nigeria, list **three** pull factors that might attract rural–urban migrants to the city. [3 marks]

> If the question says 'Using Figure 4 …' you must only take your answers from the figure and refer to things you can see.

▲ **Figure 4** Lagos, Nigeria

1.9 Find the letter X on Figure 4. It shows an industrial area on the coast of Lagos Lagoon. Using an example of an LIC or NEE city you have studied, such as Lagos, explain how urban industrial areas can be an opportunity for the city. [4 marks]

1.10 Figure 14.8 on page 213 shows a squatter settlement. Using an example of a named city in an LIC or NEE you have studied, such as Lagos, suggest reasons why managing urban growth can be challenging. [6 marks]

1.11 Study Figure 15.3 (page 225), a map showing population change for major towns and cities in the UK.

Which city in the UK saw a population decline from 2004–13? [1 mark]

> Make sure you look very carefully at the key on maps and graphs.

1.12 Suggest **two** reasons for the decline in its population. [2 marks]

1.13 Describe the pattern of the fastest growing cities in Figure 15.3. [2 marks]

1.14 Suggest **two** reasons for this pattern. [2 marks]

1.15 Using Figure 15.30 (page 238), describe the pattern of the percentage of people receiving benefits in London. [2 marks]

1.16 Suggest an impact receiving benefits may have on people's quality of life. [1 mark]

1.17 Using an example of a UK city you have studied, to what extent have social inequalities created challenges in the city? [6 marks]

Make sure you cover all the parts of this question – the UK case study, the social inequalities and the challenges. A good way to make sure you spot all the parts is to use a highlighter to pick them out.

1.18 What do you understand by the term, 'sustainable urban living'? Choose the definition from:

    A Living for the moment, so that everybody in the city has what they want now

    B Living in a city in a way that meets people's needs now without making it more difficult for future generations to meet their needs

    C Living in an environmentally-friendly way in the city, banning anything that harms the environment, such as cars

    D Carrying on living in the city in the same way that we've always lived [1 mark]

1.19 For each of the following aspects of urban living, suggest one way in which we could live more sustainably. [4 marks]

    a) Water and energy

    b) Waste

    c) Green space

    d) Transport

Look at the mark allocation – in this question there are 4 marks, one for each aspect. Take care not to write too much – only a simple statement is needed for each one.

1.20 Explain how water can be conserved in an urban environment. [4 marks]

1.21 Evaluate the success of a transport strategy in a named city you have studied in reducing traffic congestion. [9 marks] [+ 3 SPaG marks]

1.22 'Regeneration of brownfield sites is the only sustainable way to increase housing provision in the city.' Using a case study of a UK city, to what extent is this statement true? [9 marks] [+ 3 SPaG marks]

17 Economic development and quality of life

⭐ KEY LEARNING

➤ How countries are classified

➤ The world map of development

World development

How are countries classified?

When we talk about a country's level of **development**, we are describing how far it has grown economically and technologically, and the typical **quality of life** (Figure 17.1).

A country's level of development can be shown by the average wealth of its citizens, using a measurement called **gross national income (GNI)**. This is calculated by adding together:

- the total value of all the goods and services produced by its population
- the income earned from investments that its people and businesses have made overseas.

To compare the level of economic development for different countries:

1. The 'raw' GNI data are divided by the population of the country to produce a per capita (per person) figure.
2. This is then converted into US dollars to help make the comparison clearer.
3. Each figure can be adjusted for each country based on prices. In LICs, goods often cost less, meaning that wages go further than might be expected in an HIC.

Countries are classified according to their level of economic development based on GNI (Figure 17.2).

Low-income countries (LICs)	This group of around 30 countries is classified by the World Bank as having low average incomes (GNI per capita) of US$995 or below (2018 values). Agriculture still plays an important role in their economies.
Newly emerging economies (NEEs)	These are around 100 countries that have begun to experience higher rates of economic growth, usually due to rapid industrialisation. **Transnational corporations (TNCs)** invest in these NEEs, which have subgroups like **BRICs** and **MINTs** (see page 277). The NEEs roughly correspond with the World Bank's 'middle-income' group of countries. The number of NEEs has increased rapidly in recent decades: this is linked to the spread of **globalisation**.
High-income countries (HICs)	This group of around 80 countries is classified by the World Bank as having high average incomes (GNI per capita) of US$12,055 or above (2018 values). Around 40 are 'developed' countries such as North America, Japan, South Korea and countries in Europe. These are states where office work has overtaken factory employment, creating a **post-industrial economy** (Figure 17.3). The other 40 are smaller high-income countries (of roughly 1 million people or less), including Bahrain, Qatar, Liechtenstein and the Cayman Islands.

▲ Figure 17.2 The three main global groups

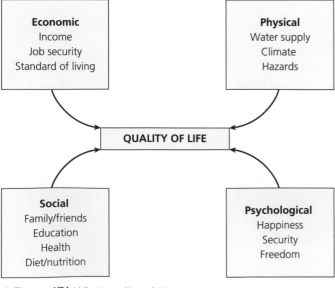

▲ Figure 17.1 What quality of life means

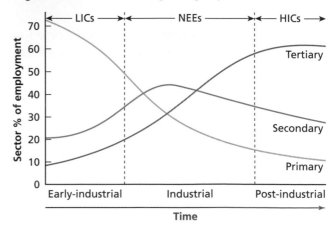

▲ Figure 17.3 The sector model of industrial change over time

What does the world map of development look like?

Figure 17.4 shows the world in 2018. The majority of the HICs lie in the northern hemisphere, with the exception of Australia and New Zealand. There are clusters of HICs in Western Europe, North America, the Middle East (including Saudi Arabia, Qatar and UAE) and East Asia (including Japan, South Korea and Singapore).

The distribution pattern for NEEs and LICs is complicated and is changing constantly. Key features are that:

- South American countries are NEEs
- Asia now has more NEEs than LICs
- Africa still has more LICs than NEEs
- Eastern European countries, including some EU members, are mainly NEEs.

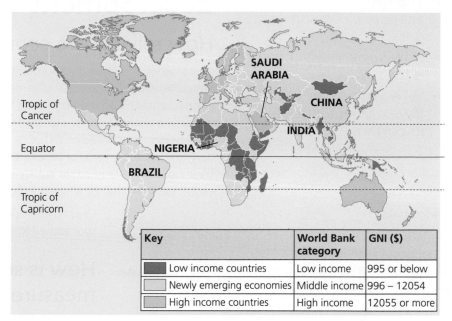

Key		World Bank category	GNI ($)
	Low income countries	Low income	995 or below
	Newly emerging economies	Middle income	996 – 12054
	High income countries	High income	12055 or more

▲ Figure 17.4 The world map of development

The global pattern of economic development has changed radically. In the 1980s, there was a clear divide between the rich 'global north' and the poor 'global south', marked by the Brandt Line (Figure 17.5). This crude division is increasingly of historical interest only, as.

- China is now the world's largest economy by one measure.
- Several of the world's highest-income countries lie south of the Brandt line (including Qatar, Kuwait and Singapore).
- The GNI per capita of some EU members, including Greece and Bulgaria, is lower than that of Malaysia and South Korea (2018 data).
- Large numbers of millionaires and billionaires can be found in every populated continent.

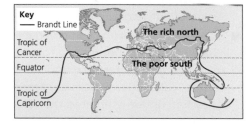

▲ Figure 17.5 How we viewed the world in the 1980s

→ Activities

1. Using Figure 17.2, describe how an LIC differs from an NEE.
2. Explain how 'raw' GNI data need to be adjusted to compare the economic development of countries.
3. Using Figure 17.4, describe the pattern of income across the world. Refer to lines of latitude and continents.
4. Using Figure 17.3 and your own understanding, explain how the importance of different sectors changes over time as a country develops.
   (The command here is 'explain', so you need to provide some reasons for the changes. In an industrialised country, machinery means fewer people need to work as farmers, for example.)
5. Using Figure 17.5 and your own understanding, explain how world development has changed since the Brandt Line was drawn. You could mention:
   - what has happened south of the line in each continent
   - the location of today's HICs
   - the variations that exist among countries north of the line.

KEY LEARNING

- The reliability of economic development data
- How social development is measured
- Why the human development index is important

Different measures of development

How reliable are economic development data?

Gross national income data sometimes provide a misleading picture of the typical level of economic development of a society, especially in LICs. The mathematical mean is a very crude way of generating a 'typical' figure. If one millionaire shares a street with 99 people who own nothing, the mean wealth of each person is counted as £10,000. Also, people in LICs and NEEs often work very hard, but the value of their efforts is not included in the GNI data – because their work consists of either **subsistence farming** (see Chapter 6) or informal sector work, neither of which is officially recognised.

Additionally:

- Data may not always be accurate: some people may lie about their earnings.
- Data may be hard to collect due to conflict or a disaster.
- The rapid **migration** of people into cities makes it hard to know how many people live in a place and what they earn.
- All GNI data are converted into US dollars, but the value of currencies changes every day.
- Errors and omissions can creep into the calculations. Some African countries like Nigeria did not include earnings from entertainment and the internet in their official calculations until very recently, meaning that they had under-estimated the value of their economy in some previous years.

Given the World Bank's categorisation of LICs, it is possible that some LICs might really be NEEs, or vice versa.

How is social development measured?

'Development' is linked with the idea of progress: there is more to this than just money! Figure 17.6 shows some of the most important social measures. There is always a strong correlation between social development measures and economic measures like GNI per capita (Figure 17.7).

Measure	Global variations	Limitations of this measure
Literacy rate (the percentage of people with basic reading and writing skills).	Most EU countries have a literacy rate of 99 per cent. In some LICs, the figure is as low as 40–50 per cent.	Carrying out surveys to determine literacy in rural populations, especially in conflict zones, or **squatter settlements** in LICs is difficult.
People per doctor (the number of people who depend on a single doctor for their health care needs).	The UK doctor-to-patient ratio is 1:350, whereas in Afghanistan it is 1:5,000.	In India and other NEEs, people in rural areas use mobile phones to get healthcare advice, but this is not taken into account by the 'people per doctor' measurement.
Access to safe water (the percentage of people who have access to water that does not carry a health risk such as cholera).	All EU citizens must have access to safe water by law. In rural Angola, around a third of people had access to safe water in 2018, in contrast.	**Water quality** can decline due to flooding or poor maintenance of pipes. Rising cost of water in cities sometimes forces poor people to start using unsafe sources. Official data may underestimate these problems.
Infant mortality rate (IMR) (the number of deaths of children under one year of age per 1,000 live births).	The UK figure is just 4 deaths per 1,000 per year but Somalia had 93 deaths per 1,000 live births in 2018.	In the world's poorest countries, not all infant deaths are recorded. Sadly, many children are buried in unmarked graves. Again, official data may be underestimated.
Life expectancy (the average number of years a person can be expected to live).	Most NEEs now have a high life expectancy of 65–75 years or more. In LICs, a figure of 50 (as in Chad, 2018) is more typical. In HICs, this is usually 75+.	In countries where infant mortality is high, the life expectancy of those who survive childhood is actually far higher than the mean life expectancy suggests. Can you see why this is the case?

▲ **Figure 17.6** Measures of social development

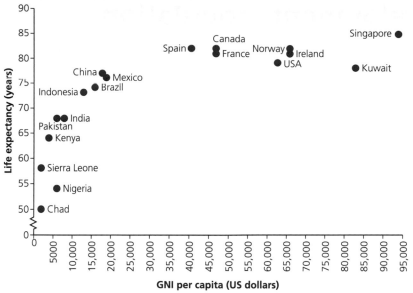

▲ Figure 17.7 Investigating the relationship between economic and social development (2018 data)

Why is the Human Development Index important?

All of the measures shown in Figure 17.6 ought to be factored in to calculate a country's overall level of development. Human rights and even happiness could also be considered. Some studies rank the world's countries according to their level of political corruption and gender inequality. With so many lists to think about, we use composite measures instead. These combine several development measures into one easy-to-use formula. The most widely used and reliable of these is the **Human Development Index** or HDI (Figure 17.8). The three 'ingredients' are processed to produce a number between 0 and 1. The world's highest- and lowest-scoring countries in 2018 are shown in Figure 17.9.

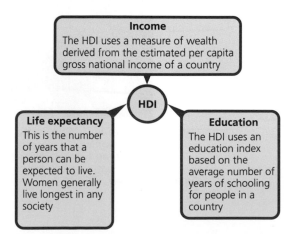

▲ Figure 17.8 The Human Development Index (HDI). In its current form, it has been used since 2010

▼ Figure 17.9 The highest and lowest GDI scores in 2018

HDI rank and score	Country
1 (0.953)	Norway
2 (0.944)	Switzerland
3 (0.939)	Australia
4 (0.938)	Ireland
5 (0.936)	Germany
185 (0.417)	Burundi
186 (0.404)	Chad
187 (0.388)	South Sudan
188 (0.367)	Central African Republic
189 (0.354)	Niger

Geographical skills

Statistical measures

1 Outline one reason why the mean average GNI per person of a country could be misleading in terms of how economically developed it is.

2 Outline one alternative statistical way of showing the 'typical' income of a population rather than calculating the mean average.

→ Activities

1 Outline two advantages of using HDI rather than GNI per capita to measure development.

2 a) Describe the relationship between the two development indicators shown in Figure 17.7.

b) Using Figure 17.7 and your own understanding, suggest reasons for the relationship shown in the graph.

c) Outline one reason why it might be misleading to draw a straight best-fit line on the scatter graph.

3 Assess the reasons why it could be difficult to collect accurate development data for different countries or regions.

■ Think about what might make data collection difficult: conflict, migration, the landscape, the vegetation.

■ Can you form an overall view of which difficulties could be hardest to overcome?

'Assess' requires you do more than just explain the reasons – you also need to 'weigh up' your ideas and make a judgement.

⭐ KEY LEARNING

➤ The population characteristics of countries with different levels of development

➤ The causes of rapid population growth in developing countries

➤ How rapid population growth can impact development

Development, population change and the demographic transition model

What are the population characteristics of countries with different levels of development?

A country's **birth rate** and **death rate** can be used to measure as social development. Over time, HICs have progressed from high rates to low rates of both. Based on the historical record of these changes, the population dynamics of LICs and NEEs can be studied according to a timeline: the **demographic transition model** or DTM (Figure 17.10).

■ LICs are in Stage 2 of the DTM. Even the world's very poorest countries have experienced a fall in death rate due to global efforts to tackle hunger and diseases like smallpox. In Sierra Leone, the death rate fell from 33 people out of every 1,000 per year in 1960s to 11 people out of every 1,000 per year in 2018. Like other Stage 2 countries, Sierra Leone's birth rate is still very high (37 births per 1,000 people per year in 2018).

■ NEEs are mostly in Stage 3 of the DTM. Compared with LICs, far fewer families in NEEs still live by subsistence means. This means that parents do not need to have large numbers of children to help farm the land. Improved healthcare means that contraception may be widely available. Independent working women in NEEs are choosing to have fewer children. In Bangladesh, the fertility rate in 2018 was just 2.1 children per woman on average. In 1970,

it was 7.1 per woman! In India, women's lives have changed similarly (Figure 17.11). This is a staggering developmental change in such a short period of time.

These rates do not always correspond with the level of development as we might expect. Occasionally, the death rate rises temporarily due to conflict or a natural disaster. It is also dependent on age. Some HICs are experiencing a rising death rate due to the growing proportion of people aged 80 and over who live there. When it comes to the birth rate, some cultures may be far more resistant to cultural change than others as they develop economically. In general, countries that promote education for women have seen a steep fall in the birth rate. Chile is an example of this.

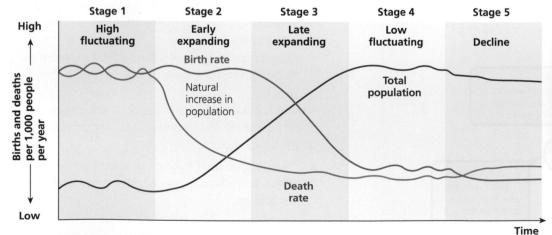

▲ Figure 17.10 The demographic transition model

▲ Figure 17.11 More young Indian women are working today than before such as in call centres in Bengaluru (Bangalore).

What causes rapid population growth in developing countries?

In an LIC where the birth rate remains high while the death rate has fallen, there is a high rate of natural increase. This is shown in Figure 17.10 by the gap that opens up between the birth rate and death rate: it is widest at the end of Stage 2, corresponding with rapid population growth. This steep gradient is the 'population explosion' that every country on the planet has experienced or is experiencing. The UK's population explosion took place in the 1800s. India experienced it between 1950 and 2000.

Global population growth is slowing as more countries gain higher levels of development. It is expected to level off by 2050, at around 9 billion people. Only a minority of the world's countries are still experiencing rapid growth – mostly sub-Saharan countries like Niger and Chad, where the fertility rate remains high, at seven children per woman. Africa's population may rise from 1 to 4 billion by 2100 due to high fertility in some countries.

How does rapid population growth impact development?

Rapid population growth can lead to **overpopulation**, where there are too many people for the available land and resources (see Figure 17.12).

▼ **Figure 17.12** Symptoms of overpopulation and their impact on the development process

Overpopulation symptom	Impact on economic and social development
Falling incomes	High unemployment, out-migration and low wages (because too many people are chasing too few jobs so employers can pay less).
Environmental degradation	**Overgrazing** of the land and water scarcity can lead to **soil erosion** in areas on the fringes of **hot deserts.**
Reduced health and happiness	Malnourishment due to insufficient food and the spread of disease (people's immune systems weaken when they are hungry).

▼ **Figure 17.13** The UK's growing population built the London Underground

Population growth is rarely the sole cause of overpopulation and its problems. When famine occurred in Ethiopia in the 1980s, population growth was a contributing factor. However, the physical cause of drought played a role. So did the ongoing civil war and lack of government assistance. To say that the cause of the **famine** was 'too many people' is a dangerous over-simplification. Most European countries did not suffer from overpopulation, despite rapid population growth in the 1800s. This was partly because they had extensive overseas empires (see page 264). As their own populations grew, they took the resources they needed from other countries.

Population growth should not be seen solely as a cost for society. In fact, every HIC and many NEEs have benefited over time from population growth. This is because people are the **human resources** that industries need.

→ Activities

1. State what is meant by natural increase.
2. Calculate the rate of natural increase in Sierra Leone.
3. a) Describe the changes in the birth rate and death rate over time shown in Figure 17.10.
   b) Using Figure 17.10 and your own understanding, explain why population begins to shrink in size in Stage 5 of the DTM.
4. Outline two reasons why life expectancy increases when a country begins to develop.
5. Explain how rapid population growth can both help and hinder the development process. Think about:
   - what happens to people's quality of life if a place with limited resources becomes overpopulated
   - the advantages of having a growing number of people who can work in different industries
   - the links between economic growth and social development.

Factors influencing development

What are the historical reasons for varying levels of national development?

Historically, colonialism harmed many countries, and sometimes created conflict which continues today. By the 1700s and 1800s, most of the 'global south' had been colonised (invaded and taken over) by European nations such as Britain, France and Spain. Their aims were to build global influence in order to better compete against rival European states and to access raw materials and labour. They also had a desire to 'civilise' indigenous people and spread Christianity. South American, Asian and African cultures were badly affected, especially those that became part of the transatlantic slave trade.

Colonialism ended, for the most part, in the twentieth century, however its legacy has lasted much longer. India and Nigeria gained independence from the UK in 1947 and 1960. But independence sometimes leads to new problems. When the Democratic Republic of the Congo (DR Congo) gained freedom from Belgium in 1960s, there were reputedly just fourteen university graduates amongst its population, so badly had the Belgians neglected the education system. Power struggles often took place in newly independent countries, especially if valuable natural resources like diamonds were at stake.

Conflict remains a major development obstacle for some LICs. Persistent political problems stem from the way that ancient African, Asian and Middle Eastern kingdoms were divided and re-assembled by competing European countries. The modern borders of many Middle Eastern and central African countries fit badly with the distribution of different cultures across these regions. Five million deaths have been linked with conflicts related to culture or ethnicity in DR Congo, Uganda and Rwanda in the 1990s. Since the conflict in Syria began in 2012, more than 12 million people have lost their homes. Many now live in refugee camps (Figure 17.14). More than half are aged under 17 and 90 per cent of them no longer receive an education.

▲ Figure 17.14 Zaatari refugee camp in Jordan. This is the world's largest camp for Syrian refugees, home to 80,000 people in 2018.

How is development affected by economic factors?

In the 1800s, European powers seized the raw materials they needed from other countries, despite local resistance. Today, their TNCs buy materials and food from LICs, but at low prices that jeopardise economic development in these countries. There are several reasons for these low prices:

■ International organisations like the World Trade Organization (WTO) have been criticised for not doing enough to establish fair terms of global **trade** for food and raw materials.

■ Sometimes, corrupt leaders of LICs have profited personally from selling resources cheaply to TNCs.

■ Food prices fluctuate wildly depending on competition and the quality of the crops. The price of cocoa beans halved in the 1990s due to overproduction, slowing down economic development for Ghana and Ivory Coast. Recently, another steep fall occurred between 2016 and 2018 (see Figure 17.15).

In contrast, NEEs have benefited from global trade. China's leaders in particular have focused on developing their own manufacturing industries, resulting in rapid economic growth in recent decades.

What role do physical factors play in the development process?

Physical factors undoubtedly play a role in hindering development. However, it is wrong to place the blame entirely on physical factors. Can you name a country with an extreme climate that suffers from frequent **earthquakes** and hurricanes? One answer could be the USA, an HIC. Physical factors alone do not explain a country's development. All the challenges in Figure 17.16 can be overcome with human ingenuity and money.

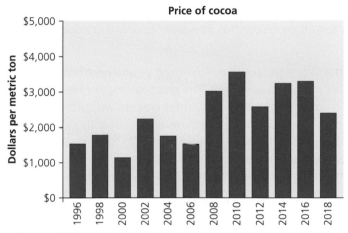

Price of cocoa

▲ Figure 17.15 World cocoa prices 1996–2018

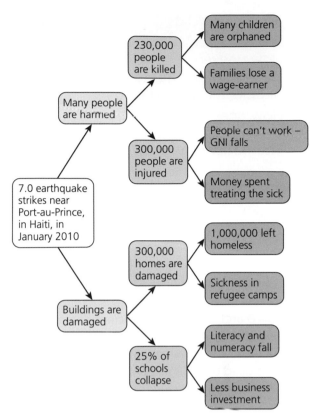

▲ Figure 17.17 How the earthquake in Haiti affected the country's rate of development

Factor	How factor influences the rate and level of development
Coastlines	There is quite a strong link between the lack of a coast and lower levels of development. With a few exceptions, the world's 45 landlocked countries are LICs or NEEs . Of the ten lowest-ranking HDI countries in 2018, seven have no coastline (including Chad, Niger and Mali). The greatest development challenge is not being able to trade goods easily without ports.
Natural hazards	In 2010, a devastating earthquake struck Port-au-Prince, capital city of the Caribbean island of Haiti. 230,000 people died. Since then, the country has struggled to develop (Figure 17.17). But many countries with a high level of development suffer from hazards too. Japan, Italy and Iceland are all HICs located on plate boundaries (see Chapter 2).
Climate	The influence of climate on development is not very clear. For every poor, hot desert country like Chad there is a rich one like Saudi Arabia. The same is true of **tropical rainforest** countries: while central Africa's equatorial climate could be viewed as a development challenge, Brazil is the world's eighth largest economy.

▲ Figure 17.16 Physical factors that can impact on development

→ Activities

1. a) State what is meant by colonialism.
   b) Outline one way in which colonialism has had a lasting economic impact on one named LIC, NEE or HIC.
2. Using Figure 17.16 and your own understanding, discuss the links between physical factors and the development of different countries. As part of your answer you could:
   - rank the three physical factors in order according to how important you think their influence is on the development process
   - give reasons for the order you have chosen.
3. a) Using Figure 17.17, describe the impact of the earthquake on Haiti's economy.
   b) Using Figure 17.17 and your own understanding, suggest how the earthquake has affected the long-term social development of Haiti.

The consequences of uneven development

How does uneven development affect the wealth and health of people in LICs and NEEs?

For some groups of people in LICs and NEEs, quality of life is not improving. In some cases it may even have worsened. Newfound trading wealth has helped Nigeria get 'promoted' from LIC to NEE status, but this wealth has not been distributed fairly. Nigeria is now one of the most uneven societies on Earth (see Chapter 19). In some local contexts, women are disproportionately affected by disparities in wealth, health and education (Figure 17.19).

These internal disparities can be studied using the Gini coefficient (Figure 17.18). This is a ratio with values between 0 and 1.0. A Gini coefficient of zero would mean that everyone in a place had exactly the same income. A score of 1.0 would mean that all the income in a place was controlled by a single person. In general, LICs and NEEs have a high Gini coefficient.

Another global-scale consequence of uneven development is how LICs have become dependent on HICs and some NEEs for aid. Many LICs have had to borrow money from the World Bank to pay for hospitals and healthcare and are now heavily in debt. The shortcomings of healthcare in some West African countries was shown by the Ebola outbreak of 2014–16, which resulted in over 11,000 people dying from the virus (see Section 19.2). Most deaths were in Sierra Leone, Guinea and Liberia. Sierra Leone's hospitals have deteriorated over time due to the country's low income, indebtedness and its civil war. This reminds us that many of the problems experienced by LICs are connected with one another.

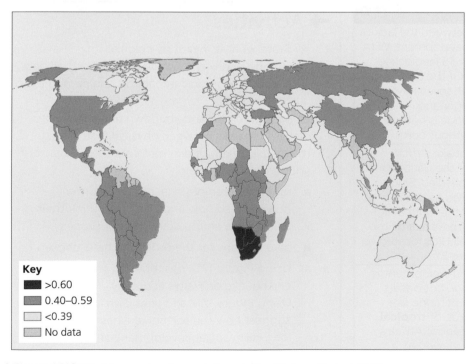

Key
■ >0.60
■ 0.40–0.59
□ <0.39
■ No data

▲ Figure 17.18 World map showing variations in the Gini coefficient, 2017

▲ Figure 17.19 Malala Yousafzai survived being shot by the Taliban when she was a schoolgirl in Pakistan. She now lives in the UK and campaigns for social development for women in LICs and NEEs.

How does uneven development lead to international migration?

At a global scale, uneven development leads to unequal flows of people between places. **Economic migrants** move voluntarily in search of a better life. Others are forced to flee persecution or disasters and are called refugees.

- International migration from poor countries reached an all-time high in 2015: a combination of poverty and conflict in places like Syria and North Africa resulted in a record 14 million people being forced from their homes. In 2018, around 260 million people lived in a country that was not where they were born.

- Also, people in LICs have become more aware of the **development gap** that exists between themselves and the NEEs and HICs. Despite their low incomes, many people in LICs are finding out about the 'bright lights' of richer places as technology spreads. In 2018, there were seven mobile phones in Africa for every ten people. Increasingly, migrants have mobile phones and can share information.

One highly visible manifestation of people from LICs on the move is the migrants who have been trying to reach Europe by boat. In 2017, 170,000 migrants came to Europe by sea, many via Italy (see page 289). Many were refugees fleeing conflict and persecution. Thousands more have drowned in the attempt. Hundreds of thousands of migrants have also tried to come to Europe by land.

It is not only poor and desperate people from LICs who cross borders. The UK employs computer engineers from India and doctors from Poland. These are highly skilled people: the countries they have left behind had invested time and money in training them. This so-called 'brain drain' of skilled human resources is sometimes given as a reason for a slow rate of an NEE's development.

However, there is a positive consequence that stems from migration caused by unequal development. Migrants from poorer countries send home **remittances**.

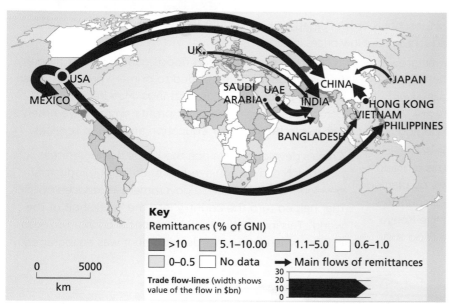

Key
Remittances (% of GNI)

■ >10	■ 5.1–10.00	■ 1.1–5.0 □ 0.6–1.0
□ 0–0.5	□ No data	→ Main flows of remittances

Trade flow-lines (width shows value of the flow in $bn)
30
20
10
0

0   5000
km

◄ Figure 17.19 Some important global remittance flows

→ Activities

1 State what is meant by the development gap.
2 a) What is the value of the remittance flow between Mexico and the USA shown in Figure 17.19?
   b) Using Figure 17.19, describe how the flow of remittances connects different countries together.
   c) Using Figure 17.19 and your own understanding, explain the links between uneven global development and flows of remittances.
3 Assess the negative impacts of uneven development on people and communities in LICs and NEEs.

- In your answer, you can explain problems linked with (i) the Ebola outbreak and (ii) the 'brain drain' caused by international migration (both issues feature on these pages). There may also be other ideas you can draw on from previous pages in the book.

- As part of your answer, try to assess which issue causes the greatest problems for poorer communities.

18 Reducing the global development gap

Industrial development and investment

Why is industrial development important for poorer countries?

For thousands of years, nations have traded to generate wealth. Not every country has the climate for bananas, cocoa or tomatoes, for instance (see Chapter 23). Some countries lack the raw materials they need for industry, like iron ore and copper, or **fossil fuels**. In theory, a country's strengths provide it with the opportunity to trade with other countries, thereby generating the income needed for economic development. In practice, countries in Africa, Asia and South America that were once colonies of rich European nations, have become trapped in an unequal global trade system.

▲ **Figure 18.1** Banana farmers

LICs which only trade in **primary products** (raw materials and agricultural produce) do not always receive a good price. This means they have insufficient money to import important manufactured products from HICs. Development goals are harder to achieve without computers for schools or specialised hospital equipment. Reasons why primary products achieve low prices include:

■ overproduction – when too many countries grow the same crop it pushes down prices globally. In years when crops are good, the problem is made even worse. In some years, prices for coffee beans, cocoa beans or bananas have fallen very low, bringing misery to producer communities (Figure 18.1).

■ import taxes – the **European Union** (EU) is a group of countries that protects its own farmers by placing import tariffs on food imports from other countries. As a result, farmers in non-EU countries like Kenya find it harder to get a good price for the food they sell to European supermarkets.

In contrast, manufactured goods can be sold at higher prices. Value has been added to primary products when they are processed to make things. Manufacturing companies often make great profits, which governments can tax to help pay for education and health services.

China's development since the 1980s demonstrates how much can be achieved by encouraging industrial development. The Chinese government introduced policies that helped turn the country into the 'workshop of the world'. This included establishing special economic zones (SEZs), where industrial development was encouraged.

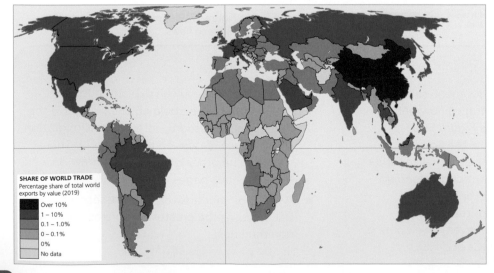

SHARE OF WORLD TRADE
Percentage share of total world
exports by value (2019)

- Over 10%
- 1 – 10%
- 0.1 – 1.0%
- 0 – 0.1%
- 0%
- No data

◄ **Figure 18.2** Map of world trade

How does investment by TNCs help NEEs and LICs to develop?

In the early 1900s, many of the world's largest companies exported their products but still produced everything in a single place. Over time, companies have grown in size to become **transnational corporations (TNCs)**. They rely less on exporting and prefer instead to produce goods and services inside the borders of many different countries.

The cash injected into other countries by TNCs is called **foreign direct investment (FDI)**. It helps the development process to take place in different ways. Local people are employed to build factories or offices. Other people will work in them. A **multiplier effect** can also develop: investment by a TNC can help other local businesses begin to thrive, creating work for even more people. Increasingly, NEEs have their own TNCs that invest globally too. Investment from Chinese and Indian companies is helping LICs in Africa to develop (Figure 18.3).

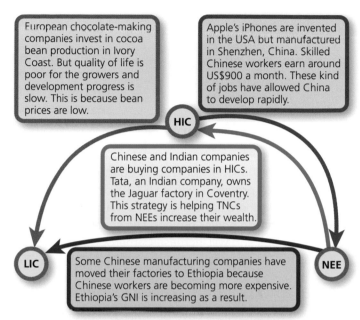

▲ Figure 18.3 How global patterns of investment can affect development and quality of life in different types of country

▼ Figure 18.4 Reasons why TNCs invest in other countries and the development this brings (shown in italics)

Reducing transport and import costs	Looking for new markets	Looking for cheap labour
By assembling their products close to the people they will be selling them to, companies can reduce transport costs and avoid import tariffs. Guinness brews beer in Nigeria. It is much cheaper to do this than to export barrels of beer all the way from Ireland. *This investment brings employment to Nigeria.*	Worldwide, over 1 billion people in NEEs now have a 'middle-class' income and lifestyle. A range of TNCs, including McDonald's, Apple and Ikea, have invested in NEEs by building retail stores there. *Globally, McDonald's has invested in over 130 countries.*	Because the cost of labour is high in HICs, many companies have relocated their operations to other countries to save money. TNCs have invested in the creation of new farms, factories and offices in many LICs and NEEs. *While this can help development, workers are sometimes exploited and may have a low quality of life.*

→ Activities

1   Using Figure 18.1 and your own understanding, suggest why many of the things sold in UK shops must be imported from other countries. Think about:

  ■ all the items in your fridge that have come from other countries

  ■ where the clothes you wear are sourced from (look at the labels).

2   State what is meant by a transnational corporation (TNC).

3   Using a case study of an NEE, explain how investment by a TNC helps that county to develop economically and also socially. Think about:

  ■ what a 'multiplier effect' is, and how it works

  ■ what workers will spend their incomes on that helps their families.

4   Using Figure 18.3 and your own understanding, explain how investment from TNCs is helping to 'bridge' different development gaps.

🌟 KEY LEARNING

➤ How international aid helps development

➤ The role of intermediate technology

➤ The importance of the work of the Fairtrade Foundation

Aid, fair trade and development

How does international aid help development?

International aid is a gift of money, goods or services to a developing country. Unlike a loan, the gift does not need to be repaid, unless it is tied-aid, which has to be spent on goods or services from the donor country (see page 286).. The donor may be a single country or a group of countries such as the European Union. Individuals in HICs give aid to poorer countries by making donations to charities like Oxfam. Most international aid is targeted at specific long-term development goals for people in LICs and some NEEs.

Economic development

Large-scale power and transport projects have been funded by international aid in some countries. In 2018, Cameroon was provided with a US$800 million grant by international donors to help it build a major new **dam** and **hydroelectric power** project. The scheme involves many new jobs and will increase Cameroon's electricity supplies by 30 per cent.

Social and political development

■ Many LICs and NEEs have benefited from gifts of money and equipment to help with education. The One Laptop per Child project was part funded by Google. It helped distribute free laptop computers to hundreds of thousands of students and teachers in South America and Africa.

■ The health of pregnant women in LICs is a spending priority for the United Nations. This issue is linked with a wider aim to improve the political rights of women across the world. A large part of

Finland's international aid budget is targeted at helping women.

We can see particular geographical patterns in the way that development aid is distributed internationally:

■ Flows of aid from the UK are directed towards **Commonwealth** countries (see Figure 20.36 on page 306). This is partly explained by the history the UK shares with its former colonies. For example, although Nigeria is an NEE, over half its population (around 100 million people) are still very poor and need help. Giving aid to Nigeria helps the UK to maintain the political influence and economic relationship with its former colony developed during colonial times.

■ India and China provide aid to LICs across Africa. India has spent over US$6 billion on education projects there. The Tazara railway that links Tanzania and Zambia was funded with international aid from China. The flow of aid from NEEs to LICs is an important new feature of the geography of development.

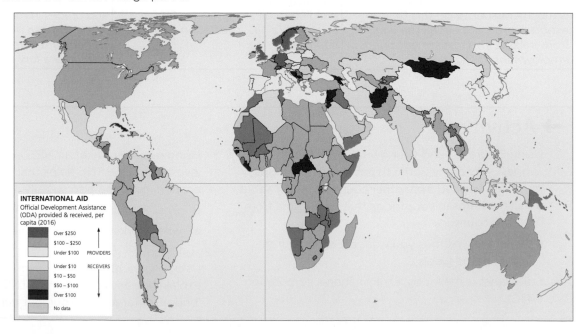

INTERNATIONAL AID
Official Development Assistance (ODA) provided & received, per capita (2016)

Over $250	
$100 – $250	
Under $100	PROVIDERS
Under $10	RECEIVERS
$10 – $50	
$50 – $100	
Over $100	
No data	

➤ Figure 18.5
International aid data

What role does intermediate technology play in the development process?

Charities and NGOs use **intermediate technology** in LICs and NEEs to help them reach their development goals, such as **sustainable energy** or more efficient ways of cooking. It is technology that the local community can take ownership of and learn to maintain. For example, WaterAid does vital work providing aid for improved water supplies in poor countries, using intermediate technology such as the Afridev hand pump (see Figure 18.6) to help provide clean water. Diarrhoea from polluted water accounts for at least 20 per cent of infant deaths in Tanzania. To help tackle this development challenge, WaterAid helped provide the community of Chessa village with a well 24 metres deep, fitted with an Afridev hand pump. The villagers can now drink safe underground water.

The Afridev hand pump is not very sophisticated, but when more advanced machines break, a specialist engineer is needed. This can leave local people without water while they wait for repairs. In contrast, an Afridev hand pump will break down fairly often but is repaired quickly by the community themselves. This is seen as the best strategy to help safeguard people's quality of life. It also contributes to the long-term development of Tanzania in two important ways:

- Life expectancy has increased due to fewer deaths from disease.
- Education has improved now students are missing fewer days of schooling due to illness.

▲ Figure 18.6 An Afridev handpump

How important is fair trade?

The work of the **Fairtrade** Foundation is very important for people in LICs and some NEEs. The aim is to give producers a better price for the goods they produce, and a price guarantee. If the global price for a particular crop like coffee collapses, Fairtrade farmers will still receive their regular income. This protects their quality of life.

Consumers in HICS have increased what they spend on Fairtrade food and goods over time. Examples of Fairtrade produce include chocolate, bananas, wine, footballs and even higher-value clothing items such as jeans. Some people are happy to pay a little more, knowing that a higher proportion than usual of the bill will find its way directly into the pay packets of some of the world's poorest people.

The benefits of the Fairtrade system are clearly shown by the experience of people living in the village of Chagelen and the city of Sialkot in the Punjab province of Pakistan. Many young adults, who did not go to school themselves, now stitch footballs under a scheme – this means their families have more money so their younger siblings can attend school, therefore breaking the cycle of poverty. Local communities have access to healthcare paid for by the Fairtrade scheme, so it is helping social development too.

However, the higher price of Fairtrade products means that many shoppers in HICs avoid buying them, especially during times of economic hardship. This means there is a limit to the number of farms or villages that can be part of the scheme.

→ Activities

1. Outline one way in which aid is different from a loan.
2. Describe the pattern on the map shown in Figure 18.5.
3. Outline two reasons why HICs give aid to LICs.
4. a) State what is meant by intermediate technology.
   b) Using Figure 18.6 and your own understanding, explain how intermediate technology can improve people's quality of life.
5. Assess how far the Fairtrade system can help all communities in LICs and NEEs to develop further. You should:
   - write about different development measures (see also pages 260–261)
   - consider whether it is possible or practical for all trade to become part of the Fairtrade system.

⊕ KEY LEARNING

➤ Why many developing countries have suffered a debt crisis

➤ How microfinance loans are helping the world's poorest people

Borrowing, debt relief and development

Why have many developing countries suffered a debt crisis?

During the period between 1960 and 1980, some HICs loaned many LICs and NEEs staggering amounts of money to develop their countries. The problem was that this money was lent under terms that would make it difficult to pay back. The **debt crisis** began in 1982, when Mexico admitted it had no way of paying back the US$80 billion it had borrowed.

The lenders were:

■ the World Bank and the International Monetary Fund (IMF) – organisations established after the Second World War to help re-stabilise the world economy. Both are based in Washington DC, USA. They lend money on a global scale to countries that apply.

■ large commercial banks – during the 1970s, US and UK banks lent large amounts of money to countries in the developing world. Levels of interest on bank loans were very high at the time.

In the most successful cases, money invested can generate enough wealth to pay back the loan and help the borrowing country to develop too. The World Bank lent Laos US$1 billion to build a dam on the Nam Theun River (Figure 18.7). Since 2010, the dam has generated hydroelectric power (HEP). Laos is projected to earn US$2 billion from 2010–35 by selling electricity to its neighbour, Thailand. This will be enough money to repay the loan and increase the GNI of Laos too.

The need for debt relief

In some cases, borrowing led to serious problems. For instance, the World Bank funded the speedy modernisation of Indonesia with large loans during the 1970s. Roads, power stations and ports were all built in order to attract investment from TNCs. In the process, a lot of money went missing: it had been taken by Indonesia's ruling family!

In DR Congo, the outcome was even more disastrous. Previous Prime Minister Joseph Mobutu kept US$4 billion that had been lent to his country. This money has never been recovered. After Mobuto's death, the World Bank decided it would be unfair to expect the people of DR Congo to pay back the stolen money. DR Congo's debt was 'written off'. As a result of pressure from charities and protests, other LICs have been offered **debt relief** too (Figure 18.8).

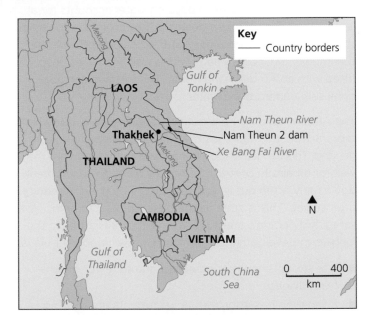

▲ Figure 18.7 The dam on the Nam Theun River

▲ Figure 18.8 A protest march supporting debt relief

Debt relief can also be achieved through '**conservation swaps**'. A richer country may agree to write off part of a poorer country's debt if that poorer country agrees to protect its physical environment. The USA agreed to let Indonesia keep US$30 million of borrowed money in exchange for increased protection of Sumatran forests, home to endangered rhinos and tigers.

How are microloans helping the world's poorest people?

At a very different scale, people in LICs and NEEs who are less likely to be loaned money by traditional banks borrow small sums of money called **microfinance loans** (Figure 18.9). The most well-known provider of microloans is the Grameen Bank in Bangladesh. It has lent money to 9 million people, 97 per cent of whom are women. In contrast to the billions of dollars lent to countries, microfinance loans involve just a few hundred dollars, but can play a crucial role in kick-starting development at a local level. The theory is that if enough villages are helped then, in time, an entire country can develop.

Microloans are needed because subsistence farmers find it hard to escape poverty (Figure 18.10). They can only grow enough food for their own needs, rather than to sell the food. The seeds they use do not always yield good enough crops. The soil they plant them in may not be fertile. This is where microfinance loans come in.

Microfinance loans provide farmers with the vital cash their families need to escape a cycle of poverty. A

microloan is not a 'free hand-out'; it must be paid back. One advantage of a small commercial loan like this, when compared with charitable aid, is that poor people feel they can stand on their own two feet instead of being dependent on others.

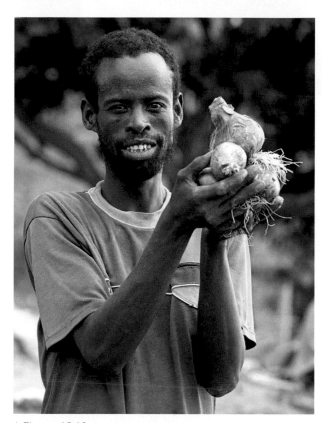

▲ Figure 18.10 A subsistence farmer in Ethiopia harvests his onion crop

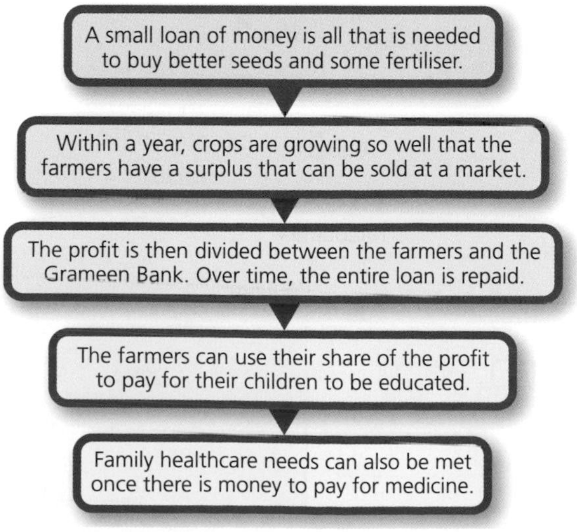

A small loan of money is all that is needed to buy better seeds and some fertiliser.

Within a year, crops are growing so well that the farmers have a surplus that can be sold at a market.

The profit is then divided between the farmers and the Grameen Bank. Over time, the entire loan is repaid.

The farmers can use their share of the profit to pay for their children to be educated.

Family healthcare needs can also be met once there is money to pay for medicine.

▲ Figure 18.9 Microloans flowchart

→ Activities

1 a) State what is meant by debt relief.
  b) Outline one reason why some countries cannot pay off their debts.
  c) Outline one way in which debt relief helps countries to develop socially.
2 a) State what is meant by microfinance loans.
  b) Using Figure 18.9 and your own understanding, discuss the importance of microfinance loans for communities in LICs. You should:
  ■ explain all the links between microloans and development
  ■ arrive at a judgement about whether microfinance loans can make a really important impact on the way a country develops, or whether they are less important than other forms of aid or trade.

273

Example

Tourism in Tunisia

Key facts

- In 2017, tourism brought 7.1 million people and around US$1 billion to Tunisia.

- This North African NEE had a GNI per capita of US$12,800 in 2018 (adjusted for prices – see page 260).

- Between 1960 and 2018, life expectancy in Tunisia has risen from 42 to 76 years showing social development.

Why has Tunisia become a popular tourist destination?

Before the 1970s, Tunisia was classified as an LIC. Agriculture still made up a very large part of its economy. A series of government reforms helped the country's economy to diversify and grow. A development strategy was introduced that promoted tourism alongside manufacturing industries. Factors that favoured the growth of Tunisia's tourist industry included:

- climate – the country's northerly coast enjoys a Mediterranean climate, with hot summers and mild, warm winters. Summer temperatures reach the high-30s in Carthage (Figure 18.12), making Tunisia a popular destination for sun-seeking Europeans.

- links with Europe – Tunisia's northern coastline is close to Europe, making it easily accessible. French colonial rule ended in Tunisia in 1956, so French is widely spoken and understood, which attracts French and French-speaking tourists.

- history and culture – Tunisia contains seven UNESCO World Heritage Sites. These include the ancient remains of the city of Carthage and the El-Jem amphitheatre built by the Romans.

- physical **landscape** – Tunisia's physical geography is diverse, ranging from Mediterranean beaches to the Dorsal Mountains and Saharan Desert. Tourists can visit their favourite film locations.

- cheap package holidays since the 1960s – the Tunisian government worked with private companies like Thomas Cook to develop the country into a tourist destination. In 2018, the industry provided jobs for 700,000 people (around 14 per cent of all employment).

The economic benefits of tourism have spread well beyond the most popular resorts due to the linked multiplier effects. In addition to money spent on hotels, tourism boosts the income of many Tunisian businesses. Visitors buy rugs and other goods in the local *souk* (market), which helps formal and informal businesses to thrive. Benefits have also spread to the agricultural sector, which supplies hotels with food, and taxi companies.

▲ Figure 18.11 Tunisia

What impact has tourism had on Tunisia's development gap?

Tunisia is now one of the wealthiest countries in Africa. Tunisian incomes quadrupled in the 1970s. Higher incomes quickly translated into longer life expectancy as diet and health improved. Tunisia's government invests almost four per cent of its annual GDP in healthcare.

Literacy rates have increased markedly over time, rising from 66 per cent to 82 per cent between 1995 and 2017. More families can now afford to send their children to school and even university. As well as job creation, tourism has helped by connecting Tunisia to other places and cultures. This may help explain changing attitudes towards girls' education. Schooling is now compulsory for girls, and women are entering higher education in increased numbers. Tunisia has made good progress towards greater equality for women – an important international development goal for any country.

Is tourism a sustainable development strategy for Tunisia?

There may be limits to how much more the development gap can be narrowed by tourism. Areas of concern include:

- **pollution** of the environment – some of Tunisia's Mediterranean beaches have been polluted with untreated sewage from hotels.
- 'leakage' of profits – companies based in other countries send holidaymakers to Tunisia but keep a large percentage of the profits. This limits how much money becomes reinvested locally and slows down the rate of economic development.
- terrorism – in 2015, there were two terrorist attacks in Tunisia aimed at tourists. One took place in the Museum Bardo in Tunis, and the second at a beach resort in Sousse. As a result, European governments said Tunisia was no longer a safe destination for their citizens. This resulted in less foreign investment, and visitors fell to 5.4 million that year. Terrorists attacked the tourist industry because of the ways in which Tunisia is developing, such as greater equality for women.
- Since 2015, tourism has recovered, but ongoing political instability in neighbouring countries, such as Libya and Nigeria, means there may be another decline.

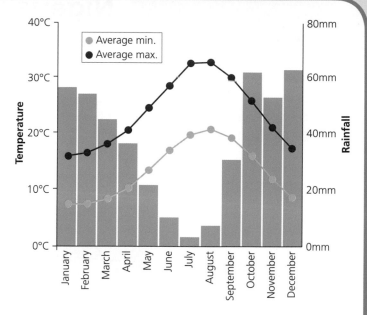

▲ Figure 18.12 The hot climate of Carthage attracts visitors from northern Europe

→ Activities

1 State what is meant by the 'development gap'.
2 Using Figure 18.12 and your own understanding, suggest how tourism has helped Tunisia to develop. Refer to both economic development and social development as part of your answer.
3 a) Using Figure 18.12 and your own understanding, suggest why tourist arrivals might vary from season to season in the city of Carthage.
   b) Outline one way in which seasonal variations in tourist numbers could impact on an NEE such as Tunisia.
4 Assess the sustainability of tourism as a development strategy for Tunisia. You should:
   - outline the success of different tourist strategies in Tunisia
   - explain why there may be limits to the growth of tourism, especially since 2015
   - judge whether the tourist industry can or cannot continue to fully support the future economic and social development of Tunisia.

19 Economic development in Nigeria

⭐ KEY LEARNING

➤ Nigeria's location and importance in Africa

➤ How Nigeria's population is growing

➤ How Nigeria's economy is growing

Nigeria's place in the world

What is Nigeria's location and importance in Africa?

Nigeria is a country in West Africa that is over three times larger than the UK. It lies just north of the Equator, with its south coast on the Gulf of Guinea, which is part of the Atlantic Ocean.

Nigeria is sometimes known as the 'Giant of Africa'. With a population of 201 million people (2019), its population is much larger than any other African country (Figure 19.1), and almost three times the size of the UK's population.

During the twenty-first century, Nigeria moved from being a low-income country (LIC) to becoming a newly emerging economy (NEE). In 2014, it overtook South Africa as the largest economy in Africa.

How is Nigeria's population growing?

Like many other LICs and NEEs, Nigeria's population is growing fast. It has a high proportion of young people and a high birth rate, and therefore, a high rate of natural increase. In 2019, Nigeria ranked seventh in the world by population, but by 2050 it is predicted to rank fourth, behind India, China and the USA (Figure 19.2).

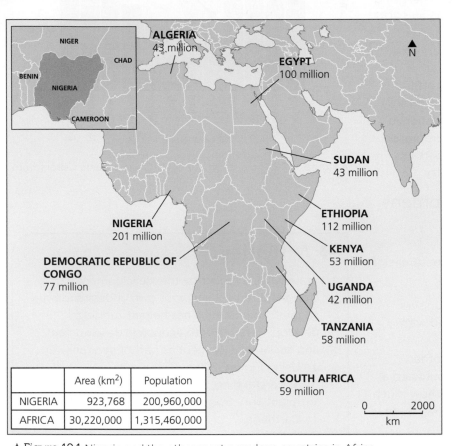

	Area (km²)	Population
NIGERIA	923,768	200,960,000
AFRICA	30,220,000	1,315,460,000

▲ Figure 19.1 Nigeria and the other most populous countries in Africa

▼ Figure 19.2 The world's top ten countries by population in 2019 and 2050

Country	Population 2019 (millions)	Country	Predicted population 2050
China	1,434	India	1,692
India	1,369	China	1,296
USA	329	USA	403
Indonesia	271	**Nigeria**	390
Brazil	215	Indonesia	293
Pakistan	218	Pakistan	275
Nigeria	201	Brazil	223
Bangladesh	163	Bangladesh	194
Russia	149	Philippines	155
Japan	129	DR Congo	149

How is Nigeria's economy growing?

In 2001, four countries with the world's fastest-growing economies were identified. They are the BRIC economies – Brazil, Russia, India and China. In 2014, four more countries, following in the footsteps of the BRICs, were also identified. These are the MINT economies – Mexico, Indonesia, Nigeria and Turkey.

Nigeria is now one of the fastest-growing economies in the world (Figure 19.3). By 2020, it is predicted to become one of the world's top twenty economies, and by 2050, it could be ahead of economies like France and Canada (see Figure 19.4).

One reason for Nigeria's predicted economic growth is its youthful population. It has a high proportion of educated young people due to start working in the next twenty years. They will provide the country with a plentiful supply of skilled labour to work in manufacturing and services.

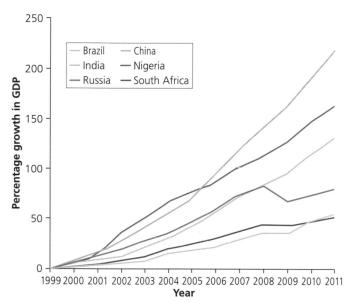

▲ Figure 19.3 Six of the world's fastest-growing economies in the twenty-first century

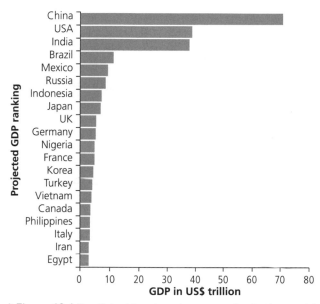

▲ Figure 19.4 Predicted top twenty economies in the world by 2050

→ Activities

1 Study Figure 19.3.
   a) Name the four BRICs on the graph.
   b) How does economic growth in Nigeria compare with that of the BRICs?
2 Study Figure 19.4.
   a) Where is Nigeria predicted to rank among world economies in 2050?
   b) Where will the other BRICs and MINTs rank?

Geographical skills

1 Study Figure 19.1.
   a) Rank the countries labelled on the map by their populations.
   b) Show this information in the form of a suitable graph.
2 a) Work out Nigeria's population density in people/km². Divide its population by the area (a calculator will help).
   b) Do the same for Africa. How does Nigeria's population density compare with population density for the whole of Africa?
3 Study Figure 19.2.
   a) Calculate Nigeria's predicted percentage population growth from 2019 to 2050.
   b) Compare Nigeria's percentage growth with other countries in the table. Will it be faster or slower?

⭐ KEY LEARNING

➤ The social and cultural context of Nigeria

➤ The environmental context

➤ How the political context is changing

Nigeria: the geographical context

What is the social and cultural context?

Modern-day Nigeria was formed in 1901 under British rule. Until then, the country was comprised of many smaller tribal kingdoms. Nigeria gained independence from Britain in 1960, but a lot of British influence remains. English is still the official language of Nigeria and is used in education, government and the media.

Nigeria has more than 500 different ethnic groups, each with its own language. However, three ethnic groups dominate – the Igbo, the Yoruba and the Hausa. The south of Nigeria, where the Igbo and Yoruba live, is predominantly Christian, while the north, where the Hausa live, is mainly Muslim (Figure 19.5).

Rapid **urbanisation** in recent years has led to a shift of population (see Chapter 13). Rural–urban migration of people from the countryside into cities has broken down some of the traditional boundaries. However, many different ethnic and cultural identities still exist, even within modern cities.

What is the environmental context?

Nigeria is located 5–12° north of the Equator in tropical Africa. Moving north from the Equator, the climate becomes drier and this determines the type of vegetation in each area. Tropical rainforest grows in the hot, humid climate in the south of Nigeria and savanna grassland in the hot, dry climate further north (Figures 19.6 and 19.7). Much of the natural vegetation in Nigeria has been replaced by agriculture. Cocoa and oil palm are grown in the south and peanuts are grown in the north.

How is the political context in Nigeria changing?

Since independence in 1960, Nigeria has progressed from civil war (1967–70), through several military dictatorships when the army ruled the country (until 1998), to a stable **democracy** today. The country now holds regular elections, when people vote to choose their government.

However, there is still conflict in Nigeria. In the north of the country, Boko Haram, an extremist organisation, wants to abolish democracy and set up its own government under its version of Sharia (Islamic) law. At least 17,000 people have been killed in the conflict since 2002 and over half a million people have fled the region.

One of the most shocking incidents happened in 2014, when Boko Haram kidnapped 276 schoolgirls as it opposes education, especially for girls. Up to now, all the girls have still not been found. The rise of Boko Haram has been blamed on inequality. Extremists exploit the growing gap between rich cities and poor rural areas in Nigeria by recruiting their fighters from among the poor.

Key
- Hausa and Fulani
- Yoruba
- Igbo

▲ Figure 19.5 The main ethnic groups and major cities in Nigeria

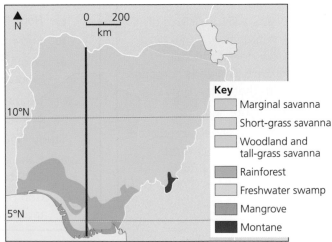

Key
- Marginal savanna
- Short-grass savanna
- Woodland and tall-grass savanna
- Rainforest
- Freshwater swamp
- Mangrove
- Montane

▲ Figure 19.6 Natural vegetation in Nigeria, including a S–N transect line (see Figure 19.7)

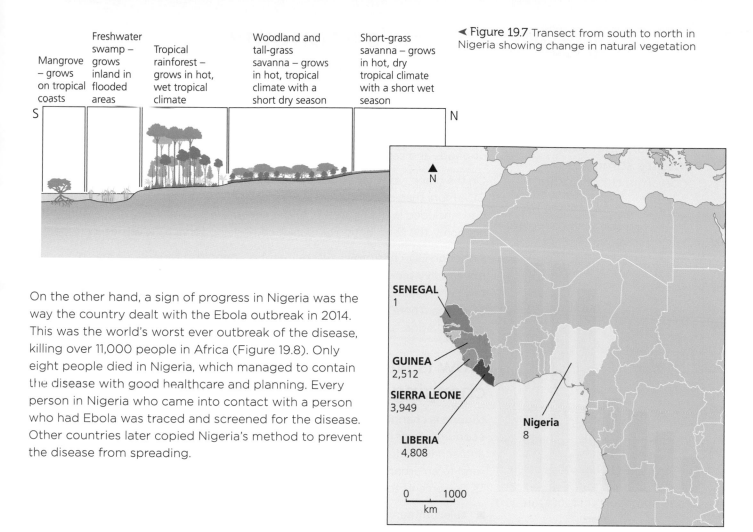

Mangrove – grows on tropical coasts

Freshwater swamp – grows inland in flooded areas

Tropical rainforest – grows in hot, wet tropical climate

Woodland and tall-grass savanna – grows in hot, tropical climate with a short dry season

Short-grass savanna – grows in hot, dry tropical climate with a short wet season

◄ Figure 19.7 Transect from south to north in Nigeria showing change in natural vegetation

S   N

SENEGAL
1

GUINEA
2,512

SIERRA LEONE
3,949

LIBERIA
4,808

Nigeria
8

0   1000
        km

▲ Figure 19.8 Numbers killed by Ebola, 2014

On the other hand, a sign of progress in Nigeria was the way the country dealt with the Ebola outbreak in 2014. This was the world's worst ever outbreak of the disease, killing over 11,000 people in Africa (Figure 19.8). Only eight people died in Nigeria, which managed to contain the disease with good healthcare and planning. Every person in Nigeria who came into contact with a person who had Ebola was traced and screened for the disease. Other countries later copied Nigeria's method to prevent the disease from spreading.

→ Activities

1 Study Figure 19.5.
   a) Describe the distribution of the three main ethnic groups in Nigeria.
   b) How might this change with urbanisation?

2 Study Figures 19.6 and 19.7.
   a) Describe how natural vegetation changes from south to north in Nigeria.
   b) Explain how this change is related to climate.

3 Look at Figure 19.8 and consider the kidnapping of the 276 schoolgirls by Boko Haram.
   a) Suggest how both events in 2014 could have been affected by politics.
   b) Should the Nigerian government take any of the blame or credit for either of these events, do you think? Explain your ideas.

4 To what extent do you think there has been political progress in Nigeria?
   a) Outline the main political events since 1960.
   b) Do each of these events demonstrate progress or not? Give reasons for your opinions.

5. Find out what has happened in Nigeria since 2014.
   ■ Are Boko Haram still active in Nigeria or has the government stopped them?
   ■ Has Ebola been eradicated or have any more outbreaks happened?
   ■ What has been done to prevent any future outbreaks of Ebola?

Case study

⭐ KEY LEARNING

➤ How Nigeria's economy is changing

➤ The industrial structure of Nigeria

➤ How manufacturing industry stimulates economic development

Nigeria's economy

How is Nigeria's economy changing?

As you know, Nigeria has the largest economy in Africa and is among the world's fastest-growing economies (see Section 19.1). Figure 19.9 shows the recent increase in Nigeria's **gross domestic product (GDP)** – the total value of goods and services the country produces.

On the graph, you may notice a dramatic change between 2012 and 2014, when GDP more than doubled. Until 2013, many of Nigeria's new industries had not been included in the figures. With the new, more accurate way of measuring the economy, the contribution of manufacturing and **service industries** increased.

Although GDP has grown, 40 per cent of people in Nigeria are still in poverty living on less than US$1 a day. There is growing inequality, with a few very wealthy people and a minority of people working in cities in well-paid jobs. There are also **regional inequalities**, with most wealth in the south, around Lagos, and greater poverty in the north and south-east.

How is Nigeria's industrial structure changing?

Nigeria is changing from a mainly agricultural economy into an industrial economy. Over half of the country's GDP now comes from manufacturing and service industries (Figure 19.10). This reflects the change from a mainly rural to an urban population, brought about by urbanisation (see Section 13.2). Some of the fastest-growing industries in Nigeria include:

- telecommunications – in 1990 there were less than a million landline telephone customers in Nigeria. Now, there are over 115 million mobile phone users.
- retail and wholesale – many small businesses that used to be in the informal sector (see Section 14.3) are now part of the **formal economy** and are included in the calculation of GDP.
- the film industry in Nigeria – known as Nollywood, it is now the third largest film industry in the world, after Hollywood (USA) and Bollywood (India).

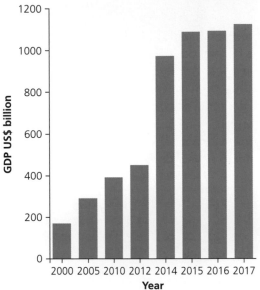

▲ Figure 19.9 Growth of Nigeria's GDP since 2000

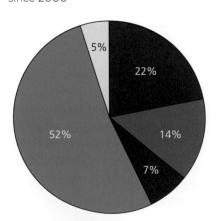

Key
- ■ Agriculture
- ■ Oil and gas
- ■ Manufacturing
- ■ Services
- ☐ Other

▲ Figure 19.10 Nigeria's GDP by economic sector

▼ Figure 19.11 Employment in different sectors of Nigeria's economy

Economic sector	% employment
Agriculture	30.5
Manufacturing (including oil and gas)	14.3
Services	
Accommodation, food, transport	12.2
Education, health, science and technology	6.3
Retail, repair and maintenance	24.9
Finance and insurance	4.2
Telecommunication, arts and entertainment	1.8
Other services	5.8

How does the manufacturing industry stimulate economic development in Nigeria?

Oil was discovered in Nigeria in the 1950s and is a vital part of the country's economy (Figure 19.12). Oil and gas account for about 14 per cent of Nigeria's GDP and 95 per cent of its export earnings. Income from oil has helped Nigeria to make the transition from low-income country to newly emerging economy.

However, the country's dependence on oil makes it vulnerable to changes in the world oil price. When oil prices fall, as they did in 2015, it damages the Nigerian economy. Additionally, because most of Nigeria's oil is exported before it is refined (or manufactured into fuels like petrol and kerosene), much of the profit is made outside Nigeria. Oil has not stimulated as much economic development in Nigeria as you might expect.

Aliko Dangote is Africa's richest person. He is a billionaire many times over and the founder of Dangote Cement, one of Nigeria's largest companies. The company has three giant cement plants in Nigeria. In a continent that is urbanising rapidly, cement is in high demand and Dangote Cement has expanded into thirteen other African countries.

New manufacturing industries, like Dangote, are increasing the pace of economic development in Nigeria in several ways:

- Improving the standard of living by the products of industries such as cement.

- Producing manufactured goods in the country reduces the need to import goods and can be cheaper.
- New industries create jobs, give people an income and contribute to the country's wealth through taxes.
- The expansion of Nigerian companies into other countries increases Nigeria's influence in the region.

▲ Figure 19.12 Drilling for oil in Nigeria

→ Activities

1   Study Figure 19.9.
    a) Describe the changes in Nigeria's GDP in the twenty-first century.
    b) Explain why an increase in GDP does not mean everybody is wealthier.

2   Study Figure 19.10.
    a) Describe the importance of four sectors in Nigeria's economy.
    b) Suggest how the importance of each sector of the economy is changing.

3   Compare Figures 19.10 and 19.11.
    a) Draw a pie chart for the data in Figure 19.11, showing employment in each sector of Nigeria's economy. Compare it with Figure 19.10.

    b) Try to explain the differences between the two pie charts for agriculture, manufacturing and services. (Hint: think about how technology is used and how much people might be paid.)

4   Look at Figure 19.12.
    a) How has the oil industry helped Nigeria's economic development?
    b) What is the problem with dependence on oil? (You will find out more on this in Section 19.4.)

5   Consider Aliko Dangote, Africa's richest person.

    Suggest why he was in the right place at the right time.

⊙ KEY LEARNING

➤ The role of transnational companies in Nigeria's oil industry

➤ The environmental impact of oil in Nigeria

➤ The advantages and disadvantages of transnational companies

The role of transnational companies

What is the role of transnational companies in Nigeria's oil industry?

The oil industry in Nigeria is located in the Niger Delta region, a vast area of wetlands on the delta of the Niger River, where it flows into the Gulf of Guinea (Figure 19.13).

The oil boom in Nigeria took off in the 1970s. It depended on the expertise and money of large transnational corporations (TNCs) based in Europe and the USA, including:

■ Royal Dutch Shell (UK, Netherlands)

■ Chevron (USA)

■ Exxon-Mobil (USA)

■ Agip (Italy)

■ Total (France).

The companies erected drilling platforms on the oil and gas fields around the Niger Delta region, linked by pipelines to export terminals in the Gulf of Guinea, where the crude oil is piped onto tankers. The oil is shipped to Europe and the USA where it is refined into petrol and other oil products. The companies make most of their profit from refined oil.

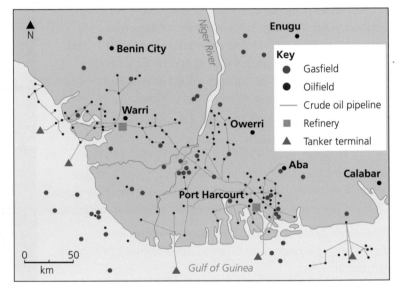

▲ Figure 19.13 The oil and gas industry in the Niger Delta region of Nigeria

When oil was discovered in Nigeria

Chief Sunday Inengite remembers the day the oilmen came hunting for oil in 1953. Members of his small community weren't sure what the men were after. Most thought they were on the lookout for palm oil, which had been exported from West Africa since 1832. Nigeria accounted for 75 per cent of this exported oil by the 1870s. But it became clear this was not the case: Inengite says, 'It wasn't until we saw what they called the oil – the black stuff – that we knew they were after something different.'

While several attempts to find oil had taken place in the early twentieth century, throughout the mid-twentieth century companies discovered numerous oil wells, many along the Niger Delta. By 1959, Royal Dutch Shell-BP had 53 wells operating in the area.

While oil quickly became a lucrative export for Nigeria, it came at a grave environmental cost. 'You see fish floating on the surface of the water, something we didn't know before,' Inengite says. 'It may be difficult to make a catch that will be enough to feed your family for one day.'

▲ Figure 19.14 A news article about oil in Nigeria (adapted from numerous sources)

What is the environmental impact of oil in Nigeria?

Some people believe, rather than being a blessing, oil has become a curse for Nigeria. In order to keep some control over the oil industry, the Nigerian government set up the Nigerian National Petroleum Corporation (NNPC) to form joint ventures with TNCs. This ensures that part of the profit from oil stays in Nigeria.

The oil industry also causes environmental damage:

- Oil spills from leaking pipelines damage farmland so crops no longer grow.
- Gas flares are used to burn off gas from the oil. Apart from being wasteful, the fumes affect people's health and contribute to global warming.
- Oil heated by the Sun becomes highly flammable and can burn out of control.
- Oil pollution, which occurs offshore from tankers, kills fish in the sea.

The Delta region contains important wetland and coastal **ecosystems**. Most people depend on the natural environment for their livelihood, either through farming or fishing.

▲ Figure 19.15 An oil spill on farmland in the Delta region

What are the advantages and disadvantages of TNCs?

Advantages of TNCs	Disadvantages of TNCs
• Bring new investment into the country's economy. • Provide jobs, often at higher wage levels than average in the local economy. • Bring expertise and new skills that the country does not have. • Have international links that bring access to world markets. • Provide new technology that helps economic development.	• Take profits out of the country to pay shareholders or to invest elsewhere. • Wage levels in LICs and NEEs are usually lower than in HICs. • Can cause environmental damage and deplete natural resources. • TNCs can withdraw their investment from a country if they wish. • They are powerful organisations and can exert political influence over the government in a country.

▲ Figure 19.16 The advantages and disadvantages of TNCs

→ Activities

1 Study Figure 19.13. Describe the distribution of oil and gas fields in Nigeria.

2 Explain the role of TNCs in the oil industry in Nigeria.

3 Study Figures 19.14 and 19.15. Imagine two conversations between Chief Sunday Inengite and an oil company executive, (a) in 1953 and (b) today.

Write a short dialogue for each conversation. What questions would Chief Sunday Inengite ask? What answers would he get?

4 Identify the main advantages and/or disadvantages of TNCs for each of these people:

- an oil worker in Nigeria
- a farmer in the Niger Delta
- a Nigerian government minister.

On balance, would each person be for or against the role of TNCs in Nigeria? Give reasons.

⭐ KEY LEARNING

➤ The relationship between Nigeria and Britain
➤ How Nigeria's trade relationships are changing
➤ The influence China now has on the Nigerian economy

Nigeria looks east

What relationship did Nigeria have with Britain?

Britain has had a trading relationship with West Africa for over 300 years. From 1650, the British traded enslaved people from countries in Africa and took them to America and the Caribbean. When slavery was made illegal in 1807, trade with West Africa turned to palm oil, used in Britain to make soap.

In the late nineteenth century, Nigeria, along with much of Africa, became part of the British Empire (Figure 19.17). The country was ruled by Britain until it gained independence in 1960. By then, a pattern of trade was established where Nigeria exported natural commodities to Britain and, in exchange, imported manufactured goods.

How are Nigeria's trade relationships changing?

Nigeria still trades with the UK, but more of its trade is now with some of the world's largest economies, which include the USA, China, India and other countries in the European Union (Figure 19.18). Since independence, oil has replaced other natural commodities as Nigeria's main export, but the country still imports manufactured goods like machinery, chemicals and transport equipment.

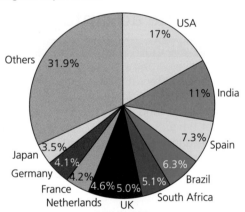

Nigeria exports to...

USA 17%
India 11%
Spain 7.3%
Brazil 6.3%
South Africa 5.1%
UK 5.0%
Netherlands 4.6%
France 4.2%
Germany 4.1%
Japan 3.5%
Others 31.9%

▲ **Figure 19.17** Much of Africa was ruled by Britain as part of the British Empire. In this 1892 cartoon, the British imperialist Cecil Rhodes is pictured ruling over Africa, from Cape Town to Cairo.

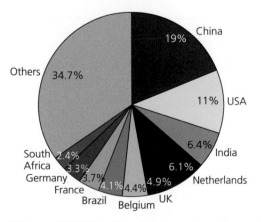

Nigeria imports from...

China 19%
USA 11%
India 6.4%
Netherlands 6.1%
UK 4.9%
Belgium 4.4%
Brazil 4.1%
France 3.7%
Germany 3.3%
South Africa 2.4%
Others 34.7%

▲ **Figure 19.18** Nigeria's main export and import partners

What influence does China now have on Nigeria's economy?

Nigeria's main import partner for manufactured goods is now China. But China's influence on Nigeria's economy goes beyond the goods it sells. There is also growing Chinese investment in Nigeria and other African countries (Figure 19.19). Both China and Nigeria benefit from this relationship.

■ Nigeria needs huge investment in **infrastructure**, particularly its transport network and power supply (see Chapter 14). China, with recent experience in building its own infrastructure, is now able to take that expertise to other countries. In 2014, the China Railway Construction Corporation won a US$12 billion contract to build a new 1,400 kilometre railway along the coast of Nigeria.

■ China's fast-growing economy needs more resources than the country can provide for itself. It can find these resources abroad, in countries like Nigeria. In 2014, another Chinese corporation agreed to invest $10 billion in exploration and drilling in a new oilfield in Nigeria.

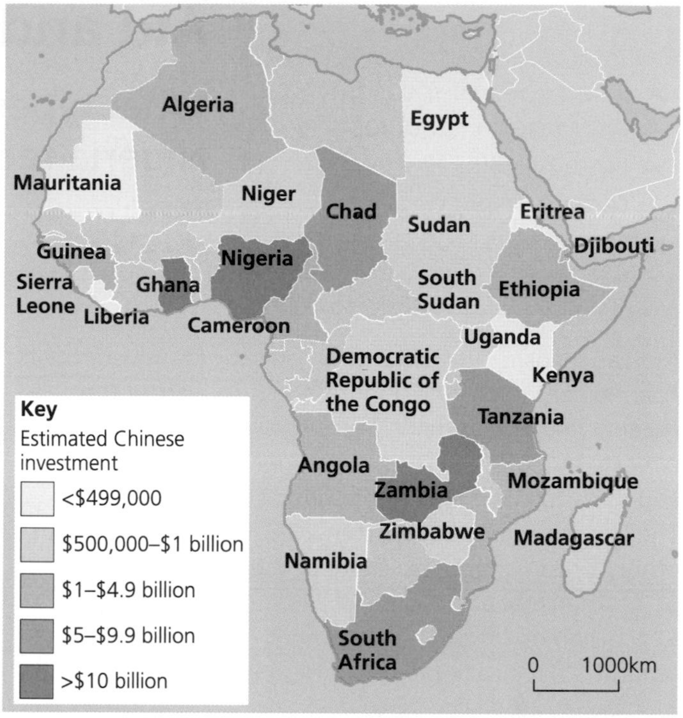

Key
Estimated Chinese investment
- <$499,000
- $500,000–$1 billion
- $1–$4.9 billion
- $5–$9.9 billion
- >$10 billion

0   1000km

▲ Figure 19.19 New Chinese investment in Africa, 2010–2014

→ Activities

1 Study Figure 19.17. What does the image suggest about Britain's relationship with Africa during the time of the British Empire?

2 Study Figure 19.19. Describe the pattern of Chinese investment around Africa. Name the countries with the most investment.

3 Suggest how a new railway along the coast of Nigeria could support economic development. Mention Lagos and the Niger Delta.

Geographical skills

Study Figure 19.18. Complete a map to show Nigeria's trading partners. On an outline world map, draw arrows from and to Nigeria linked to its trading partners, showing (a) exports and (b) imports. Use proportionate arrows to show the relative importance of trade with each country.

▲ Figure 19.20 A Chinese company will be building bullet trains in Nigeria

⭐ KEY LEARNING

➤ The types of international aid Nigeria receives

➤ How Nigeria got in to (and out of) debt

➤ Whether Nigeria still needs aid or not

Aid and debt

What types of international aid does Nigeria receive?

International aid or 'aid' is help given by one country to another in the form of money, food, technology or advice. Usually the help is from high-income countries (HICs) like the UK to low-income countries (LICs) like those in Africa.

Aid can involve international organisations, governments, charities or individuals (Figure 19.21). As part of the Millennium Development Goals to eradicate extreme poverty by 2015, the UN set a target for high-income countries to commit 0.7 per cent of their GDP as aid. The UK achieved this target in 2013. Nigeria, along with other African countries, is one of the main recipients of international aid (Figure 19.22). In 2016, it received $2.5 billion of Official Development Assistance (ODA), around $12 for each person in Nigeria.

Nigeria has one of the highest death rates from malaria in the world. Malaria is a disease transmitted by a mosquito bite that can cause death or long-term health problems. It can easily be prevented by sleeping under a mosquito net at night (Figure 19.23). Each net costs as little as £2.

From 2009 to 2013, 60 million mosquito nets were distributed to households across Nigeria as part of an international aid project funded by the World Bank, IMF and USA government.

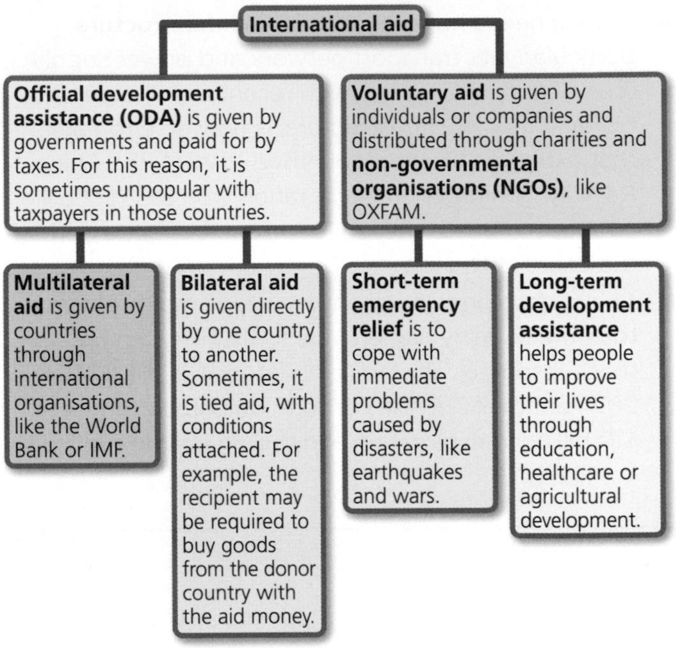

International aid

Official development assistance (ODA) is given by governments and paid for by taxes. For this reason, it is sometimes unpopular with taxpayers in those countries.

Voluntary aid is given by individuals or companies and distributed through charities and **non-governmental organisations (NGOs)**, like OXFAM.

Multilateral aid is given by countries through international organisations, like the World Bank or IMF.

Bilateral aid is given directly by one country to another. Sometimes, it is tied aid, with conditions attached. For example, the recipient may be required to buy goods from the donor country with the aid money.

Short-term emergency relief is to cope with immediate problems caused by disasters, like earthquakes and wars.

Long-term development assistance helps people to improve their lives through education, healthcare or agricultural development.

▲ Figure 19.21 Types of aid

▼ Figure 19.22 Top ten African countries receiving ODA in 2016

	Country	ODA in $ million	Population in millions
1	Ethiopia	4,074	112
2	Nigeria	2,501	201
3	Tanzania	2,318	58
4	Kenya	2,189	53
5	Egypt	2,130	100
6	Democratic Republic of Congo	2,107	87
7	Morocco	1,992	36
8	Uganda	1,757	42
9	South Sudan	1,590	11
10	Mozambique	1,531	30

▲ Figure 19.23 Aid paid for this mosquito net in Nigeria

How did Nigeria get into (and out of) debt?

During the 1980s and 1990s, many low-income countries like Nigeria faced a debt crisis. They were unable to repay their debts without cutting essential government spending (see Figure 19.24). Spending cuts in low-income countries can be a matter of life or death.

It was impossible for the low-income countries to escape from the cycle of debt repayment. Although millions of US dollars were repaid, the total amount would have been impossible to ever repay. So, debt continued to increase. In 2005, leaders of the world's richest countries finally agreed to **debt relief** for 39 of the world's most highly indebted poor countries (HIPCs), which included Nigeria. Some, or all, of the low-income countries' debt was cancelled, so it no longer had to be repaid.

Does Nigeria still need aid or not?

Since 2005, the Nigerian economy, like the economies of many other African countries, has been growing. Nigeria moved from being a low-income country (LIC) to a newly emerging economy (NEE). Given this achievement, does Nigeria still need aid?

In 2018, the UK gave £300 million in aid to Nigeria but this figure is declining. Critics point out that Nigeria now funds its own space programme so, surely, it does not need aid. The UK government's response is that Nigeria's space programme is about investment in weather satellites that will help to improve food production. While Nigeria's economy is growing, it should not be forgotten that 60 million people still live below the poverty line on less than US$1.25 a day The largest chunk of UK aid went to development programmes in North East Nigeria to help counteract the influence of Boko Haram in the region.

Timeline (Figure 19.24):

- **1960** — Nigeria gains independence
- The economy is booming. Nigeria invests in large infrastructure projects, like roads
- **1970**
- An economic slump means Nigeria borrows money to pay for public services. The country gets into debt
- **1980**
- Debt begins to spiral out of control. Public services have to be cut
- **1990**
- Nigeria becomes Africa's most indebted country, with $36 billion of debt
- The UN sets Millennium Development Goals to eradicate extreme poverty by 2015
- **2000**
- Leaders of the world's rich countries agree to cancel the debt of HIPCs like Nigeria
- Nigeria's economy, along with other African countries is growing again. It is now the largest African economy
- **2010**

▲ Figure 19.24 Nigeria's debt timeline

→ Activities

1 Study Figures 19.21 and 19.23.
   a) Into which category of aid does the mosquito net project fall? Give reasons.
   b) Apart from saving lives, what other benefits could the project have for Nigeria? Make a list.

2 Study Figure 19.24. Explain the terms (a) debt crisis and (b) debt relief.

3 Which do you think is the most effective way of helping the Nigerian economy to grow – aid or debt relief? Give reasons to justify your answer.

4 Does Nigeria still need aid? Argue the case, using evidence either for or against giving aid to Nigeria.

Geographical skills

Study Figure 19.22. It lists the top ten African countries in order of the ODA they receive.

1 Calculate the amount of ODA per person for each country. To do this, divide ODA by population. Write a new list of countries in order of ODA per person. Which country is top of the list? Where does Nigeria come?

2 Find out more about the country at the top of your list. Why might it receive more ODA? Why might Nigeria receive less?

⭐ KEY LEARNING

➤ How quality of life in Nigeria has improved

➤ How the improvements are connected with economic development

➤ Why economic migrants risk their lives to leave Nigeria

Nigeria: development for all?

How has quality of life in Nigeria improved?

Quality of life can be measured using the UN's Human Development Index (HDI) (see Section 17.2). It is a combined measure of three important aspects of people's quality of life:

■ life expectancy as a measure of health
■ years of schooling as a measure of access to education
■ gross national income (GNI) per capita as a measure of wealth (Figure 19.25).

Nigeria is ranked 152 out of 187 countries by its HDI, which puts it in the low category of human development. However, over a longer period, since 1980, there have been significant improvements in life expectancy, years of schooling and GNI per capita in Nigeria.

below the average. There are large differences between:

■ north and south of the country
■ urban and rural areas
■ educated and uneducated people.

How are the improvements connected with economic development?

Nigeria's improved quality of life is connected with the country's economic development. With the new jobs that come with development, people are able to earn more money to pay for the things they need. The government also earns more money through taxes (Figure 19.26).

However, the benefits of economic development are not equally shared. The data in Figure 19.25 are averages. What they do not tell us is how many people are above or

▼ Figure 19.25 Changes in quality of life in Nigeria since 1980

Year	Life expectancy	Expected years of schooling	GNI per capita	HDI
1980	45.6	6.7	4,259	
1985	46.4	8.6	3,202	
1990	46.1	6.7	2,668	
1995	46.1	7.2	2,594	
2000	48.6	8.0	2,711	
2005	48.7	9.0	3,830	0.466
2010	51.3	9.0	4,716	0.492
2015	53.0	10.0	5,527	0.527
2017	53.9	10.0	5,231	0.532

▼ Figure 19.26 The benefits of economic development

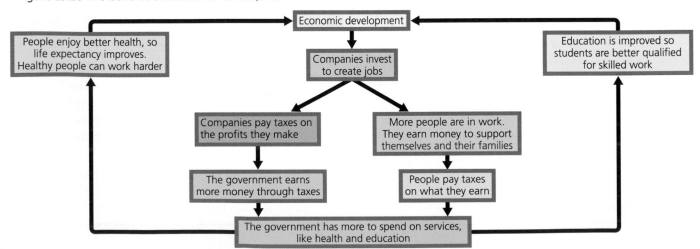

Why do economic migrants risk their lives to leave Nigeria?

In recent years, over a million migrants from Africa and the Middle East have crossed the Mediterranean Sea in makeshift boats to reach Europe; tens of thousands of them came from Nigeria. Nigerian migrants are driven by several factors, including **climate change**, extreme poverty, **food insecurity** and the actions of Boko Haram in the north. Some migrants are simply economic migrants, seeking a better quality of life in Europe.

▲ Figure 19.27 The route for migrants from Nigeria to Europe

The journey from Nigeria to Europe is long and hazardous (Figure 19.27). It involves a journey across the Sahara Desert to Libya, a country that has recently been in a civil war. In Libya, migrants pay hundreds of pounds to sail across the Mediterranean in boats that are often unfit for the voyage (Figure 19.28). Since 2014, over 3,000 migrants have died each year making the crossing.

The aim for most Nigerian migrants is to earn enough money in Europe to send home to support their families and, perhaps, one day to return to Nigeria to enjoy a better life with the money they have earned.

▲ Figure 19.28 Migrants on a boat in the Mediterranean Sea

Geographical skills

1 Study Figure 19.25. Draw suitable graphs to show changes in quality of life in Nigeria, using the data in the table. Draw graphs to show:
   a) life expectancy
   b) years of schooling
   c) GNI/capita
   d) HDI.

2 a) Write a sentence about each graph to describe what it shows about changes in quality of life in Nigeria.
   b) Explain why HDI in Nigeria increased from 2005.

→ Activities

1 Study Figure 19.26.
   a) Explain how economic development can lead to improvements in quality of life.
   b) Which groups are most likely to benefit from economic development in Nigeria:
      ■ north or south
      ■ urban or rural
      ■ educated or uneducated?
      In each case, give reasons.

2 Study Figures 19.27 and 19.28.
   a) What are the risks of a migrant's journey from Nigeria to Europe?
   b) Suggest why many Nigerian migrants believe it is worth the risk.
   c) What does this tell you about economic development in Nigeria?

3 The issue of migration from Africa to Europe is never far from the news. Find out what has happened recently, in the last year.
   ■ Do people still migrate across the Mediterranean?
   ■ Where are they from? How many are Nigerian?
   ■ Why are they migrating?
   ■ What measures have been taken to reduce migration and/or make it safer?

20 Economic change in the UK

➤ How the industrial structure of the UK has changed
➤ The effects of globalisation on the UK economy

The changing UK economy

How has the industrial structure of the UK changed?

The type of work you will do when you leave school or university is likely to be different from the work your parents do. And the jobs they do are probably different from what your grandparents did. The **industrial structure** of the UK – the types of work people do – is always changing (Figure 20.1).

Back in 1841, at the height of the Industrial Revolution in the UK, more people worked in manufacturing than services. Almost a quarter of the workforce still worked in primary industry.

During the twentieth century, the UK industrial structure changed.

■ The primary sector (agriculture, mining and fishing) declined mainly due to the increased use of machinery.

■ Since the 1960s, the manufacturing sector also declined, due to mechanisation and competition from abroad.

■ The service sector increased with the growth of public services and financial services.

■ Since the 1980s, the new knowledge sector (research and development) has grown.

By 2011, 80 per cent worked in services, 9 per cent in manufacturing and just 1 per cent in agriculture. However, industrial structure varies around the country, with areas specialising in different industries (Figure 20.2).

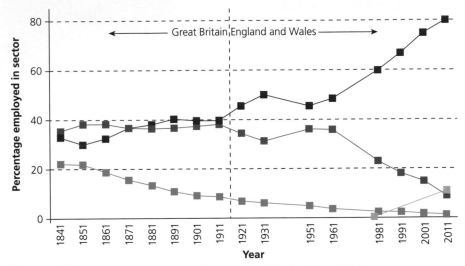

▲ Figure 20.1 The changing industrial structure of England and Wales

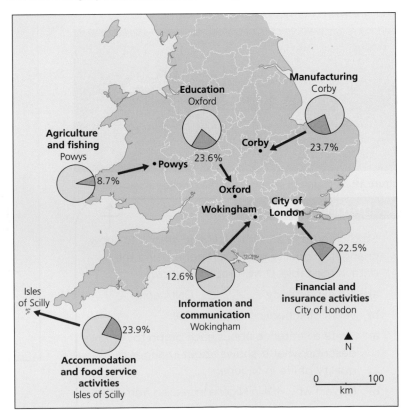

➤ Figure 20.2 Places around the UK specialise in different industries

What impacts does globalisation have on the UK economy?

The UK economy is changing due to globalisation (see Figure 20.3). This is the way business, ideas and lifestyles spread rapidly around the world. For example, more businesses in the UK are now owned by companies, based in other countries while, in the same way, more British companies own businesses in other countries.

It would be almost impossible for the UK to be isolated from the global economy. We have been trading with the rest of the world for centuries. For the UK economy to thrive, we need to be part of the global economy.

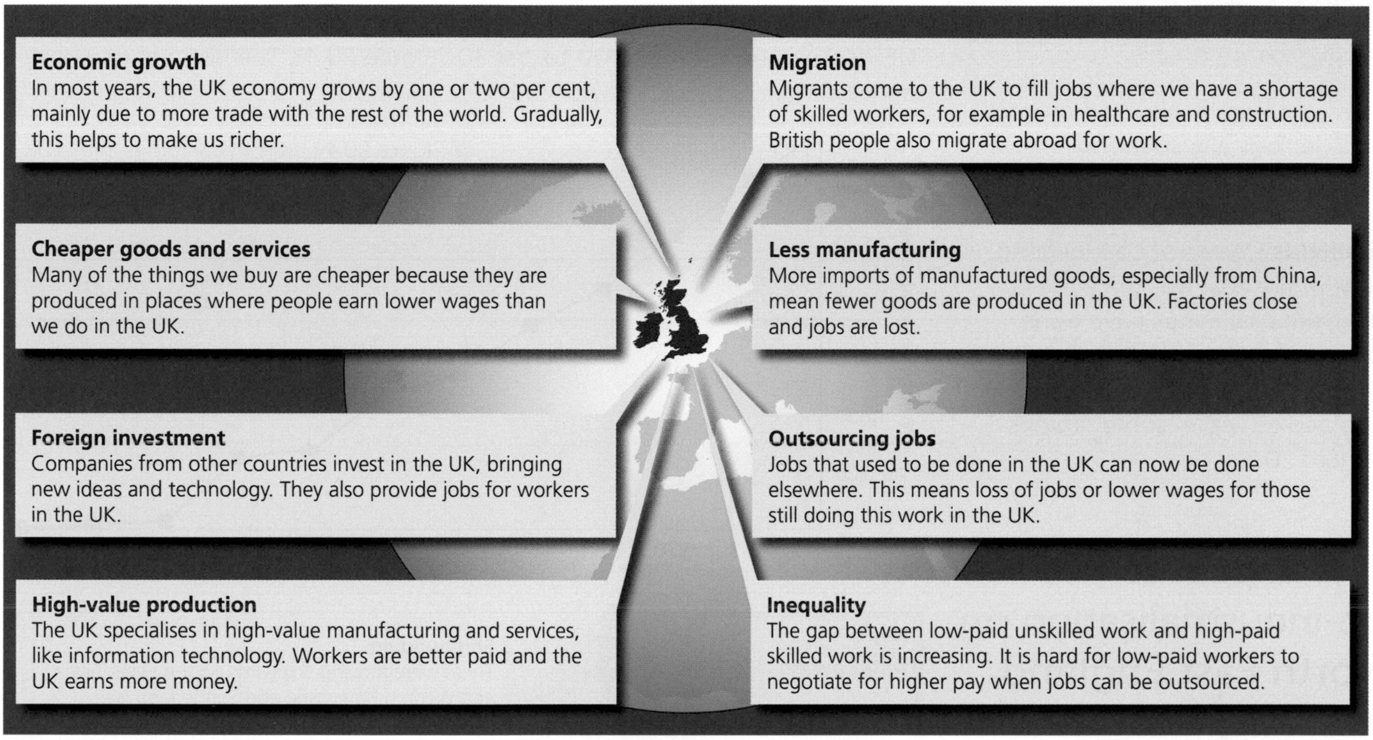

Economic growth
In most years, the UK economy grows by one or two per cent, mainly due to more trade with the rest of the world. Gradually, this helps to make us richer.

Migration
Migrants come to the UK to fill jobs where we have a shortage of skilled workers, for example in healthcare and construction. British people also migrate abroad for work.

Cheaper goods and services
Many of the things we buy are cheaper because they are produced in places where people earn lower wages than we do in the UK.

Less manufacturing
More imports of manufactured goods, especially from China, mean fewer goods are produced in the UK. Factories close and jobs are lost.

Foreign investment
Companies from other countries invest in the UK, bringing new ideas and technology. They also provide jobs for workers in the UK.

Outsourcing jobs
Jobs that used to be done in the UK can now be done elsewhere. This means loss of jobs or lower wages for those still doing this work in the UK.

High-value production
The UK specialises in high-value manufacturing and services, like information technology. Workers are better paid and the UK earns more money.

Inequality
The gap between low-paid unskilled work and high-paid skilled work is increasing. It is hard for low-paid workers to negotiate for higher pay when jobs can be outsourced.

▲ Figure 20.3 The impacts of globalisation in the UK

→ Activities

1 Think about jobs in your family. Classify them into primary, secondary or tertiary jobs.
- What job do you hope to do?
- What jobs do the adults in your family do?
- What jobs did your grandparents do?

From the three generations in your family, is there any evidence that jobs in the UK are changing? You could carry out a class survey to get a bigger sample.

2 Study Figure 20.1.
a) What percentage of people worked in agriculture, manufacturing and services in 1841?
b) Describe the change in industrial structure from 1841 to 2011.

3 Study Figure 20.2. Most people in the UK now work in services.
a) Make a list of the service jobs people do in different parts of the UK.
b) Think of at least five more service jobs to add to your list. You could include jobs in your area or look at London in Figure 15.22 on page 233 for more ideas.

4 Study Figure 20.3.
a) Classify the impacts of globalisation into benefits and problems. Draw a table to list them.
b) Overall, do you think globalisation is good or bad for the UK? Give reasons for your opinion.

⭐ KEY LEARNING

➤ How traditional industries have declined in the UK

➤ How government policies have affected the UK economy

➤ How the government has responded to de-industrialisation in the north east

They say 'Britain was built on coal', and, geologically speaking, it is almost literally true. All around the country, from Kent to Cumbria, are **coalfields** – areas of coal-bearing rock where coal was once mined. Now, the coal mines have gone. Other manufacturing industries, like shipbuilding, textiles and steel have declined as well in some regions (Figure 20.5).

The impact of de-industrialisation on North East England

North East England was one of the first industrial regions in the UK at the start of the Industrial Revolution. It also became one of the first to experience de-industrialisation with the closure of the coal mines and shipyards.

The impact on towns like Easington Colliery ('colliery' is another name for a mine) has been devastating. The town of Easington grew around its coal mine (Figure 20.6), and when the mine closed in 1993, over a thousand men were left unemployed. Over twenty years later, the town has not recovered. Unemployment is still high and people are on low incomes, so businesses in the town struggle to survive.

➤ **Figure 20.5** Industry in North East England in the 1960s

De-industrialisation and government policies

How have traditional industries declined in the UK?

De-industrialisation is the process of decline of traditional heavy industries. Nothing illustrates the story of de-industrialisation in the UK better than coal mining. During the twentieth century, the number of coal mines in the UK declined from over 3,000 to just 30 (Figure 20.4). The last working coal mines in the UK closed down in 2015.

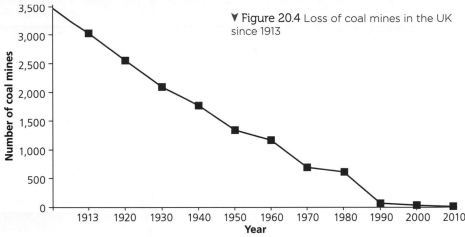

▼ **Figure 20.4** Loss of coal mines in the UK since 1913

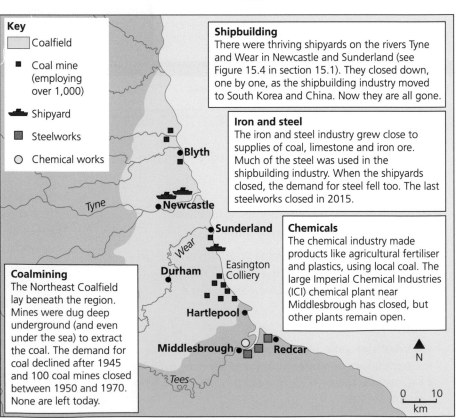

Key
- ▨ Coalfield
- ■ Coal mine (employing over 1,000)
- ⚓ Shipyard
- ▧ Steelworks
- ○ Chemical works

Shipbuilding
There were thriving shipyards on the rivers Tyne and Wear in Newcastle and Sunderland (see Figure 15.4 in section 15.1). They closed down, one by one, as the shipbuilding industry moved to South Korea and China. Now they are all gone.

Iron and steel
The iron and steel industry grew close to supplies of coal, limestone and iron ore. Much of the steel was used in the shipbuilding industry. When the shipyards closed, the demand for steel fell too. The last steelworks closed in 2015.

Chemicals
The chemical industry made products like agricultural fertiliser and plastics, using local coal. The large Imperial Chemical Industries (ICI) chemical plant near Middlesbrough has closed, but other plants remain open.

Coalmining
The Northeast Coalfield lay beneath the region. Mines were dug deep underground (and even under the sea) to extract the coal. The demand for coal declined after 1945 and 100 coal mines closed between 1950 and 1970. None are left today.

How have government policies affected the UK economy?

The government has played an important role in shaping the UK economy since 1945, responding to both globalisation and de-industrialisation:

- **1945–79** – the government **nationalised** many industries, creating new, state-run companies such as the British Steel Corporation. This helped to support declining industries and protect jobs. However, low productivity and competition from abroad eventually led to factory closures and loss of jobs. During the 1970s, there were workers' strikes and unrest in protest at the closures.

- **1979–2010** – the government **privatised** many of the nationalised industries, believing that private companies could run them better. Companies tried to increase productivity, leading to even more factory closures and job losses. Joint investment between government and the private sector led to the regeneration of former industrial areas, such as London's Docklands.

- **2010 to present day** – the government has tried to rebalance the economy to reverse the loss of manufacturing jobs, particularly with jobs in new, hi-tech industries. They have invested in new transport infrastructure such as London's Crossrail and HS2 (see Section 20.7) to encourage more **private investment**.

▲ Figure 20.6 Easington Colliery when its mine was still open

How has the government responded to de-industrialisation in the north east?

Successive governments have tried different strategies to revitalise North East England, such as:

- investment in nekitna time lagegew infrastructure, including roads and industrial estates

- encouraging investment from large, transnational companies. Nissan, the Japanese car manufacturer opened a new car plant near Sunderland in the northeast in 1986. It now employs 7,000 people

- setting up a regional development agency in 1999, which was replaced in 2012 by a local enterprise partnership which supports businesses, improves skills and plans for economic growth.

→ Activities

1 Define these terms in your own words:
   - de-industrialisation
   - globalisation.

2 Study Figure 20.4. Describe how the number of coal mines in the UK changed from 1913 to 2010. Include numbers and dates from the graph.

3 Study Figure 20.5.
   a) Outline the main industries in the north east in the 1960s. Explain how the industries were linked with each other.
   b) To what extent has the north east become de-industrialised? Which industries have gone and which remain?

4 Read the information about Easington Colliery and study Figure 20.6. Create a flow diagram to show the impact of de-industrialisation on Easington.

a) Write a list of the problems the town faces, arranged on a page in your book, like this:

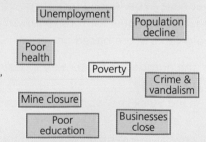

b) Draw lines to link problems that are related to each other, for example mine closure and unemployment. Try to link all the problems to at least one other.

c) Explain the links by annotating on the lines you have drawn (for example, people were made unemployed by the mine closure).

5 Suggest how people in Easington would have been affected by the three phases of government policy from 1945 to the present day.

⭐ KEY LEARNING

➤ What a post-industrial economy looks like

➤ Where most economic growth is found in the UK

➤ How the M4 corridor contributes to the economy

Towards a post-industrial economy

What is a post-industrial economy?

A **post-industrial economy** is one where manufacturing industry has been replaced by tertiary and quaternary jobs. A new sector of the UK economy that is growing rapidly in the twenty-first century is the **quaternary sector**.

The quaternary sector is sometimes described as the 'knowledge economy' because it involves research and the development of new ideas. This includes **information technology**, biotechnology and new creative industries.

The development of information technologies has transformed people's working lives and economic development in the UK.

- Computers can store and process vast amounts of information very quickly.
- Mobile devices, such as smartphones, enable information to be accessed almost anywhere.
- Satellites and the internet promote the flow of information.
- The internet and computers enable people to work from home and be self-employed.
- The UK is one of the top IT countries in the world, attracting investment from overseas companies.

▼ Figure 20.7 Ten UK cities with the most potential for growth

Rank	Cities (outside London)	Potential growth score
1	Cambridge	175
2	Reading	146
3	Manchester	131
4	Bristol	129
5	Oxford	128
6	Brighton & Hove	127
7	Milton Keynes	123
8	Leeds	114
9	Warrington	111
10	Nottingham	107

Where is most economic growth found?

Economists have identified a list of cities outside London that have experienced recent growth and have the greatest potential for future growth (Figure 20.7). Cities are given a score based on:

- the number of quaternary industries with potential for growth
- a highly skilled workforce, educated to degree level or above
- new, start-up businesses with the potential to grow larger
- good transport connections, including road, rail and air.

These cities are often the focus of **growth corridors**, following major transport routes, where the fastest economic growth is happening (Figure 20.8). While this map shows England, similar corridors could be identified in other parts of the UK, around Glasgow, Edinburgh, Cardiff and Belfast.

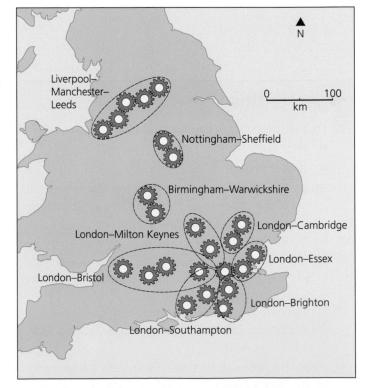

▲ Figure 20.8 Growth corridors in England

How does the M4 corridor contribute to the economy?

The M4 corridor, from London to Bristol, has become home to **hi-tech industry** over the past 30 years (Figure 20.9). Many companies, like Microsoft, Sony and Vodaphone, are based there, usually in **business parks** (Figure 20.10). It is estimated that the M4 corridor produces eight per cent of the UK's economic output, as much as Manchester and Birmingham combined.

Recently, businesses in the M4 corridor have begun to be sucked into London. Vodaphone moved in 2009 and Google has now opted to move too. The factors drawing companies to London include:

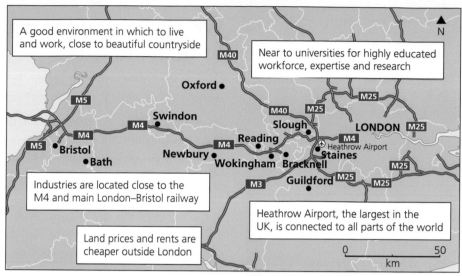

▲ Figure 20.9 The M4 corridor from London to Bristol

- the attraction of urban living for a young workforce
- the proximity of similar companies to swap ideas and workers
- new businesses require less space than the first generation of hi-tech industry.

→ Activities

1 What do you understand by the term, 'post-industrial economy'? Try to use your own words.

2 Choose a workplace you know something about. It could be your school! Write a list of all the ways in which information technology has changed the way people work in the past 20 years.

3 Study Figure 20.7.
   a) Which cities have universities that you have heard of? Write a list.
   b) Explain why quaternary industries are often linked with university cities.

4 Study Figure 20.8.
   a) Describe the distribution of growth corridors in England.
   b) Suggest why London is so dominant in a post-industrial economy. (Hint: look back at Sections 15.4 and 15.5.)

5 Study Figures 20.9 and 20.10. Design an advert for a business park in the M4 corridor to attract new business. Think particularly about what your location offers that London cannot provide.

▲ Figure 20.10 A business park near Reading

Fieldwork: Get out there!

How are the economic activities in this area changing?

How would you investigate changing jobs in your area? You will need:

- **primary data** (that you collect yourself through fieldwork) about the workplaces that are there now
- **secondary data** (that you obtain from another source) about workplaces that were there in the past.

You could classify jobs, both past and present, into primary, secondary, tertiary and quaternary sectors. How has the proportion changed?

Cambridge: a hi-tech hub

Why is Cambridge growing as a hub for hi-tech industry?

Cambridge is probably best known for having one of the top universities in the world. Until recently, it was less well known for its industry, but this is changing.

Cambridge is fast emerging as one the UK's main hubs for hi-tech industry. Over 1,500 information technology and **biotechnology** companies are now based there. The city lies about 80 kilometres north of London, close to the M11 (Figure 20.11), in one of the UK's growth corridors (see Section 20.3).

Hi-tech industry in Cambridge began with the Cambridge Science Park, in the north of the city (Figure 20.12, square 4661), which was opened in 1970 by Trinity College. Trinity was the first of several Cambridge University colleges to make links with industry.

▼ Figure 20.11 The London–Cambridge growth corridor on the M11

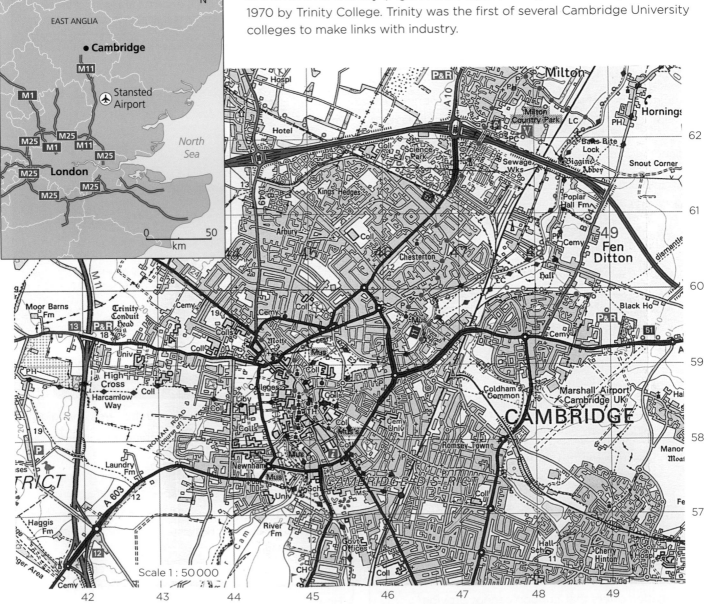

Scale 1 : 50 000

▲ Figure 20.12 OS map extract of Cambridge, 1:50,000. © Crown copyright and database rights 2020 Hodder Education under licence to OS

What are the advantages and disadvantages of Cambridge as a location for industry?

Many of the hi-tech companies in Cambridge began as small start-up businesses, formed by university graduates who wanted to stay in the city when they had finished their degrees. Some of these businesses, like the biotech company Abcam, have grown into successful companies.

Abcam (named for 'antibodies Cambridge') is based at Cambridge Science Park. It produces antibodies that are used in the treatment of diseases. The company is now worth £1 billion and employs 200 staff with PhD degrees – more than some universities!

▼ Figure 20.13 Advantages and disadvantages of Cambridge as a location for industry

Advantages	Disadvantages
● Good transport links, including the M11 motorway to London and Stansted Airport ● Graduates from the university provide a highly educated workforce ● There are few traditional industries to compete for space, so rents are lower ● The city offers a good quality of life, with plenty of shops and open spaces ● There are good links between colleges and industry, helping to develop new business ideas	● The city is overcrowded and congested, making it difficult to drive or park ● House prices are high and still rising, making it expensive to live there ● Road and rail routes need to be improved to speed up connections to other cities apart from London

Geographical skills

1. Study Figure 20.12. Find Cambridge Science Park in square 4661. Draw a sketch map to show the location of the science park. Include the following features and label them on your map: Cambridge (urban area), M11, A10, ring road (A14), city centre, Cambridge Science Park.

2. Find these grid references on the map in Figure 20.12: 470621, 452583, 487585, 480623. For each location:
   a) State what you find there.
   b) Explain why it might be a factor for a company to choose Cambridge as a location for their business.

→ Activities

1. Study Figures 20.11 and 20.12.
   a) Describe the location of Cambridge Science Park in relation to:
      ■ the built-up area of Cambridge
      ■ the transport network.
   b) Explain why this would be a good location for a science park.
2. Study Figure 20.13. What would be the main advantages and disadvantages for:
   a) a university graduate looking for a job in Cambridge
   b) a large biotech company thinking of moving to Cambridge.

Fieldwork: Get out there!

Where is the best location for a new business in your area?

■ Choose a type of business that might want to locate in your area. (It does not need to be a hi-tech firm. For example, you could choose a café.)

■ Think of the main criteria your business would have for finding the best location. For example, the number of people walking past.

■ Select two or three possible locations where the business could be located.

■ Carry out fieldwork at each location to find the best one for your business. For example, you could count the number of people walking past at different times of day.

Example

➤ The impacts of the car industry on the environment

➤ How the car industry can be more environmentally sustainable

Making industry more sustainable

What impacts does the car industry have on the environment?

The car industry is one of the few large-scale manufacturing industries left in the UK. More than 1.5 million new cars are made in the UK every year and most of them at just seven giant manufacturing plants. All of these are owned by TNCs from other countries such as Nissan, Toyota and BMW.

The car industry does not enjoy the best reputation due to its impact on the environment; for example, most people know that in cities emissions from cars is one of the main causes of air pollution (see Section 15.10). Less well known are the other **environmental impacts** cars have through their lifetime, from the resources used in their manufacture to their disposal at the end of the car's life (Figure 20.14).

Fuel consumption

Most cars run on petrol or diesel, which are both obtained from oil; the cause of many environmental problems:

• Drilling for oil uses energy and can endanger ecosystems.

• Shipping oil can cause oil spills.

• As oil is used up, new sources are harder to obtain and can cause more problems.

Resources

Cars are made from a range of resources including steel, rubber, glass, plastic, paint and fabric. Manufacturing and transporting these resources also uses energy.

Air pollution

Burning petrol or diesel in cars is a major cause of air pollution. The main pollutants are carbon dioxide (the main greenhouse gas), nitrogen dioxide (a cause of breathing problems) and particulates (tiny particles, which also cause breathing problems).

Manufacture

Cars consume a lot of energy even before they are driven. It is estimated that manufacturing a car uses as much energy as the car will consume in its lifetime on the road.

Disposal

Cars end up on the scrap heap at the end of their lives. Some components like plastic are hard to recycle. Others, like the acid in batteries, can leak into the environment.

▲ **Figure 20.14** Environmental impacts of the car industry

How can the car industry be more environmentally sustainable?

You may have noticed electric or hybrid (combined electric and petrol-powered) cars on the road (Figure 20.15). They are a sign that the car industry is becoming more environmentally aware. Car companies are responding to demand from consumers for more sustainable cars and to stricter government regulations.

By most measures, the car industry in the UK is becoming more sustainable (Figure 20.16). The amount of energy and water used in the production process has declined. There has been a dramatic fall in the amount of waste going to **landfill sites** at the end of a car's life. And, importantly, the average CO_2 emissions from new cars are falling.

▲ **Figure 20.15** The Nissan LEAF, an electric car produced in the UK

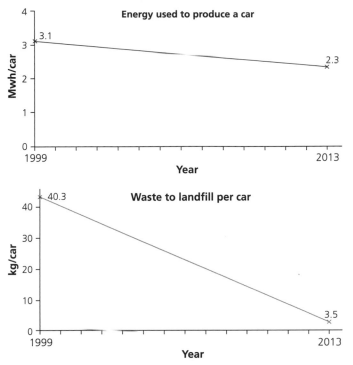

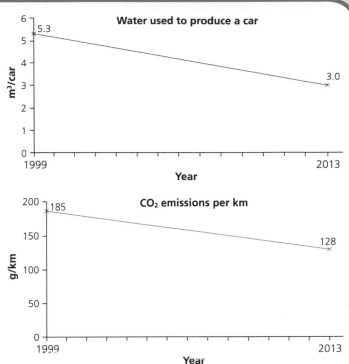

▲ Figure 20.16 How the car industry has become more sustainable

Production begins at Nissan: 1986

Number of people employed: 7,000

Estimated number of jobs created in the UK by Nissan's car plant: 28,000

Number of cars produced in a year: 500,000 (one-third of all cars produced in the UK)

Models produced at the plant: Nissan Note, Nissan Qashqai, Nissan LEAF, Compact Infiniti

Amount of energy generated by wind turbines: 7%

▲ Figure 20.17 The Nissan car manufacturing plant near Sunderland. Note the wind turbines (top left)

→ Activities

1 Study Figure 20.14 and the rest of the information on pages 298–99. Which of the five environmental impacts of the car industry would be:
   a) easiest for the car manufacturer to reduce
   b) easiest for the car owner to reduce
   c) most difficult to reduce?

   In each case, give reasons for your answer.

2 Look at Figure 20.15. The Nissan LEAF is an example of a more sustainable car. In what ways:
   a) does it have a reduced environmental impact
   b) does it still have a harmful environmental impact?

3 Study Figure 20.16. Describe the four ways in which the UK car industry has become more sustainable. Write four sentences, including data from each of the graphs.

4 Study Figure 20.17. Is it possible for a car plant both to contribute to the economic development of an area and to be more sustainable? Give evidence from Nissan's car plant to support your opinion.

➤ How rural areas are changing
➤ What changes in an area of population growth
➤ What changes in an area of population decline

Rural changes

How are rural areas changing?

Most people in the UK live in urban areas, but 19 per cent still live in rural areas. Although they might not look crowded, the population of most rural areas in the UK is actually growing (Figure 20.18) as a result of **counter-urbanisation**. People leave cities to live in the countryside for a better quality of life. The population of urban areas is growing faster, but this is mainly the result of natural increase and immigration.

Around major cities in the UK is the greenbelt – green open space in which further building development is not allowed (see Section 15.9). Within the greenbelt, and just beyond, are towns and villages that are desirable places to live and **commute** to work in the city. Without the greenbelt these areas might have experienced even greater population growth (Figure 20.19).

While the greenbelt has been successful in preserving rural areas, it has also limited the amount of land available for building new homes. As there is a housing shortage in the UK, there is pressure on the government to allow more building in the greenbelt.

Even **sparsely populated** rural areas furthest from cities have experienced population growth. Some of these areas are popular with tourists and second home owners, particularly in **national parks** like the Lake District.

There is high demand to live in both the greenbelt and national park areas. This has pushed up house prices, making homes for local people unaffordable. People are forced to rent locally or to move away.

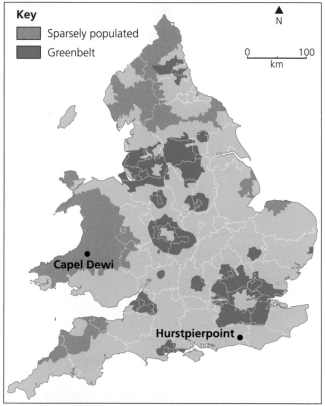

▲ Figure 20.19 Sparsely populated and greenbelt areas in England and Wales

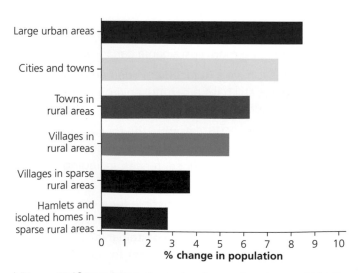

▲ Figure 20.18 Population change in urban and rural areas, 2001–11

▲ Figure 20.20 A house in the village of Hurstpierpoint, Sussex

What changes in an area of population growth?

The greatest population pressure on rural areas is in South East England, where many people want to live in rural surroundings but work in London (Figure 20.22). Population growth in these areas results in social and economic changes.

> The large village of Hurstpierpoint has a high street full of shops and a good pub. Trains from the nearest station, which is two miles away, take 50 minutes to journey to London. An annual season ticket costs £3,504. Properties range from two-bedroom cottages valued at £230,000, to modern four-bedroom houses selling for over £500,000 and period family houses from £650,000.

▲ Figure 20.22 An advert for property in the village of Hurstpierpoint, Sussex

What changes in an area of population decline?

The village of Capel Dewi is in a sparsely populated area of mid-Wales. During the twentieth century, its population declined as young people moved away to find work, leaving older people to continue farming. One by one, shops in the village closed as the number of customers fell until, eventually, none were left.

In 2012, the local community got together to open a new shop. It is a convenience store for the village run by volunteers. It saves people making long car journeys to supermarkets in town and is most useful for older people who do not drive.

▼ Figure 20.21 Social and economic changes due to population growth

Social changes	Economic changes
• Population growth helps to maintain demand for rural services, like shops or schools that otherwise might close. • If a village has a high proportion of commuters it may feel dead during the day, losing its sense of community. • Car-owning commuters do not need public transport so services may be reduced, affecting local people. • Local people may feel resentful towards newcomers.	• Population growth brings new energy. Newcomers are more likely to start their own business and employment opportunities may rise. • Some services such as shops may close if commuters do not use them, or only use them on weekends. • Newcomers are often wealthy and this helps to push up house prices. • House prices may be unattainable for most local people.

▲ Figure 20.23 Capel Dewi

→ Activities

1 Study Figure 20.18.
   a) Describe the different rates of population growth in urban and rural areas.
   b) Most rural areas of the UK have growing populations. Explain why.

2 Study Figure 20.19. Describe the distribution of (a) greenbelt and (b) sparsely populated areas.

3 Locate Capel Dewi and Hurstpierpoint on the map in Figure 20.19.
   a) Compare the locations of the two villages.

   b) How do their locations help to explain the characteristics of the two villages?

4 Study Figure 20.22. What would be (a) the advantages and (b) the disadvantages for someone moving to the village and commuting to London. Think of at least three of each.

5 Create your own table of social and economic changes in rural areas with population decline, like Capel Dewi.

➤ How government investment in transport is changing

➤ The arguments for high-speed rail

➤ Supporters of and objectors to high-speed rail

Improvements in transport infrastructure

How is government investment in transport changing?

There are over 35 million vehicles on the roads in the UK and this grows each year. Despite government investment over many years to improve the road network, **traffic congestion** remains a serious problem and journey times become slower rather than quicker (Figure 20.24).

In the twenty-first century, government investment in an older form of transport – the railways – has increased to relieve congestion on the roads. In particular, there are plans for a new high-speed rail network in the UK. Already, High Speed 1 runs from London to Kent on the same route as the Eurostar from London to Paris. Now, there are plans for High Speed 2 (HS2) from London to Birmingham, then on to Manchester and Leeds (Figure 20.25). The government's most recent road investment strategy, in 2014, is to increase road capacity with over 100 new road schemes by 2020 and over 100 miles of new lanes added to motorways. Additionally, there are plans to improve the M4 motorway by making it a 'smart motorway'. This involves

▲ Figure 20.24 Congestion on the M25 motorway

helping reduce congestion, for example by varying the speed limits to keep traffic moving smoothly.

What are the arguments for high-speed rail?

The main arguments in favour of high-speed rail, including HS2, are:

■ It will take the pressure off the existing road and rail networks, encouraging more people to travel by rail.

■ It will reduce journey times between cities (Figure 20.26), so people spend less time travelling.

■ It will bring economic benefits to the Midlands and northern England where de-industrialisation has led to a loss of jobs.

Even though HS2 is not planned to start running until 2031, there are already ideas for High Speed North; a rail route from Manchester to Leeds, to link cities in northern England.

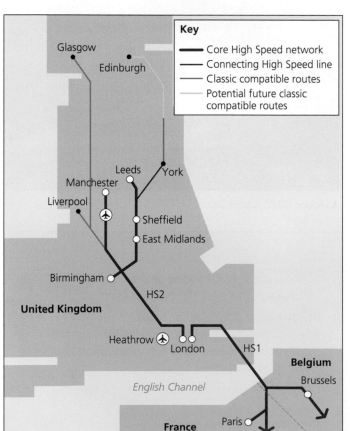

Key
- ▬▬ Core High Speed network
- ▬ Connecting High Speed line
- ▬ Classic compatible routes
- ▬ Potential future classic compatible routes

Glasgow
Edinburgh
Leeds · York
Manchester
Liverpool ✈
Sheffield
East Midlands
Birmingham
HS2
United Kingdom
Heathrow ✈ London   HS1
English Channel
Belgium
Brussels
France   Paris

▲ Figure 20.25 The proposed route for HS2

▼ Figure 20.26 Journey times for HS2 compared with current journey times

Journey	Current rail journey time	HS2 journey time
London – Birmingham	1hr 24 mins	49 mins
Birmingham – Manchester	1hr 8 mins	49 mins
Birmingham – Leeds	2 hrs	57 mins
London – Manchester	2hrs 8 mins	1hr 8 mins
London – Leeds	2hrs 12 mins	1hr 22 mins
London – Glasgow	4hrs 30 mins	3hrs 30 mins
London – Edinburgh	4hrs 30 mins	3hrs 30 mins

Who are the supporters of and objectors to high-speed rail?

The plan for HS2 has proved to be controversial (Figure 20.27).

Supporters of the plan include:

- the main UK political parties
- large cities, including Birmingham, Manchester and Leeds
- businesses in those cities
- the Scottish government.

Objectors to the plan include:

- county councils on the route, like Buckinghamshire and Oxfordshire
- residents living close to the route
- environmental organisations and the Green Party
- taxpayers' groups.

In 2019, the government called for a review of the whole project, to decide whether it will go ahead or not, even though billions of pounds have already been spent.

What supporters say	What objectors say
• It will create thousands of jobs in the Midlands and northern England. • It is estimated HS2 will help to generate £40 billion for the UK economy. • It will increase the number of rail passengers and make transport more sustainable. • It will also reduce the number of people who fly between UK cities. • It will be a faster way to travel between cities. • It will be carbon-neutral because it will reduce journeys that use other transport.	• It is more likely to create jobs in London and people will commute there instead. • The cost of HS2 is estimated at over £80 billion and it is difficult to predict how much money it will generate. • Existing rail routes could be improved to increase the number of passengers. • The number of people flying within the UK is already falling. • People do not want to travel any faster. Intercity routes are already fast. • It will increase carbon emissions because high-speed trains use more power.

▲ Figure 20.27 Supporters of and objectors to HS2

→ Activities

1 Look at Figure 20.24. Should the government invest more money in roads or railways? Give reasons for your answer.

2 a) Rank the reasons for investing in high-speed rail in order of importance.
   b) Explain why you ranked them in this order.

3 Study Figures 20.25 and 20.26. Draw your own map of the HS2 route. Label HS2 journey times on the correct sections of the route.

4 Study Figure 20.27. Give reasons why each of the following either support or object to the plan for HS2.
   a) Birmingham City Council
   b) A business in Birmingham
   c) Oxfordshire County Council
   d) A resident living near the route

5 Either write a short newspaper article or present a short TV report about the HS2 controversy. Give arguments on both sides, for and against HS2.

KEY LEARNING

- Where the main ports and airports in the UK are located
- Why a new port has opened on the Thames Estuary near London
- Whether Heathrow airport should expand or not

Ports and airports

Where are the main ports and airports in the UK located?

The locations of ports and airports in the UK are very different, as you might expect, because they serve different purposes (Figure 20.28).

Ports are found at coastal and **estuary** locations all around the UK. They are used mainly for the import and export of bulky raw materials as well as manufactured goods, usually in large metal containers. These are then transported by road or rail around the UK. Some ports are also for passengers travelling on ferries or cruise ships.

Airports are located close to major cities, especially London, and are used mainly by passengers travelling on international flights. By far the largest airport in the UK, London Heathrow also serves as a hub airport, used by passengers in transit from one country to another rather than staying in the UK. Airports are also used for the transport of less bulky high-value goods.

N
Key
Port
Airport
Forth
Edinburgh
Glasgow
Newcastle
Tees and Hartlepool
Belfast
Manchester
Liverpool
Grimsby and Immingham
Birmingham
London Luton
Felixstowe
London Heathrow
London Stansted
Milford Haven
Bristol
London
Dover
Southampton
London Gatwick
0
200
km

Figure 20.28 The main ports and airports in the UK

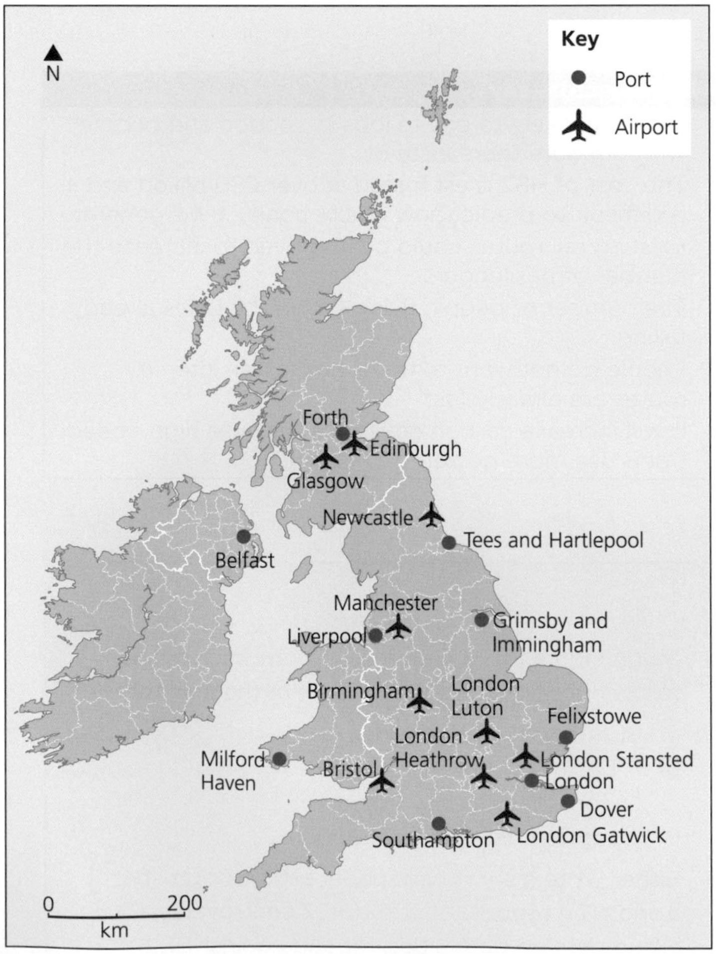

Why has a new port opened on the Thames Estuary near London?

The London Gateway on the Thames Estuary opened in 2013 (see Figure 20.29). It is the first port in London to open since the old docks closed in the 1970s (see Section 15.5) and there are plans for it to expand. Unlike the docks, which were too small for large container ships, London Gateway can accommodate the largest container ships in the world, up to 400-metres long and carrying up to 18,000 containers.

Once it is complete, London Gateway will employ 2,000 people, with another 6,000 employed at the new logistics park next to it. This is where many companies will base their distribution centres.

One advantage of London Gateway is that it brings the largest ships closer to London, the biggest market for consumer goods in the UK. It will reduce the distance lorries need to travel and help to cut carbon emissions.

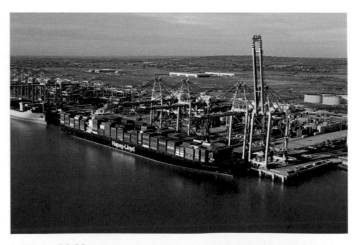

Figure 20.29 London Gateway, a new container port close to London

Should Heathrow Airport expand or not?

UK airports handle almost 200 million passengers per year. They are important for the UK's economy. Heathrow is already by far the largest airport in the UK (Figure 20.30). By 2030, it could expand even further. A new runway would be built at an estimated cost of £18.6 billion. Heathrow currently operates at almost full capacity, with 480,000 flights a year. It would be impossible to increase the number of flights on its two existing runways.

There were alternatives to expanding Heathrow (Figure 20.31). Some people thought that Manchester Airport should be expanded to boost the economy in northern England and help reduce the North–South divide (see section 20.9). However, Heathrow supporters pointed out that, unless Heathrow is allowed to expand, London would be in danger of losing its position as a leading world city.

▼ Figure 20.30 Passenger numbers at UK airports (2013)

Airport	No. of passenger (millions)
London Heathrow	73.4
London Gatwick	38.1
Manchester	21.9
London Stansted	19.9
London Luton	10.5
Edinburgh	10.2
Birmingham	9.7
Glasgow	7.7
Bristol	6.3
Newcastle	4.5

▼ Figure 20.31 Expansion of Heathrow or Manchester?

	Heathrow expansion	Manchester expansion
For	• It will help London to compete with rivals like New York and Paris. • The airport employs 76,000 people and supports a similar number of jobs in London. • Expansion would boost the UK economy by over £200 billion.	• The airport is further from the built-up area so fewer people will be affected by noise. • 22 million people live within a two-hour drive and HS2 will improve connections. • It would boost the economy of northern England.
Against	• It is already the largest emitter of CO_2 in the UK. This will increase when the airport expands. • Noise pollution will get worse for one million people who live below the flight path. • One village will be demolished and two others are threatened.	• The boost to the UK economy as a whole would not be as great as expanding Heathrow. • London would be less able to compete with rival cities. • CO_2 emissions would increase by 50 per cent if the runways double from one to two.

→ Activities

1 Study Figure 20.28.
   a) Describe the location of (i) ports and (ii) airports in the UK.
   b) Explain the different locations of port and airports.
2 Look at Figure 20.29. Suggest what benefits the new port has for (a) London and (b) the environment.
3 Study Figure 20.31. Make the case for the expansion of either Heathrow or Manchester Airport. Give points in favour of your preferred airport and points against the alternative.

Geographical skills

1 Study Figure 20.30. Complete a map to show the number of passengers at UK airports.
   a) Mark and label each airport at the correct location on an outline map of the UK.
   b) Draw proportional bars at each airport on the map to show the number of passengers.

➤ How is there a North–South divide in the UK

➤ Exceptions to the North–South divide

➤ Strategies to reduce differences between North and South

The UK's North–South divide

Is there a North–South divide in the UK?

In people's minds there has long been a **North–South divide** in the UK. Depending on where you live, you will often hear people talk about 'up north' or 'down south'. But, is there a real North–South divide in the UK, and, if so, does it matter?

Geographers have drawn a line on the UK map to show the North–South divide (Figure 20.32). North of the line:

■ are the hills and mountains of upland Britain

■ is where most manufacturing industry was located until de-industrialisation began

■ there are higher unemployment levels (Figure 20.33)

■ population is growing more slowly as people move south to find work

■ house prices are lower because there is less demand for housing.

South of the line:

■ is the flat, fertile farmland of lowland Britain

■ there was less manufacturing, so de-industrialisation has been less of an issue

■ higher employment levels are found

■ population is growing more quickly as people move here to find work

■ house prices are higher because there is more demand.

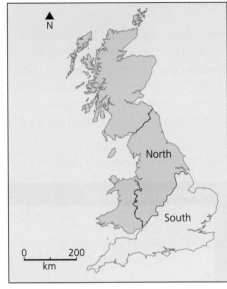

▲ Figure 20.32 The North–South divide in the UK

What are the exceptions to the North–South divide?

The North–South divide is not as simple as the map shows. There are exceptions. For example, although London has a booming economy (see Section 15.5), it still has higher levels of unemployment than other regions in the south (Figure 20.33). There are also inequalities within London (see Section 15.8).

Scotland is in the north but it has lower unemployment than other regions in the north of England. This is partly due to wealth from North Sea oil, but Scotland also has its own government, which has powers to raise and spend more money than the rest of the UK.

There are also individual cities that are exceptions to the North–South divide (Figure 20.34).

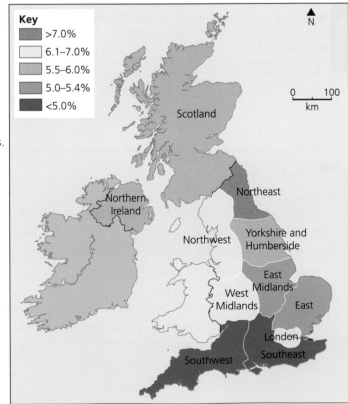

Key
- >7.0%
- 6.1–7.0%
- 5.5–6.0%
- 5.0–5.4%
- <5.0%

➤ Figure 20.33 Unemployment in UK regions, 2015

▼ Figure 20.34 Cities with the highest and lowest employment levels in the UK (UK average: 71.9)

Rank	City	Employed (%)	Rank	City	Employed (%)
1	Warrington	79.8	55	Bradford	66.4
2	Cambridge	78.9	56	Swansea	65.8
3	Swindon	78.0	57	Hull	64.8
4	Aldershot	78.0	58	Birmingham	64.2
5	Reading	77.2	59	Coventry	63.6
6	Aberdeen	77.1	60	Rochdale	62.8
7	Gloucester	76.8	61	Blackburn	62.6
8	Crawley	76.3	62	Liverpool	62.3
9	Brighton	75.5	63	Burnley	62.1
10	Ipswich	75.2	64	Dundee	61.9

What strategies can be used to reduce differences between the north and south?

Over the years, the UK government has attempted to reduce the North–South divide without much success. In the twenty-first century, the gap between the north and south has widened, with most economic growth happening in the south.

The strategies the government is trying, or hopes to try, include:

■ identifying areas of the UK that need special help, called assisted areas, to provide money for new business (Figure 20.35)
■ improving the transport infrastructure, linking cities in the north, including improvements to the M62 motorway and a proposed new high-speed rail link
■ giving more power to individual cities to take decisions on how to raise and spend their own money (as Scotland does)
■ designating 24 new enterprise zones to encourage new businesses, including low rates, simpler planning regulations and super-fast broadband.

Geographical skills

1 Study Figure 20.34.
   a) Mark and label the cities in the correct location on an outline map of the UK. You can use an atlas to help you. Use two colours to mark the cities – one for the top ten and another for the bottom ten.
   b) What pattern do you notice on the map?

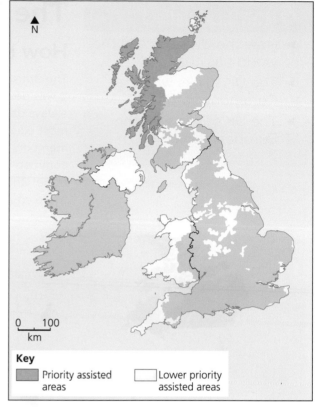

Key
Priority assisted areas
Lower priority assisted areas

▲ Figure 20.35 Assisted areas in the UK

→ Activities

1 Look at Figure 20.32. For each of the following cities, say whether they are in the north or the south: Manchester, Sheffield, Bristol, Norwich, Leeds, Birmingham, Plymouth, Cardiff, Milton Keynes. You can check in an atlas.
2 Study Figure 20.33.
   a) Describe the regional pattern of unemployment in the UK. Mention differences between the north and the south and any exceptions to the pattern.
   b) Explain the pattern. (Hint: look back at the previous Sections 20.1 to 20.3.)
3 Study Figure 20.35.
   a) Describe the pattern of assisted areas on the map.
   b) To what extent does the map support or not support the idea of a North–South divide in the UK? Give evidence from the map.
4 Does the North–South divide matter, and does it matter that most wealth and economic activity are concentrated in the south? Give reasons to support your answer.

⭐ KEY LEARNING

➤ How the UK's place in the world has changed
➤ The UK's links with the Commonweath
➤ The UK's links with the European Union (EU)

The UK's place in the world

How has the UK's place in the world changed?

The British Empire once covered about one-third of the world's land surface. It was described as 'the Empire on which the Sun never sets' because it was always daytime somewhere in the Empire. During the twentieth century, most countries in the Empire gained independence from the UK and became members of the Commonwealth (Figure 20.36). All these countries share common values, including the promotion of democracy, human rights and trade. One consequence, amongst many others, of Britain's historical role in these countries is that the English language is often used.

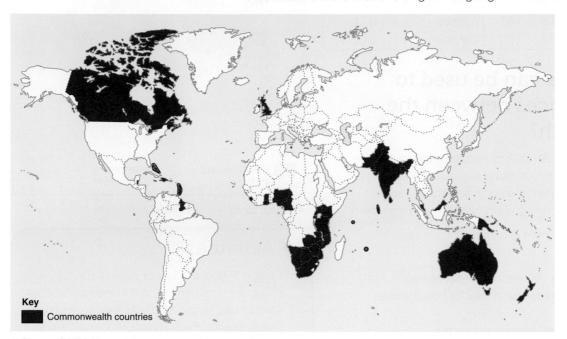

Key
■ Commonwealth countries

▲ Figure 20.36 Map of Commonwealth countries

Commonwealth countries

Africa Botswana, Cameroon, Ghana, Kenya, Lesotho, Malawi, Mauritius, Mozambique, Namibia, Nigeria, Seychelles, Sierra Leone, South Africa, Swaziland, Uganda, Tanzania, Zambia

Asia Bangladesh, Brunei, India, Malaysia, Pakistan, Singapore, Sri Lanka

The Caribbean and America Antigua and Barbuda, Bahamas, Barbados, Belize, Canada, Dominica, Grenada, Guyana, Jamaica, St Lucia, St Kitts and Nevis, St Vincent and the Grenadines, Trinidad and Tobago

Europe Cyprus, Malta, United Kingdom

Pacific Australia, Fiji, Kiribati, Nauru, New Zealand, Papua New Guinea, Samoa, Solomon Islands, Tonga, Tuvalu, Vanuatu

What are the UK's links with the Commonwealth?

The UK maintains its links with the Commonwealth through trade, culture and also migration. Many people of British descent now live in Commonwealth countries like Australia, Canada and New Zealand.

And, of course, there are many people of Asian, African and Caribbean descent now living in the UK. Migration from these countries grew in the second half of the twentieth century and still continues, often filling gaps in the UK workforce as our population gets older.

There are strong cultural and sporting links between Commonwealth countries, including the Commonwealth Games every four years. Now the UK has left the **European Union (EU)**, some people expect these links to become stronger.

What are the UK's links with the European Union?

In 1973, the UK joined the European Union (EU) (Figure 20.37). It left in 2020. Now comprised of 27 countries, the EU is one of the world's major trading blocs, with considerable political and economic influence worldwide. The EU also has its own currency – the Euro – shared by 19 countries. In 2016, people in the UK voted in a referendum to leave the EU – a decision that came to be known as 'Brexit'. Until it left the EU, the UK benefited from its support while keeping to its rules.

EU membership had many effects on the UK.

- Goods, services, capital (money) and labour could move freely between countries. Many EU citizens lived here, while many UK citizens lived there (Figure 20.38).
- European funds helped to provide support for some of the poorest regions in the UK, such as the assisted regions (see Section 20.9).
- Hundreds of thousands of people from Eastern Europe came to the UK to work in agriculture and service industries on relatively low wages. This was a key factor in the Brexit vote.
- Many EU laws and regulations affected working practices, product standards and environmental guidelines in the UK.
- The UK contributed billions of pounds each year to the EU budget.

UK citizens in other EU countries (2012)	
Spain	761,000
Ireland	291,000
France	200,000
Germany	115,000

EU citizens in the UK (2012)	
Poland	646,000
Ireland	403,000
Germany	304,000
France	136,000
Italy	133,000
Lithuania	130,000
Romania	101,000

▲ Figure 20.38 Free movement of people in the EU, allowed UK citizens to live in Europe and EU citizens to live in the UK

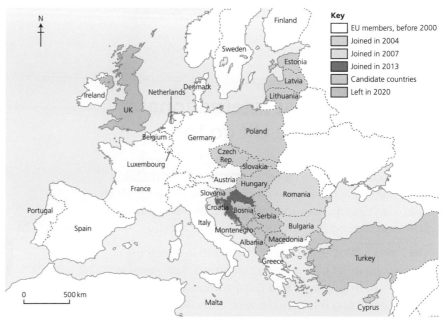

▲ Figure 20.37 Map of EU countries

Key
- EU members, before 2000
- Joined in 2004
- Joined in 2007
- Joined in 2013
- Candidate countries
- Left in 2020

Geographical skills

Study Figure 20.38. Create a map of Europe to show the movement of people between the UK and other EU countries. Use proportional arrows to show the direction and numbers of people who moved to and from the UK.

→ Activities

1 Study Figure 20.36.
   a) Describe the distribution of Commonwealth countries around the world. Include the names of continents and oceans where they are located.
   b) 'The Empire on which the Sun never sets.' Why was this a good description of the British Empire?

2 Study Figure 20.37.
   a) Describe how the EU has grown in the twenty-first century.
   b) Explain the impact this had on migration to the UK.

3 How might each of the following people have felt about the UK's membership of the EU? Would they have wanted to leave or remain? In each case, give a reason for their view.
   - A farmer in East Anglia
   - An unemployed factory worker in South Wales
   - An employee in financial services in London
   - A government minister

➤ Which countries are the UK's main trading partners

➤ Where most international flights from the UK go to

➤ The impact of the internet on our global links

The UK's global links

Which countries are the UK's main trading partners?

Most of the UK's trade has been with other countries in Europe (Figure 20.39). This will probably continue, even though the UK has left the EU, as:

■ European countries are geographically close to the UK, so transport costs are cheaper.

■ Also, European countries are among the world's wealthiest economies, so the volume of trade is greater.

However, the USA and China are also major trading partners for the UK.

Where do most international flights from the UK go to?

Heathrow is the largest UK airport, with the most international flights (Figure 20.40). The flight routes from Heathrow reflect the parts of the world with which people in the UK have the most links, as a result of business, visiting family and friends, and holidays.

They also reflect the parts of the world in which the UK has the most cultural and trade links. For example, festivals such as Diwali and Eid are known in the UK due to its strong relationship with India, Pakistan and Bangladesh.

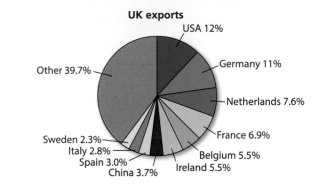

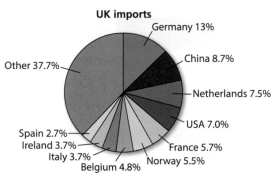

▲ Figure 20.39 The UK's main trading partners

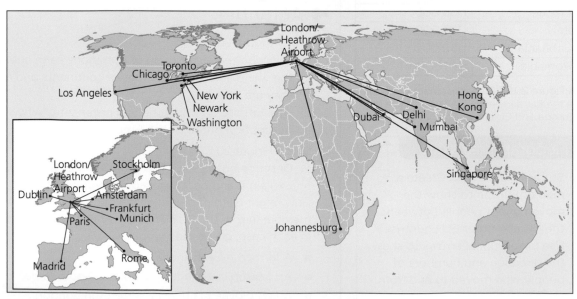

▲ Figure 20.40 Most popular international routes from Heathrow Airport

What is the impact of the internet on our global links?

The biggest change to our global links in the twenty-first century has come through the internet. Globally, the growth of the internet has become an almost unstoppable force. In 2018, there were 3.7 billion global mobile internet users, almost half of the world's population (Figure 20.41).

On average, 269 billion emails are sent and received each day. Of course, this figure will be out of date when you read this, because it increases all the time. Social media is one of the internet's biggest success stories. Facebook alone has over a billion users, meaning that if it were a country, it would be the third largest in the world.

The UK is one of the world's most connected countries. Ninety per cent of people in the UK use the internet now, compared to just 27 per cent in 2000.

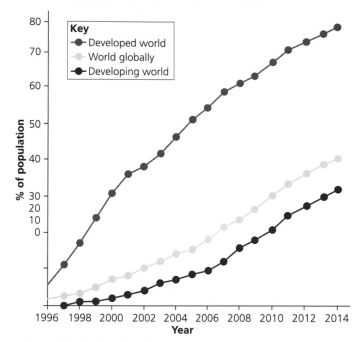

▲ Figure 20.41 The global growth of the internet

→ Activities

1 Map your personal global links on a world map. Include all the links you can think of, including;
- family and friends – where they live outside the UK
- clothing and other items – where they were made
- holidays – where you have been
- the internet – where you have linked with people by email or social media.

Mark and label each country you are linked to. Give the map a key to show different types of links.

2 Study Figure 20.39. Map the UK's exports and imports on a world map.
a) Mark and label the countries.
b) Use arrows in two colours to indicate exports and imports, from and to the UK.

3 Study Figure 20.40. Choose at least three cities around the world that are linked with the UK. From what you know of each city, suggest reasons for that link as a result of one or more of:
- business
- family and friends
- tourism.

4 In 2016, people in the UK voted in a referendum to leave the EU. People under 18 were not allowed to vote. But imagine they were. Would it have made any difference to our decision to stay in the EU or not? These are some of the main issues to consider:
- the influence of the UK within the EU
- the influence of the UK on the rest of the world
- the amount of trade the UK can do, inside and outside the EU
- the number of jobs created or lost in the UK
- how easy it is for UK residents to travel or live abroad
- how easy it is for EU residents to live and work in the UK.

a) Find out more about these issues. Do you think being a member of the EU made them better or worse?
b) Write a short speech for or against the UK staying in the EU. You could give your speech as part of a class debate. Which way would your class vote? If people under 18 voted in the referendum, would it have made any difference to the result?

2.1 Describe the distribution of LICs on Figure 1. [2 marks]

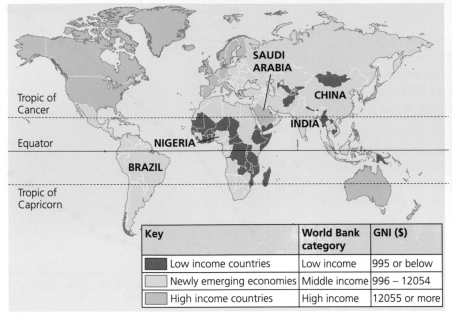

To describe a distribution, focus on where the main areas are and then look to see if there are any outside of this which are anomalies. For example, the main areas of HICs are in the northern hemisphere, such as Europe and North America but there are also some in the southern hemisphere, such as Australia and New Zealand.

Key	World Bank category	GNI ($)
Low income countries	Low income	995 or below
Newly emerging economies	Middle income	996 – 12054
High income countries	High income	12055 or more

▲ **Figure 1** The world map of development

2.2 Look at Figure 2. Compare the GNI values in Indonesia and Singapore. [2 marks]

'Compare' means to identify similarities and differences.

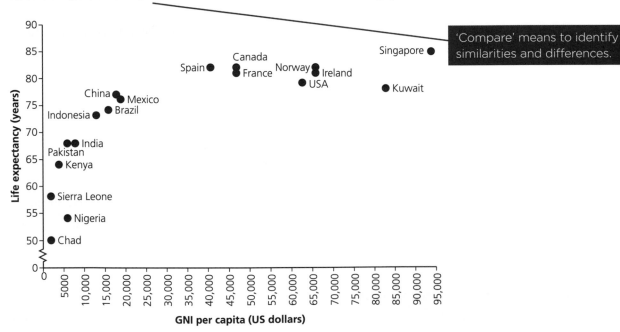

▲ **Figure 2** Investigating the relationship between economic and social development

2.3 Outline **one** disadvantage of using a single measure of development such as life expectancy. [2 marks]

2.4 Explain how the indicators of development in Figures 2 and 3 show the differences in the quality of life between Sierra Leone, Norway and the USA.

HDI rank and score	Country	HDI rank and score	Country
1 (0.944)	Norway	183 (0.374)	Sierra Leone
2 (0.933)	Australia	184 (0.372)	Chad
3 (0.917)	Switzerland	185 (0.341)	Central African Republic
4 (0.915)	Netherlands	186 (0.338)	DR Congo
5 (0.914)	USA	187 (0.337)	Niger

[4 marks]

▲ **Figure 3** The highest and lowest HDI scores in 2014

2.5 Study Figure 4.

Calculate the difference in the price of cocoa from 1996 to 2010. [1 mark]

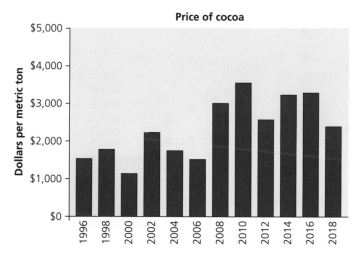

▲ **Figure 4** World cocoa prices 1996–2018

2.6 Suggest a reason for the increase in cocoa prices since 2006. [1 mark]

2.7 Using an example of an LIC or NEE you have studied, explain how tourism [4 marks] can improve the quality of life for the people who live there.

2.8 Outline **one** way that intermediate technology can help LICs and NEEs. [2 marks]

2.9 'Microfinance loans are more successful than international aid in reducing the development gap.' To what extent do you agree with this statement? Use evidence to support your answer. [9 marks]

Using evidence means you need to provide information to either prove or disprove your decision.

2.10 What is the definition of the term 'de-industrialisation'? Choose **one** answer from:

A The action of government to privatise state-run industries

B The process of decline of traditional heavy industries

C The action of government to nationalise industries owned by private companies

D The loss of employment in manufacturing industry as a result of mechanisation

[1 mark]

2.11 Study Figure 5.

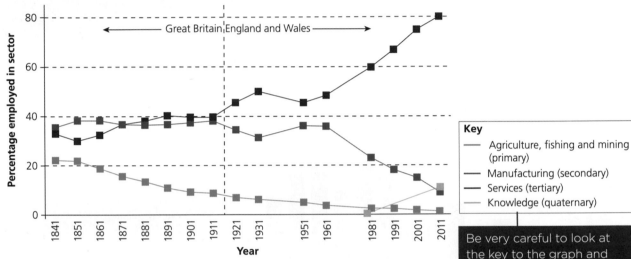

▲ Figure 5 Change in the industrial structure of England and Wales, 1841–2011

Key
— Agriculture, fishing and mining (primary)
— Manufacturing (secondary)
— Services (tertiary)
— Knowledge (quaternary)

Be very careful to look at the key to the graph and make sure you are looking at the correct line.

Complete the passage below to describe the changes in employment structure between 1921 and 2011. Use the correct words from the list to fill the gaps:

fell sharply   fell slowly   declined steadily   rapid increase
increased slightly   decreased slightly

Employment in UK manufacturing _____ between 1961 and 2011. From 1961 to 2011 there was a _____ in employment in the service sector. Since 1921 the number of people working in primary industries has _____.

[3 marks]

2.12 Explain how UK government policy could have affected the changes in manufacturing employment:

a) from 1978 to 2010? [2 marks]

b) since 2010? [2 marks]

2.13 Using Figure 20.12 (page 296), identify the feature found at grid reference 470621. [1 mark]

2.14 Explain why this feature might be considered a benefit for a company that is considering choosing Cambridge as the location for its headquarters. [2 marks]

2.15 Using Figure 20.25 (page 302) and your own understanding, explain how improving transport links can help reduce the North–South divide. [4 marks]

2.16 'The UK's most important global links are economic.' To what extent do you agree with this statement? [6 marks]

2.17 Using Figure 20.17 (page 299), state **two** ways the UK car industry is becoming more sustainable. [2 marks]

2.18 On Figure 6 annotate two impacts this industry is having on the physical environment. [2 marks]

> Make sure you link your annotation to the exact point on the figure you are talking about with a label line.

▲ Figure 6 Drax Power Station in Yorkshire, UK

2.19 Using an example of an LIC or NEE such as Nigeria, evaluate to what extent economic development has improved the quality of life for people. [9 marks]

21 Global resource management

▲ Figure 21.1 Key resources

▼ Figure 21.2 Daily calorie guidelines

Category	Calories
Women	2,000
Men	2,500
Child, 5–10	1,800
Girl, 11–14	1,850
Boy, 11–14	2,200

Essential resources

What are the key resources needed for economic and social well-being?

Food, water and energy resources are essential for our **social well-being** and **economic well-being**. Social well-being involves a person's relationships with others and whether they have a sense of belonging. Economic well-being is a person's or family's standard of living, based on how well they are doing financially and materially. For example, it can include your income and what you can afford to buy.

How are food, water and energy significant for our well-being?

If people have a good supply of food, water and energy, their quality of life, as well as their standard of living, improves. Where availability is relatively scarce, a small decrease can substantially reduce human well-being.

Food

The need for food is obvious. 'Calories in' (fuel for our bodies) are needed to work and enjoy ourselves, which equals 'calories out'. The calories needed per day depend on how active you are, your age and gender. Average figures can be seen in Figure 21.2.

Water

Water has wide-ranging uses in our current society. We need it to drink to survive, but we also need it to wash, to dispose of waste, to both grow and process our food, and in industrial manufacturing processes. The average person in the UK today uses 150 litres of water daily at home, of which only four per cent is used for drinking (see Figure 21.3). Nearly 75 per cent of the water used in the UK is used in industry. This includes the production of items as varied as cakes and cars.

▲ Figure 21.3 Domestic water use in the UK

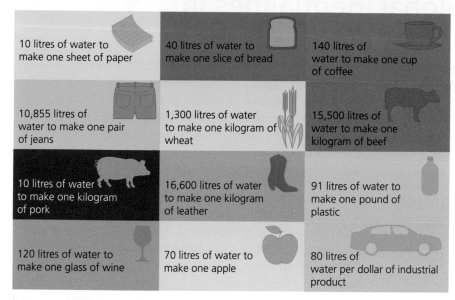

▲ Figure 21.4 Volume of water it takes to make common products

The cells in the figure read:
- 10 litres of water to make one sheet of paper
- 40 litres of water to make one slice of bread
- 140 litres of water to make one cup of coffee
- 10,855 litres of water to make one pair of jeans
- 1,300 litres of water to make one kilogram of wheat
- 15,500 litres of water to make one kilogram of beef
- 10 litres of water to make one kilogram of pork
- 16,600 litres of water to make one kilogram of leather
- 91 litres of water to make one pound of plastic
- 120 litres of water to make one glass of wine
- 70 litres of water to make one apple
- 80 litres of water per dollar of industrial product

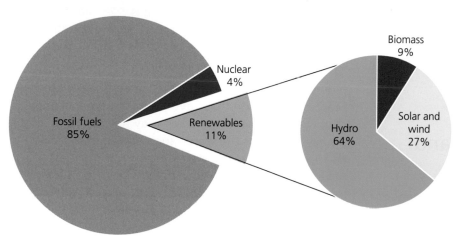

▲ Figure 21.5 World energy consumption by source (2017)

Pie chart labels: Fossil fuels 85%, Nuclear 4%, Renewables 11%. Renewables breakdown: Biomass 9%, Solar and wind 27%, Hydro 64%.

Energy

Energy is used to make the bricks for our houses, to heat our homes, transport us, power machinery and process food.

The amount and type of energy used depends on a variety of factors, including where people live and how wealthy they are. Traditionally, energy has come from burning naturally occurring fuels such as wood and coal. However, today there is more and more energy from renewable sources such as wind, waves and solar power.

As the world's population grows and becomes wealthier, so does the pressure on the supply of resources. The rate of growth can cause huge problems, as the supply of resources struggles to keep up with demand. Technology cannot change or improve fast enough to provide the essential resources needed. One of the major problems is meeting the demand for these essential resources, and, in particular, solving the problems caused by their unequal distribution and consumption.

→ Activities

1 What are the three main domestic uses of water in the UK?

2 Using Figure 21.4, choose one of the products and suggest ways in which the water may be used to produce it. Think about what each product is made from and how its raw materials are produced and processed.

3 To what extent is energy important to the economic well-being of people?

    Think about the uses which energy has which helps towards economy.

Geographical skills

1 Use Figure 21.3 to calculate the percentage of water people use in washing and cleaning in the UK.

2 Using Figure 21.2, draw a bar graph to demonstrate the advised average calorie intake per person.

3 Using Figure 21.5 describe the global consumption of energy by source. Start with overall energy consumption, then look at individual renewable types, and use evidence.

➤ Global inequalities in food
➤ Global inequalities in water
➤ Global inequalities in energy
➤ The growing demand for essential resources

Global inequalities in essential resources

The consumption of resources varies greatly throughout the world. Generally, HICs consume more resources than LICs. The problem we face is not that we do not have enough of these essential resources, but that they are unevenly distributed. As the wealth of LIC s grows, so does the demand for resources. This demand for resources, combined with the growth in population, leads to shortages or scarcity of these essential resources. Inability to access essential resources can impact on social and economic well-being.

What are the global inequalities in the supply and consumption of food?

The average calorie consumption in a country such as the UK is 3,440 per person, while in a country such as the Democratic Republic of Congo it is 1,590 per person.

Figure 21.6 shows some correlation between the areas of greatest population growth (see Figure 23.3 on page 335) and the areas which have the highest levels of **undernourishment**.

What are the global inequalities in the supply and consumption of water?

The global supply of freshwater is limited and unequally distributed. The **water footprint** of countries can be calculated to compare consumption. This is the amount of water used throughout the day, for example, from a tap for drinking or showering. It also includes the water it takes to produce food, products, energy and even the water saved when products are recycled. This virtual water may not be seen, but it makes up the majority of a country's water footprint.

The global average water footprint is 3,287 litres per person. The water footprint of the USA is 7,786 litres per person. The water footprint of the Democratic Republic of Congo is just 1,500 litres per person per day.

Figure 21.7 shows the areas that suffer from water scarcity and those that have water to spare. Many countries may have water but may not have the money to access the water, such as Sudan. This is known as **economic water scarcity**. Others may not have as much water due to the physical conditions such as climate, for example, Saudi Arabia (**physical water scarcity**).

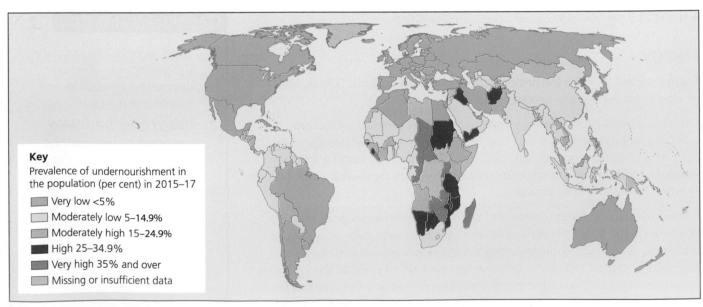

Key
Prevalence of undernourishment in the population (per cent) in 2015–17

- Very low <5%
- Moderately low 5–14.9%
- Moderately high 15–24.9%
- High 25–34.9%
- Very high 35% and over
- Missing or insufficient data

▲ Figure 21.6 World hunger

What are the global inequalities in the supply and consumption of energy?

Energy consumption varies considerably in different countries (Figure 21.8). The richest 1 billion people in the world actually consume 50 per cent of the world's energy, while the poorest 1 billion people consume only 4 per cent of the world's energy.

Why is the demand for essential resources growing?

The demand for essential resources has grown over time as we develop new processes, new products and change our way of life. As LICs and NEEs develop industrially and economically, their demand for resources has grown too. For example, as industry has grown in China, energy consumption has increased with it. Between 2003 and 2011, China saw an increase of 53 per cent in its consumption of energy.

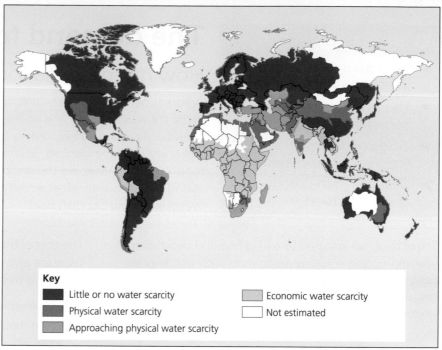

Key

- ■ Little or no water scarcity
- ■ Physical water scarcity
- ■ Approaching physical water scarcity
- ■ Economic water scarcity
- □ Not estimated

▲ Figure 21.7 Map to show global water scarcity

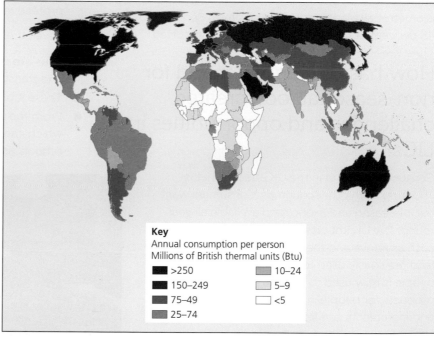

Key
Annual consumption per person
Millions of British thermal units (Btu)

- ■ >250
- ■ 150–249
- ■ 75–49
- ■ 25–74
- ■ 10–24
- ■ 5–9
- □ <5

▲ Figure 21.8 Energy consumption per person by country

→ Activities

1. Explain the difference between physical water scarcity and economic water scarcity.

2. Using the data from Figure 21.8, explain why some countries use more energy than others.

3. Assess the extent to which population growth will affect global water scarcity.

Geographical skills

1. Using Figure 21.6, describe the global pattern of undernourishment.

2. Using Figure 21.7, describe how global water scarcity links with the global pattern of undernourishment.

22 Resources in the UK

The demand for food in the UK

How has demand for food in the UK changed?

Before supermarkets were commonplace, the majority of the food eaten in the UK was seasonal and sourced in the UK. Fruit and vegetables were grown, sold and eaten according to the seasons; for example, lettuce and strawberries in the summer, and cabbage and parsnips in the winter. More food was also preserved (frozen, bottled or made into jam and pickles) for eating out of season. Meat would also have been produced in the UK, such as Welsh lamb and Scottish beef.

These days, we are used to eating fruit and vegetables all year round, and enjoying exotic fruits such as mangoes (which cannot be grown in the UK due to the climate). However, even seasonal fruit and vegetables in the UK are often imported from other countries as they can be grown more cheaply elsewhere. In September, you would expect to find seasonal fruits and vegetables, such as apples and onions, in your local supermarket, but you will also find apples from South Africa, imported from over 8,000 kilometres away, along with onions from Spain. In 2017, 45 per cent of the UK's food supply was imported.

How has the high demand for non-seasonal food created challenges and opportunities in LICs?

Consumer demand in the UK affects what is imported from other countries. Consumers want out-of-season and exotic food available all year round. The UK imports food like this from places such as Kenya and the Caribbean. This means that land previously used to produce food for local people is now used to provide high-value food products for people in the UK. **High-value foods** and ingredients can fetch retail prices that are up to five times those of similar products. The high value may be due to the product itself, such as Madagascan vanilla, specialist honeys and gourmet coffees, or because they are luxury items available out of season that are in high demand.

The cost of these products to the UK consumer is high, but there are also challenges for the people in the LICs where they are grown:

■ Less land is available for locals to grow food to eat.
■ Often these crops need huge amounts of water in areas where the water supply is unreliable or poor.
■ Sometimes the people growing the crops are exposed to chemicals such as **pesticides** without protective clothing.

However, there are opportunities for people in LICs too:

■ Jobs are created, for example in farming, packaging and transport.
■ These jobs supply wages for local people.
■ From the wages, taxes are paid to the government, which can then fund facilities for the country such as schools and hospitals.

▲ Figure 22.1 Green beans being packed for export in Kenya

How and why has the demand for organic produce changed?

Another change to the UK's eating habits has been the increasing demand for **organic produce**. Organic produce, including meat, fruit and vegetables, is produced by organic farming, a type of farming which does not include the use of chemicals such as pesticides and fertilisers.

- The aim is to protect the environment and wildlife by using natural predators to control pests, for example using ladybirds to eat blackfly.
- Farmers maintain the fertility of the soil by rotating crops and using a variety of natural fertilisers, including green manure and compost.
- Weeds are controlled by mechanical weeding rather than using chemical weed killers.
- Animals are farmed without the use of antibiotics and the regular use of drugs such as hormones to increase growth.

The demand for organic products has been rising steadily since the early 1990s, as people are increasingly concerned about the effect of what they eat on their health. Organic food is believed to be healthier than non-organic food. In a government survey of households in 2014, the main reasons for choosing organic products were:

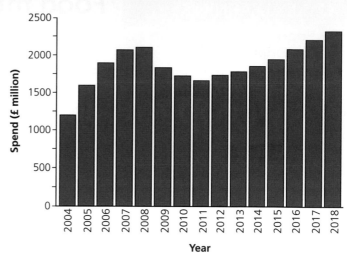

▲ Figure 22.2 Sales of organic products in the UK, 2004–18

- 'It contains fewer chemicals and pesticides.'
- 'It's natural and unprocessed.'
- 'It's healthier for me and my family.'

A closer examination of the figures will show a decline in sales between 2009–11. This was due to the global recession and the reduction in incomes for many families, making expensive organic produce a luxury item. It is expensive because yields from organic farming tend to be lower, but many people claim that organic food tastes better. They are prepared to pay extra for this and the reduced impact on the environment.

Today, all the major supermarkets sell organic produce, providing about 75 per cent of all organic food sold. Other sources of organic food are local farmers' markets and vegetable box schemes, where households receive a box of seasonal organic fruit and vegetables weekly or monthly, usually delivered to their home. The highest-selling organic products are dairy produce including milk, cheese and yoghurt.

→ Activities

1 Why is organic produce more expensive than non-organic produce?

2 Make a list of the challenges and opportunities provided by:
   a) the increased demand for fruit and vegetables all year round
   b) organic products.

   Include the challenges and opportunities to:
   a) the consumers
   b) the producers.

3 Suggest what challenges importing seasonal food causes farmers in the UK.

4 Evaluate the impact of the increased demand for non seasonal products on global food production. Think about the impacts of importing non seasonal foods on both the UK and the countries they are being imported from.

Geographical skills

1 Using Figure 22.2:
   a) Describe the pattern of sales of organic products between 2004 and 2018.
   b) Calculate the change in organic sales between 2008 and 2014 in pounds.

⭐ KEY LEARNING

➤ Food miles and carbon footprints

➤ The impact of importing food on the UK's carbon footprint

➤ Alternatives to importing food

➤ The trend towards agribusiness

Food miles and carbon footprints

What are food miles and carbon footprints?

Food miles are the distance that food travels from producer to consumer. For example, green beans from Kenya travel 6,818 kilometres to reach the UK. This does not include the distance the beans travel from the airport to the supermarket, or from the supermarket to your home. In the UK, food travels over 30 billion kilometres every year. This includes transport by air, ship, train and road.

A **carbon footprint** is the measure of the impact that human activities have on the environment in terms of the amount of greenhouse gases they produce (see Section 4.3). It is possible to calculate an individual's carbon footprint as well as that for a country or a business.

How does importing food increase the UK's carbon footprint?

The transport used to import food into the UK adds over 19 million tonnes of carbon dioxide to the atmosphere every year, which increases the UK's carbon footprint.

In theory, the further a product has travelled, the higher the food miles and the higher its emissions. However, there are other aspects of food production which add to the UK's carbon footprint. Carbon dioxide is also produced when food is grown and harvested for example, when farm machines harvest the crops or when greenhouses are heated. Figure 22.3 shows how the production and transportation of food contributes to the UK's carbon footprint. Food contributes at least 17 per cent of the total UK carbon dioxide emissions, but only 11 per cent of this is linked to transport.

Emissions

The emissions created by producing a food product in the UK can sometimes create higher emissions than those imported from overseas. An example of this is tomatoes. Even including the transport emissions from the aircraft bringing them to the UK, the carbon footprint of Spanish tomatoes is smaller because growing tomatoes in the UK requires heated greenhouses whereas Spain's warmer climate means no additional heat is needed.

The actual emissions produced by different forms of transport used in transporting food also need to be taken into consideration when looking at food miles and carbon footprints. Food transported by plane generates around 100 times the amount of emissions of food transported by boat. This means that bananas from Ecuador which travel by ship are much more carbon friendly than avocados flown from Mexico, although both are a similar distance away from the UK. In general, food products which are perishable and have a high value relative to their weight are transported by plane, while others are sent by sea, which takes a lot longer.

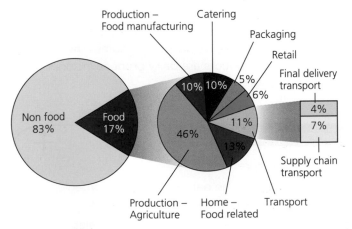

▲ Figure 22.3 The contribution of food to the UK's carbon footprint

▼ Figure 22.4 Transport emissions, kilogram CO_2 equivalents per tonne kilometres

Type of transport	Carbon emissions (kg CO_2/t.km)
Long-haul air, e.g. Australia–UK	1.762
Short-haul air, e.g. Spain–UK	0.733
Road	0.410
Rail	0.037
Water	0.038

What are the alternatives to importing food?

To reduce the amount of carbon emissions, we need to reduce the amount of food products that are flown into the UK. There are several ways to do this:

- eating seasonal produce grown in the UK
- limiting imported foods to those we cannot grow in the UK, and limiting those that are transported by air. Many restaurants now label the origins of food and work in close collaboration with the food producers. This helps the farmers and producers and allows consumers to make informed choices. Some supermarkets such as the Co-operative have also started to stock only British meat, and all supermarkets now use the Red Tractor scheme (Figure 22.5)
- eating locally produced food, which reduces the amount of food miles that the food travels within the UK, and also supports local farmers and producers. There has been an increase in local farmers' markets or farm shops in the UK. To qualify to sell at a farmers' market the products must be produced within a specific local area.

- Growing food at home or on an allotment. The move towards growing our own food has increased in the last five years, with over a third of people now growing some of their own fruit and vegetables.

Why is there a trend towards agribusiness?

Agribusiness refers to treating food production from farms like a large industrial business, making it a large scale, capital-intensive, commercial activity. This has meant increasing the size of farms by:

- removing hedgerows
- increasing field sizes and combining smaller family farms
- using modern production methods (see Chapter 23)
- increased mechanisation
- using the latest technology, better seeds, and increased use of chemicals such as pesticides and fertilisers.

Today, the vertical integration of food production, commonly known as 'from farm to fork', is increasingly common. Large agribusinesses now often own not only the farms where food is grown, but also the processing factories, the transport and the retail outlets. Food-processing companies and supermarkets will often buy the crops before they are planted. Agribusiness has some huge advantages for the increased production of food, but can have considerable impacts on the environment and **local food** production (see Chapter 23) . This trend can be seen clearly in East Anglia, where farm sizes have increased dramatically over the last 40 years. This has led to a decline in agricultural employment, especially in more isolated areas.

◄ Figure 22.5 The Red Tractor label, which assures consumers that the source of food is British and has been inspected for safety, welfare and environmental impacts

---

→ Activities

1   Suggest two ways of reducing food miles.

2   Give three advantages of buying local food products.

3   Assess the extent to which the method used to transport imported food can affect the carbon footprint of a food item. Include reasons why food items are transported by different methods.

4   a)   Choose your favourite meal and work out the basic ingredients needed to cook it.

    b)   Find out where these products come from by looking at the labels or by using a supermarket website. The origin of every product should be on the label.

    c)   Use the following website to find out how far each part of your meal has travelled: www.foodmiles.com.

    d)   Plot the locations on a world map and add up how far your 'dinner' has travelled.

    e)   Using the food miles website to help you, suggest an alternative meal that would have lower food miles.

Geographical skills

1   Using Figure 22.4, draw a bar graph to show the different emissions produced by different methods of transport.

2   Using Figure 22.3, describe the contribution food makes to the UK's carbon footprint.

Keeping it clean

What are the causes of water pollution?

Today, rivers, lakes and coastal waters in the UK are cleaner than they have ever been since before the Industrial Revolution. The improvement in water quality has seen wildlife, including salmon, otters and birds, returning to live in these habitats. However, water **pollution** still exists. In the UK, only 35 per cent of our water is classified as being of a 'good status' under the EU Water Framework Directive, which looks at **water quality** throughout the EU. This means that 73 per cent of our water in the UK could be polluted or of poor quality. Rivers, lakes and coastal waters are polluted by a wide variety of items as seen on Figure 22.9.

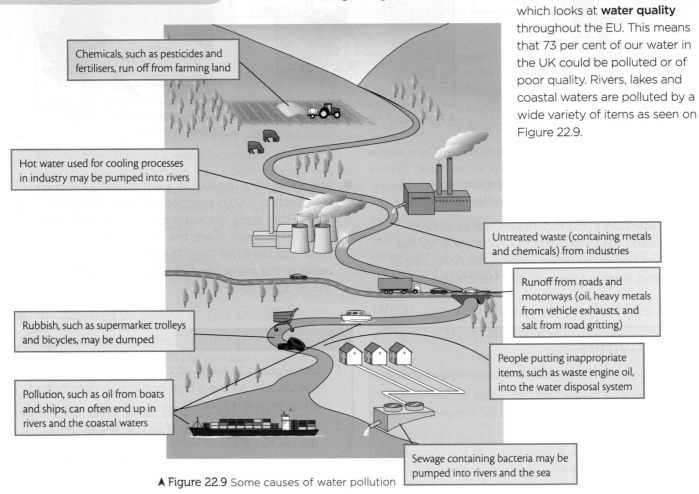

Chemicals, such as pesticides and fertilisers, run off from farming land

Hot water used for cooling processes in industry may be pumped into rivers

Untreated waste (containing metals and chemicals) from industries

Runoff from roads and motorways (oil, heavy metals from vehicle exhausts, and salt from road gritting)

Rubbish, such as supermarket trolleys and bicycles, may be dumped

People putting inappropriate items, such as waste engine oil, into the water disposal system

Pollution, such as oil from boats and ships, can often end up in rivers and the coastal waters

Sewage containing bacteria may be pumped into rivers and the sea

▲ Figure 22.9 Some causes of water pollution

How does water pollution affect the UK?

Water pollution can have a number of serious effects, with long-term and short-term consequences.

- Toxic waste can poison wildlife. Sometimes the toxins can be transferred to humans if they eat the shellfish or fish, leading to birth defects and, in some cases, cancer.
- The supply of drinking water can be poisoned.
- Increased water temperatures can lead to the death of wildlife and the disruption of habitats.
- Increased fertilisers can increase nutrients in the water, speeding up the growth of algae and leading to **eutrophication**. This means there may not be

sufficient oxygen in the water, so other wildlife will also die. The increased algae may also block the sunlight to other water plants.

- Pesticides can kill important parts of the **ecosystem**.
- The microbacteria in sewage can cause the spread of infectious diseases in aquatic life, animals and humans.
- People whose livelihoods depend on a clean water supply, for example fishermen or workers in the tourist industry, may suffer.

How is water quality managed in the UK?

Legislation – The UK and EU have strict laws which ensure that factories and farms are limited in the amount and type of discharge they put into rivers. Water companies which provide our drinking water and sewage systems have very clear regulations and penalties.

Education campaigns – These inform the public about the damage caused by putting inappropriate items into the sewage systems, such as engine oil and baby wipes, and advise how to dispose of them correctly. A good example of this is the Yellow Fish campaign by the **Environment Agency**. It encourages people not to put anything down the street drains except water, with the slogan 'only rain down the drain' and yellow fish stencilled on the roads next to drains (Figure 22.10).

Waste water treatments – Local water treatment plants remove suspended solids such as silt and soil, bacteria, algae, chemicals and minerals, to produce clean water for human consumption. They use a number of processes to do this.

Building better treatment plants and investing in new infrastructure – Better sewers and water mains can prevent spills and accidents, but can lead to higher water and sewage bills to pay for the investment. For example, Thames Water in London is investing heavily in its sewage works and new tunnels to prevent the overflow of the current sewers.

Pollution traps – For example, when new roads and motorways are built close to rivers and watercourses, pollution traps such as reed beds are often installed to 'catch' and filter out the pollution.

Green roofs and walls – In cities, new buildings often have green roofs, which naturally filter out the pollutants in rainwater. Green roofs also offer excellent sustainable water management, reducing the risk of flooding by reducing runoff from the roof. Green roofs can also help to combat **climate change** by increasing the absorption of CO_2 from the atmosphere (Figure 22.11).

▲ Figure 22.10 A yellow fish next to a drain – part of the Yellow Fish campaign

▲ Figure 22.11 Green roofs and walls filter out pollutants reduce funoff and absorb CO_2

→ Activities

1 Describe three causes of water pollution.

2 Explain the impact of runoff from agriculture on rivers and lakes.

3 To what extent are education campaigns successful in managing water quality in the UK? Use the information on this page to think about how successful campaigns are compared to other methods in managing water pollution. Consider the advantages and disadvantages of each method before reaching a final judgement.

KEY LEARNING

➤ How the demand for energy in the UK is changing

➤ The UK's energy mix

➤ How the UK's energy mix has changed

Energy resources in the UK

How is the demand for energy changing in the UK?

Energy is important in everything we do. It powers our cars and other transport; it heats our homes, schools and offices; it powers the machines that produce our clothes and food; and it provides the electricity we use to watch TV and use computers.

Today, we actually consume less energy than we did in 1970, despite there being more than an extra 9.1 million people living in the UK. The average household uses 12 per cent less energy while the decline of heavy industry has led to this sector using 60 per cent less. There has, however, been an increase in the amount used by the transport sector. The number of cars on the road has increased dramatically – in 1970 there was 10 million but today there is over 37.5 million. The increase in air travel has also contributed to this rise.

This reduction in domestic energy consumption can be explained by:

■ the introduction of energy-efficient devices, such as light bulbs and washing machines
■ the increasing awareness of the public that they must save energy
■ the increased cost of energy leading to lower demand.

What is the UK's energy mix?

The **energy mix** of the UK refers to the different sources of energy used by households, industry and other commercial users, such as shops and offices. Most of the energy we use in our homes is in the form of electricity. Electricity can be generated by burning **fossil fuels**, such as oil and coal, or by using renewable energy sources, such as wind or water.

Fossil fuels

Coal, oil and gas make up the three fossil fuels used today in the UK. These were formed over many thousands of years and, as they take so long to be replaced, they are regarded as non-renewable. In other words, they will run out. When burnt, fossil fuels release carbon dioxide, along with other greenhouse gases. Fossil fuels can be burnt directly to produce heat. They can also be used to generate electricity in power stations and, finally, they can be used to power vehicles and machinery.

Nuclear power

Nuclear power uses uranium to produce heat in a nuclear reactor. This heat is then used to drive a turbine to make electricity. It is not a fossil fuel, but is considered non-renewable as the supplies of uranium are finite.

Renewable energy

Renewable energy sources are sources that can be used again and again, and will not run out. They include the Sun, wind, waves, the tides, running water in rivers and **geothermal** heat created deep underground. Methane produced from **landfill sites** is also burnt to generate electricity from vegetation (**biomass**). The main problems with these energy sources are the cost of the technology and the relatively small amounts of energy produced. However, they are seen as 'clean' and non-polluting.

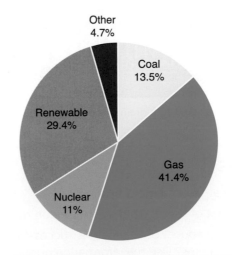

▲ Figure 22.12 The UK's energy mix in 2017

How is the UK's energy mix changing?

Until recently, the UK produced enough energy to power homes and industry. However, a reduction in oil and gas reserves and in the production of coal has led to an increasing reliance on imported fossil fuels.

The production of coal, gas and oil has declined for many reasons (see Figure 22.12) but there are still supplies that could be exploited, often in less accessible areas. One such area is the Mariner oil field, 150 km east of the Shetland Islands, which started production in 2016.

In addition, policies introduced both nationally and internationally can have an effect on production and the mix of energy used. For example, the use of coal increased in 2011 as older coal-powered stations worked to full capacity, in the knowledge they were soon to be closed in 2025 due to EU regulations on emissions.

To reduce the reliance on imported fuels and the carbon emissions generated by burning fossil fuels, the British Government is encouraging investment in renewable energy sources such as **wind** and **solar energy**.

Figure 22.13 shows the reduction in the use of fossil fuels in the UK to generate electricity in recent years. It also shows the small percentage of energy production from renewable

sources. The reliance on fossil fuels to produce our energy will continue into the future mainly due to the cost and unreliable nature of renewable sources. However, the energy production from renewable sources is growing year on year, as the number of wind and solar farms increases.

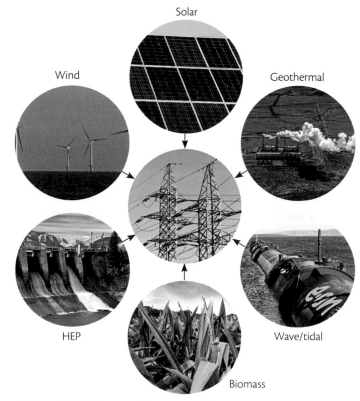

▲ Figure 22.14 Renewable sources of energy

Key
- Coal
- Gas
- Nuclear
- Renewable
- Other

1cm = 10%

Year

▲ Figure 22.13 UK electricity generation by source

→ **Activities**

1 Make a list of:
   a) non-renewable sources of energy
   b) renewable sources of energy.
2 Explain why the consumption of energy in the UK in recent years has reduced.
3 Why is the energy mix of the UK changing?
4 'The UK will be able to supply most of its power from renewable sources in the near future.' To what extent do you agree with this statement?

Geographical skills

Study Figure 22.13.

1 Calculate the increase in the percentage of electricity generated by renewable energy sources between 2006 and 2017.
2 Give two reasons for this increase.

Economic and environmental issues of energy production

▼ Figure 22.20 Challenges and opportunities for different types of energy

Fossil fuels

Economic challenges	Environmental challenges
• Much of the remaining coal is in hard-to-access areas, often deep underground, which is very expensive to mine. • With the UK's last coal mine closed in 2015, coal must be imported from countries like South Africa. • Mining coal causes environmental problems, such as waste or spoil heaps, which are expensive to clean up. • Miners often suffer from diseases related to their jobs, which will incur a cost to the health service. Emissions from fossil fuels can also cause respiratory diseases, again incurring a cost. • Costs of exploring more remote and inaccessible areas in the North Sea, or costs of drilling in heavily populated areas (Sussex) or sensitive areas (Dorset). • Costs of climate change, for example, increased flooding requiring flood defences (see Chapter 11).	• The burning of fossil fuels creates greenhouse gases which may contribute to climate change (Chapter 4) and causes acid rain. • Waste heaps from coal mining can create visual pollution. • Opencast coal mines are unsightly and create dust and noise, disturbing local people and wildlife. They also use huge areas of land. • Access roads and support industries for all of these can destroy wildlife habitats and impact on land visually. • Gas and oil coastal terminals take up space and can mean the digging up of large areas of land for pipelines to transfer to the power stations. • Fracking (see opposite) presents its own challenges.
Economic opportunities	**Environmental opportunities**
• Creation of jobs directly, in support industries and in the manufacture of equipment. It can bring money and jobs in to an area – a multiplier effect.	• Carbon captive storage (CCS) is more efficient but expensive (see Section 4.5).

Nuclear

Economic challenges	Environmental challenges
• The costs of building nuclear power stations are huge. • There are enormous costs to store and transport nuclear waste. It is expensive to **decommission** power stations.	• The waste from nuclear power stations must be stored safely for many years to avoid contamination. • The environment can be considerably more dangerous if an accident occurs. Nuclear accidents can lead to the release of radiation into the atmosphere which can have a long-term detrimental impact on wildlife and people.
Economic opportunities	**Environmental opportunities**
• Creates jobs in research and development for new technologies in the nuclear power industry. • After the initial investment, energy generated by nuclear power is seen to be cheaper.	• Nuclear power is considered cleaner and less polluting than energy generated by fossil fuels.

Renewables

Economic challenges	Environmental challenges
• High set-up costs of renewable energy sources, such as wind turbines, solar farms and **tidal power** stations, especially in the remote areas suitable for this type of energy generation. • The impact on the visual environment can affect tourism and reduce income and jobs. • Low profitability is also a concern.	• Evidence shows that wind turbines can affect bird migration patterns and bat life in the area. Turbines located at sea are believed to impact on sea currents and on fish and birdlife. • Many people consider wind turbines ugly. • Wind turbines and the associated access roads can impact on untouched land such as the Highlands of Scotland. • Turbines are also noisy and can disturb people and wildlife living nearby. They can also block TV and phone signals.
Economic opportunities	**Environmental opportunities**
• Many jobs are created in the manufacture of solar panels and wind turbines along with jobs in research and development.	• They produce much lower carbon emissions. • Land used for siting wind turbines can also support other uses, such as farming and leisure activities. • Offshore wind turbines can act as an artificial reef, creating habitats for marine wildlife.

Fracking: a new energy production issue in the UK

Fracking or hydraulic fracturing was being explored in the UK as a means to extract gas that is locked in rocks thousands of metres below the Earth's surface.

A hole is drilled deep into the rock and a mixture of sand, water and chemicals are injected into it at high pressure, which splits the rock and releases the gas. This technique has been used in the North Sea oil and gas fields for many years.

Areas where this was being considered included Lancashire, Yorkshire, the East Midlands and some areas in southern England, such as Sussex and Surrey.

In November 2019, the UK Government stopped all fracking indefinitely due to fears of earthquakes. This may be a temporary ban.

The economic and environmental challenges and opportunities of fracking

- Fracking for shale gas can lead to pollution of ground water, which in turn can lead to contamination of drinking water with hydraulic fracturing fluids.
- It also requires the use of large quantities of water and can impact water supplies in some areas.
- Additionally, the process of fracking has been linked to low-level **earthquakes**.
- However, fracking can bring economic benefits in terms of more government revenues and more jobs for people.

▲ Figure 22.21 Shale gas sites in the UK

> ### → Activities
>
> 1 Using Figure 22.21, describe the pattern of areas where fracking may be carried out.
>
> 2 Using Figure 22.22, draw a flowchart to explain the sequence of actions used to 'frack' for gas.
>
> 3 'Fracking for shale gas has huge advantages for the UK's energy supply.' To what extent do you agree with this statement? Use the information on this page in your answer, and make sure you reach a final judgement.

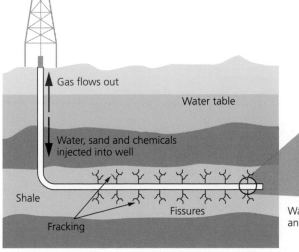

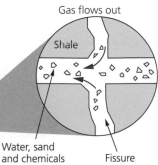

◄ Figure 22.22 The fracking process

23 Food

Global patterns of food consumption

What are the differences in global calorie intake and food supply?

Food security was defined by the World Food Summit in 1996 as 'when all people at all times have access to sufficient, safe, nutritious food to maintain a healthy and active life'. Many people do not have food security (as you may remember from Figure 21.6 on page 318) and are severely undernourished. There is a great difference in calorie intake in different parts of the world (Figure 23.1).

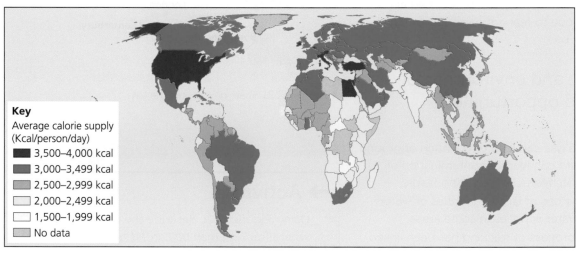

Key
Average calorie supply (Kcal/person/day)
- 3,500–4,000 kcal
- 3,000–3,499 kcal
- 2,500–2,999 kcal
- 2,000–2,499 kcal
- 1,500–1,999 kcal
- No data

▲ Figure 23.1 Average daily supply of calories per person, 2013

The world as a whole produces enough food for everyone, but not everyone has equal access. If we distributed the world's food evenly, there would be enough for every person to receive around 2,831 calories per day, plenty for everyone to live a healthy life. As Figure 23.1 shows, the countries where people have the lowest calorie supply are almost all in sub-Saharan Africa. Another issue is the composition of the average diet:

- In HICs, over a quarter of the diet is made up from meat, fish, eggs, milk and cheese and approximately a quarter from cereals.

- In the countries in Africa and other LICs which have the worst **food insecurity**, approximately half the diet is made up from cereals and a further 20 per cent from tubers such as yams. Cereals and tubers provide food energy (calories) but are low in other nutrients, which leads to undernourishment.

Why is demand growing for food?

The demand for food has grown over time, as we develop new processes and new products, and change our way of life. As LICs and NEEs develop industrially and economically, their demand for different food products increases as well.

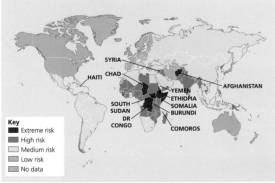

Key
- Extreme risk
- High risk
- Medium risk
- Low risk
- No data

▲ Figure 23.2 Global food security risk

Population growth

Global population growth puts a lot of pressure on our essential resources such as food. In the past ten minutes, while you have been reading this, the world's population has grown by approximately 1,690 people, so, in the past hour, that's around 10,100 people.

Where are the areas of fastest population growth?

The problem for food supply is that the growth of population is not even. Parts of the world are growing at a faster rate than others (Figure 23.4). There is a huge difference in the rates of population growth. The population in Africa is growing by 2.52 per cent per year, while the population of Europe is growing by 0.1 per cent per year.

Figure 21.6 on page 318 shows the areas of the world where there is a real risk of hunger due to lack of food. If you link this to the graph of population growth shown in Figure 23.4, and Figure 23.1 showing calorie supply, there is some correlation between the areas of greatest population growth and the areas which have the highest levels of undernourishment and lowest calorie intake per person.

Economic development

As people become richer, their diets change. They begin to eat more meat instead of grains such as rice. Grain is fed to animals to produce the meat for people to eat, rather than grain being eaten by people. This means more food resources are needed to feed the people of the country. Additionally, as wealth grows, so does the demand for highly processed and convenience foods, which can also increase the calorie intake. China provides a good example of this, as shown in Figure 23.5.

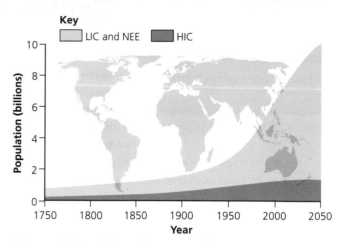

▲ **Figure 23.3** World population growth

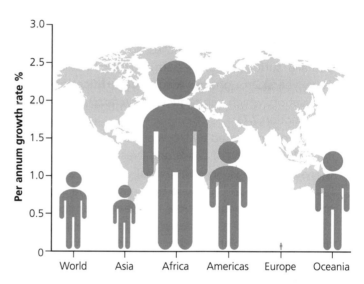

▲ **Figure 23.4** Population growth by continent

▼ **Figure 23.5** China in 1981 and 2011

Year	Average calorie intake	Percentage of meat in diet	GDP (US$) per capita
1981	2,165	6	193.30
2011	3,073	17	5,574.20

Geographical skills

1 Use Figure 23.1 to describe the global pattern of countries that are affected by food insecurity. Use the information in Figure 21.2 (page 316) to help you establish the average calories per person needed for a healthy diet and to have food security.

2 Use Figure 23.4 to calculate the difference in the percentage population growth between Europe and Africa.

→ Activities

1 What is meant by the term 'food security'?
2 Explain how population growth can affect food security.
3 Using Figure 23.5 and your own understanding, explain how changes in GDP can affect diet.

⭐ KEY LEARNING

➤ The impacts of food insecurity on people

➤ The impacts of food insecurity on the environment

The impacts of food insecurity

How does food insecurity affect people?

The most obvious problem associated with food insecurity for people is the lack of food.

Famine and undernutrition

Famine is the clearest indicator that people in an area do not have sufficient food. Famine is described as 'a widespread scarcity of food'. Famine:

■ not only causes death, but also **undernutrition**, which weakens immunity and makes people more vulnerable to diseases

■ may also lead to deficiency diseases such as beriberi or anaemia

■ can also hinder the physical and cognitive development of children.

The UN Food and Agriculture Organisation estimates that, in 2019, over 820 million people of the approximately 7.7 billion people in the world were chronically undernourished. Almost all of these people live in LICs.

Rising prices

A shortage of food can lead to an increase in prices. Population growth can increase demand, while poor grain harvests can reduce supply – both of which will increase prices as demand exceeds supply. More recently, the increased use of grains for **biofuels** has reduced the supply available for food, as higher prices can be gained by selling the grain for fuel.

Food shortages can cause prices of basic food stuffs to rise out of reach for the average family. When they are not facing famine, people remain undernourished which leads to diseases associated with poor diet as they cannot afford access to nutritious food.

Conflict and social unrest

The increased competition for scarce food resources can lead to conflict in both local and international communities.

The need for water for farming can lead to international disputes over ownership of the water sources (see page 357). For example, the River Nile runs through several countries and all of them need the water from the river to help feed their population. There have been long

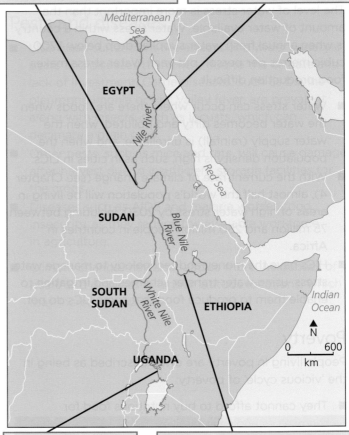

The Sudanese government would like to use more of the water for irrigation to help grow more food.

Egyptian farmers use the water to grow vast amounts of food, such as avocados and fruit, for the global and domestic markets.

The Ugandan government wants to dam the river, which will restrict the flow downstream.

The Ethiopian government currently uses the water from the Nile for irrigation to support its coffee industry, which brings in much-needed income.

▲ Figure 23.8 The international demands on water in the River Nile

negotiations over how the water is shared. Who has the right to the water?

Additionally, social unrest in local communities is often related to rising prices of staple food stuffs. Between 2008 and 2011, 60 riots linked to rising food prices or food shortages were recorded around the world, especially in North Africa and the Middle East. There were five days of rioting in Algeria in 2011, when flour and cooking oil prices doubled.

How does food insecurity affect the environment?

In LICs, the best land is often used to grow cash crops for export to HICs, such as flowers and green beans in Kenya. This leaves the less suitable land for growing food to feed the local population of the country. This marginal land often does not have sufficient nutrients or water to produce a good harvest and will quickly become infertile. Infertile land cannot support plant growth, which leaves the soil exposed and prone to erosion through wind and water.

Soil erosion and water shortages

Overgrazing by cattle and other farm animals also leaves the soil exposed. Once the soil has no vegetation to hold it in place, it can be blown or washed away, so causing **soil erosion**.

Where cash crops are grown in both LICs and HICs, the increased use of pesticides and fertilisers to produce a large healthy crop can cause water pollution. It can also increase the demand for water for irrigation and lead to water shortages. Water shortages and pollution can also have an impact on the indigenous wildlife habitat.

▲ Figure 23.9 A protest against rising prices in Algeria, 2011

▲ Figure 23.10 Overgrazing and soil erosion

→ Activities

1 Outline the meaning of the term 'undernutrition'.
2 Study Figure 23.10. Use it to draw a flow chart to show the causes and sequence of soil erosion in LICs.
3 'The greatest impact of food insecurity is environmental.' Discuss this statement.

   Discuss means you must look at both points of view. Use the information on these pages to help you do this using evidence to support your points. Extra research may be needed.

⭐ KEY LEARNING

➤ The advantages and disadvantages of a large scale agricultural development

Almería, Spain: a large-scale agricultural development

The southeast of Spain near Almería has always been an arid area and receives only an average of 200 millimetres of rainfall per year. In the past, film production companies used the region to film cowboy films, as the **landscape** is so similar to the deserts in the USA.

In the last 35 years, this area has developed the largest concentration of greenhouses in the world, covering over 31,000 hectares. The greenhouses are owned and operated by a mixture of large businesses and individual farmers. Most of the UK's out-of-season crops, such as tomatoes, lettuce, melons, courgettes, cucumbers and peppers, are grown here. The scheme brings in over US$1.5 billion per year in income, as it delivers over half of Europe's fruit and vegetables. This large-scale **commercial farming** has developed for several reasons:

■ changes in people's diets to eat more fruit and vegetables

■ the development of suitable plastic to build the greenhouses

■ new and fast transport methods, which have lowered shipping costs

■ the average temperature in the region is 20 °C, with about 3,000 hours of sun per year, meaning that the crops can be grown in the winter without artificial heating, unlike other areas in Europe (the Netherlands), so reducing costs

■ low labour costs from immigrants

■ funding from the EU and the Spanish government.

The greenhouses are so successful that they have covered the plain of Dalías, and have moved up the valleys of the nearby Alpujarra hills, one of Spain's most unspoilt areas. Almost all the plants are grown using hydroponics. Figure 23.18 shows the impacts on the local economy in Almería.

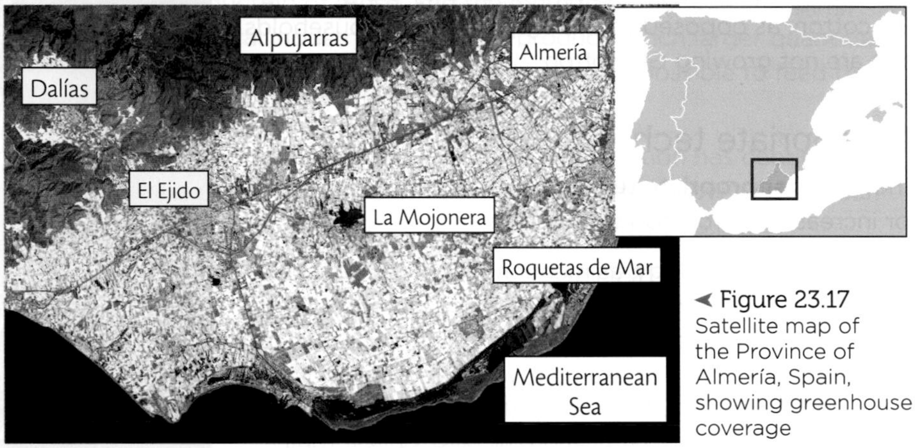

◀ **Figure 23.17** Satellite map of the Province of Almería, Spain, showing greenhouse coverage

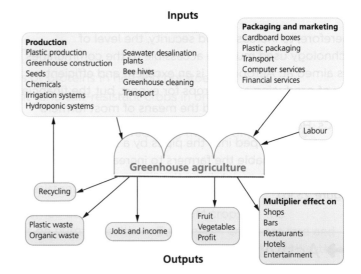

▲ Figure 23.18 The impact of the scheme on Almería

What are the advantages and disadvantages of the scheme?

Advantages

- Large amounts of cheap, temporary labour from North Africa, Eastern Europe, and Central and South America
- The advance of hydroponic growing techniques
- Less water used due to drip irrigation and hydroponics
- A new **desalination** plant supplying fresh water from sea water to the region
- Low energy costs due to the all-year-round warmer temperatures
- Additional jobs created in packing plants
- Factories producing and recycling the plastic for the greenhouses also provide jobs
- Relatively cheap fresh fruit and vegetables provided all year
- New scientific agribusiness companies have located in the area, providing high-skilled jobs in research and development
- Strict UK regulations on quality have reduced levels of chemicals used and raised production standards

Disadvantages

- The immigrant labour is paid very low wages and often live and work in poor conditions
- There are often clashes between immigrants from different countries
- Many immigrants are working illegally and so have little control over their working conditions
- The local environment has been badly affected – large areas of land have been covered with plastic, destroying the natural ecosystems
- Large amounts of litter are left around, including chemical containers and plastic sheeting
- Plastic is dumped into the sea and is affecting marine ecosystems
- The increased use of pesticides in the area has led to increased health risks for those who work or live near the greenhouses
- The natural water sources (aquifers) in the area are drying up
- The greenhouses reflect sunlight back into the atmosphere and have contributed to the cooling of the area. Average temperatures have increased in the rest of Spain since 1983, but in Almería they have dropped by 0.3 °C in ten years.

◄ Figure 23.19 Greenhouses in Almería

→ Activities

1. Use Figure 23.18 to list how the greenhouse agriculture in Almería contributes to the local economy.
2. Why is Almería suitable for the intensive production of salad crops?
3. Evaluate the success of the greenhouse industry in Almería in increasing food production. Think about the social, economic and environmental impacts and whether the industry is successful or not.

⊛ KEY LEARNING
➤ Sustainable water supply in Ethiopia

Hitosa Ethiopia: a local sustainable water supply scheme

How has Hitosa used a sustainable water supply scheme?

Hitosa is a largely rural area located 160 kilometres south of Addis Ababa, the capital city of Ethiopia (Figure 24.22). Ethiopia, in northeast Africa, is one of the poorest countries in the world with a **GNI** of $790 per capita (2018).

The plains are hot and very dry, with annual rainfall averaging around 122 cm per year, but have been extensively farmed for wheat, barley and oil-producing crops. Prior to the water scheme, the people collected their water from a few shallow, largely seasonal rivers and one spring.

The gravity-fed water scheme began in the 1990s. It involves taking water from permanent springs high on the slopes of Mount Bada, a mountain reaching to over 4,000 metres above sea level. Under gravity, the spring water flows through 140 kilometres of pipeline to over 100 public water points (known as tap stands) and nearly 150 private connections largely related to agriculture.

The cost of the project was £1,084,213, funded by WaterAid (54 per cent), The Overseas Development Agency (17 per cent) and the Ethiopian Red Cross (4 per cent).

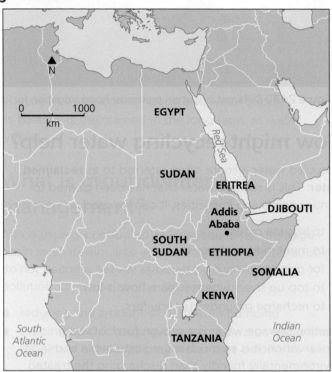

▲ Figure 24.22 The location of Ethiopia in Africa

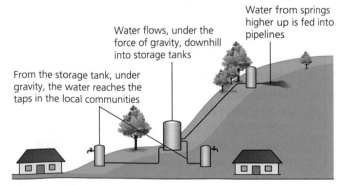

▲ Figure 24.24 A gravity-fed water system

Water from springs higher up is fed into pipelines

Water flows, under the force of gravity, downhill into storage tanks

From the storage tank, under gravity, the water reaches the taps in the local communities

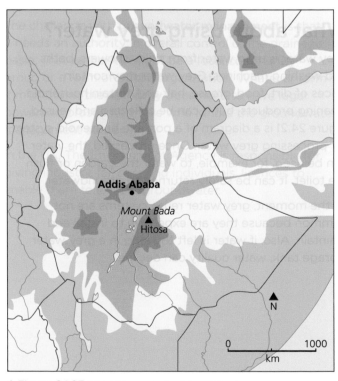

▲ Figure 24.23 The Hitosa project in Ethiopia

Successes

- Construction was completed on time and within budget. WaterAid largely designed and supervised the project.
- Twenty years on, it continues to provide a reliable supply of water to Hitosa. Over 65,000 people are each supplied with 25 litres of water a day. It reaches 32 villages and 3 small towns along the 140 kilometre pipeline.
- The project is completely managed by local communities.
- People are charged a small amount for the water and the money raised is used to maintain the physical infrastructure.
- There has been no misuse of funds.
- The economic benefits include new businesses attracted to the towns and an increase in farmers involved in cattle fattening.

Problems

- The pipeline, supplied from the UK, may be too costly to replace after its expected lifetime of 30 years.
- The scheme did not include any accompanying education about hygiene and **sanitation** initially and has only been implemented after the project was completed. This led to poor hygiene around the tap stands so increasing the risk of disease.
- It has been argued that agriculture is using too much of the water, leading to disputes.
- The availability of water has encouraged migration into the area, which means that the scheme is now required to meet the water needs of well over 70,000 people, threatening the sustainability of the project.

Conclusion

The scheme showed that community management of a large-scale project such as this is possible, with local people operating and managing the scheme without specialist skills. The availability of water has had direct economic benefits. Cattle fattening has become one of a number of new businesses in the area. In addition, the time spent collecting water from rivers has been vastly reduced, improving access to education particularly for girls. However, conflicts have arisen as people living close to the springs felt that other people were taking their water.

▲ Figure 24.25 Local people working on the construction of the pipeline from Mount Bada to Hitosa

▲ Figure 24.26 Women collecting water at a tap stand

→ Activities

1. Study Figure 24.24. Use it to draw a flow chart to show how a gravity-fed water scheme works.
2. Using Figures 24.2 (page 352), 24.21 and 24.22, explain why this scheme is needed.
3. Explain how a reliable supply of water will improve the quality of life for people in the Hitosa region of Ethiopia.
4. To what extent is the gravity-fed water scheme in the Hitosa region of Ethiopia sustainable?

⊛ KEY LEARNING

➤ The occurrence, extraction and use of natural gas

➤ The advantages and disadvantages of natural gas

The dash for natural gas

What should we know about natural gas?

Natural gas provides 24 per cent of the world's energy supply. It was around the middle of the twentieth century that oil took over from coal as the main source of energy (Figure 25.4). It is only since the 1970s that natural gas has begun to challenge the other two fossil fuels.

Natural gas is formed from decaying animal and plant matter that lived millions of years ago. Today, it is found underground, mainly trapped in deep shale rock formations. Wells are drilled and the gas comes to the surface, either under its own pressure or forced up by pumped water. Once at the surface, the gas is pumped through pipelines to where it is used. This transfer often involves pumping the natural gas to ports, where it is shipped in huge tankers. At some point in its transmission, the gas will be refined to remove unwanted chemical impurities.

Figure 25.12 shows the 15 countries that between them hold over 80 per cent of the world's proven reserves of natural gas. The rankings are dominated by Russia and some Middle Eastern and Asian countries. The USA is well placed, but Russia is the only European country shown. The largest gas field in Western Europe is at Groningen in the Netherlands.

Figure 25.13 shows that natural gas has a wide range of uses. Most natural gas is burnt to create heat that is either used directly in factories, the home, commercial premises and transport or is converted into electricity.

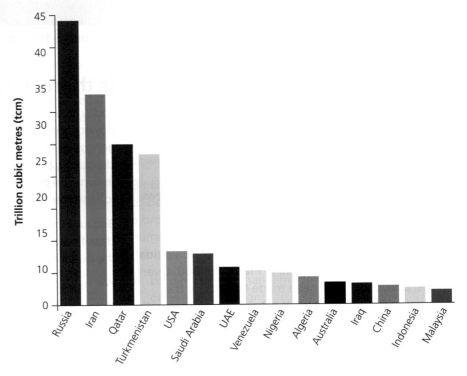

▲ Figure 25.12 The world's largest natural gas reserves

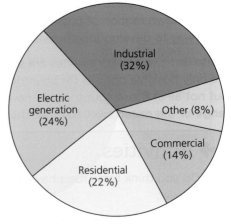

▲ Figure 25.13 Natural gas use by sector

What are the advantages and disadvantages of natural gas?

Figure 25.14 summarises the main advantages and disadvantages of natural gas.

	Advantages	Disadvantages
Environmental	<ul><li>A cleaner fuel producing less carbon emissions – 45 per cent less than coal and 30 per cent less than oil.</li><li>Does not produce waste, such as coal ash.</li><li>The extraction **infrastructure** causes less damage to the ground surface.</li><li>It is lighter than air and therefore disperses quickly in the case of leakages.</li></ul>	<ul><li>Leakages can be very dangerous, causing explosions and fire. If inhaled, the gas is very toxic.</li><li>Burning releases greenhouse gases into the atmosphere, but less so than coal or oil.</li><li>Ground subsidence and earthquakes are caused by pumping of gas and fracking.</li></ul>
Practical	<ul><li>Can be used for many different purposes.</li><li>An economic and instant fuel for heating water and large areas, as well as for cooking.</li><li>Ideal because it allows precise control and quick results.</li><li>More abundant than other fossil fuels with large proven reserves.</li><li>Easy to distribute via pipelines.</li></ul>	<ul><li>Is odourless and leaks cannot be detected unless some odorant is added to the gas.</li></ul>
Economic	<ul><li>Cheaper than electricity.</li><li>Also produces competitively priced electricity.</li></ul>	<ul><li>The infrastructure for extraction and distribution is fairly expensive.</li><li>As a motorvehicle fuel, it gives less mileage than petroleum (refined oil).</li></ul>

▲ Figure 25.14 The advantages and disadvantages of natural gas

What is the overall verdict on natural gas?

Natural gas is not a perfect source of energy. There is not one known to us at the moment. But the above evaluation shows that there is much to commend it: the large proven reserves; its relative cleanness and its versatility. Overall, it certainly compares favourably with the other two major fossil fuels, coal and oil, but of course there are issues with fracking as a method of extraction (see page 333).

→ Activities

1 Explain why natural gas is challenging coal as a leading energy source.

2 Study Figure 25.13. Suggest how natural gas is used in each of these sectors, including the 'other' sector.

3 Overall, do you think natural gas is a good source of energy? Weigh up all the advantages and disadvantages.

4 Find out whether the UK has any natural gas sources of its own and where they are located.

⭐ KEY LEARNING

➤ What a sustainable energy future involves
➤ Ways of achieving energy conservation
➤ Making the use of fossil fuels more efficient

- Switch off lights, power sockets, phone chargers and televisions when not in use
- Use energy-efficient light bulbs and rechargeable batteries
- Only use the washing machine or dishwasher when they have a full load
- Use curtains and blinds to provide insulation – from heat in summer and from heat loss in winter
- Wear warm clothing indoors in winter and turn down the central heating
- Walk and cycle more and become less reliant on transport over short distances
- Spend less time on the internet, playing electronic games and texting friends

▲ Figure 25.15 Some ways of reducing personal energy use

Towards a sustainable energy future

What does a sustainable energy future involve?

We need to make sure that today's sources of energy supply will also be available for future generations to use. As global energy consumption increases and even greater pressure is put on energy sources, three priorities have become clear:

■ Reliance on fossil fuels must be reduced and every effort made to reduce emissions of carbon dioxide to a minimum.
■ More use must be made of renewable sources of energy. Research into new sources of renewable energy should also be encouraged.
■ Energy efficiency must be improved. This means getting more from the energy we use and eliminating wastage.

What can we do by way of energy conservation?

Energy conservation is all about minimising the wastage of energy and using it as efficiently as possible. The following includes some actions we can take to do this.

Reducing our individual use of energy and our carbon footprints

Energy conservation begins with us as individuals, both in our homes and in our lifestyles. If we all contribute, then the impact on energy conservation can be massive. Figure 25.15 details some simple actions that we all should take.

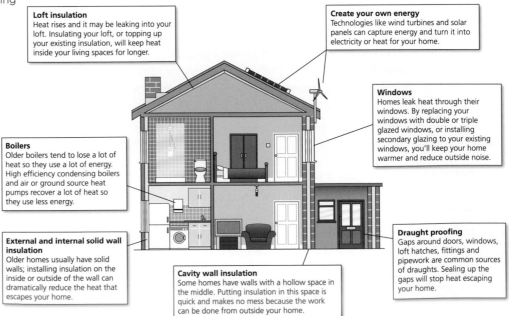

Loft insulation
Heat rises and it may be leaking into your loft. Insulating your loft, or topping up your existing insulation, will keep heat inside your living spaces for longer.

Create your own energy
Technologies like wind turbines and solar panels can capture energy and turn it into electricity or heat for your home.

Windows
Homes leak heat through their windows. By replacing your windows with double or triple glazed windows, or installing secondary glazing to your existing windows, you'll keep your home warmer and reduce outside noise.

Boilers
Older boilers tend to lose a lot of heat so they use a lot of energy. High efficiency condensing boilers and air or ground source heat pumps recover a lot of heat so they use less energy.

External and internal solid wall insulation
Older homes usually have solid walls; installing insulation on the inside or outside of the wall can dramatically reduce the heat that escapes your home.

Cavity wall insulation
Some homes have walls with a hollow space in the middle. Putting insulation in this space is quick and makes no mess because the work can be done from outside your home.

Draught proofing
Gaps around doors, windows, loft hatches, fittings and pipework are common sources of draughts. Sealing up the gaps will stop heat escaping your home.

➤ Figure 25.16 Making the home more energy efficient

In the UK, as in many other countries, most of our energy still comes from the burning of fossil fuels. This means that any saving in our consumption of energy will mean a reduction in our carbon footprint.

Designing homes and workplaces

There are a number of things that can be done to make both existing and new buildings more energy efficient. Figure 25.16 shows some design features that would make homes more energy efficient. The same features should also be adopted in the design of new workplaces. Figure 25.17 has some suggestions for energy-saving actions in the workplace (including schools and colleges).

Transport

In this age of **globalisation**, it does not look as if there is much hope of reducing either the use of transport or its consumption of energy. But as responsible citizens, there are some actions that we can take that collectively would contribute to a more sustainable use of energy (Figure 25.18).

Demand reduction

The energy conservation actions just described all have the same outcome. They will, to varying degrees, reduce the demand for and consumption of energy. Any such, reduction is an important step towards energy conservation. It means that there will be more energy left for use tomorrow.

Can modern technology help us to use fossil fuels more efficiently?

Carbon emissions from the burning of fossil fuels continue to threaten global warming and climate change. Can modern technology help at all?

Chapter 3 mentions carbon capture as one technology (see page 52). Another is combined heat and power (CHP, see page 251). This generates heat and electricity from a single fuel source, most often oil or natural gas. It uses the heat created during the generation of electricity to provide a supply of hot water for the heating and air conditioning of housing schemes, shopping centres and hospitals. Its overall fuel efficiency is 85 per cent, compared to 52 per cent for a traditional thermal power station. This better efficiency means that consumption of fossil fuels and carbon dioxide emissions are much less. These are two steps in the right direction. But who can predict what better solutions technology might devise in the future?

- Ensure temperatures are set at no more than 19 °C in offices and classrooms.
- Keep doors and windows closed when the heating is on.
- Avoid heating unused spaces such as corridors and storerooms.
- Don't leave electronic equipment (computers and printers) in standby mode.
- Ensure that inbuilt energy-saving software is activated.
- Use daylight where possible.
- Use low-wattage, compact fluorescent bulbs rather than tungsten ones.

▲ Figure 25.17 Some ways of energy saving in the workplace

- Use public transport rather than private cars.
- Use smaller, more energy-efficient hybrid cars.
- Use alternative, cleaner fuels, such as electricity.
- Car-share when commuting to school or work.
- Reduce the number of aircraft journeys taken, especially short-haul flights.
- Cut down on the number of holidays taken abroad.

▲ Figure 25.18 Some ways of energy saving on transport

→ Activities

1 Explain what is meant by energy conservation.
2 Read all the ideas about energy conservation on these pages. Draw a large table with three columns and categorise the ideas into these groups:
   - things you could do yourself to conserve energy
   - things that your family could do (you couldn't do on your own)
   - things that we could only do with government help.
3 Can you suggest other energy-conserving actions?
4 Investigate ways in which the UK government is encouraging a more efficient use of energy.

⭐ KEY LEARNING

➤ The energy situation in Nepal

➤ Micro-hydro plants in Nepal

➤ Other possible sustainable sources of energy in Nepal

A local scheme in an LIC: sustainable energy in Nepal

What is the present energy situation in Nepal?

In the Himalayan kingdom of Nepal (Figure 25.19), the present demand for energy is relatively small, but the demand is growing as the country tries to develop and its people aim for a better quality of life. For centuries, wood was the traditional source of energy for heating and cooking. This led to widespread **deforestation**. Crop and animal waste continue to be used as a fuel for cooking.

The wish to supply electricity to its population of 28 million is held back because Nepal has no significant deposits of coal, oil or natural gas of its own. As Nepal is a land-locked and mountainous country, importing fossil fuels is difficult. There is an electric grid system covering part of the country. But power cuts lasting an average of ten hours a day are common. They are so common that the Nepal Electricity Authority even publishes a timetable of power cuts!

What are micro-hydro plants?

The government of Nepal, with support from the World Bank, is helping to create micro-hydro plants across rural Nepal. The plants are built and run by local communities. They are sustainable and bring much-needed electricity for use by local industry, agriculture and commerce, as well in the home.

The micro-hydro schemes in Nepal, as elsewhere, are of the 'run-of-the-river' type (Figure 25.20). They do not need a dam or reservoir to be built, but instead divert water from a stream or river. This water is then channelled to a forebay tank. This is a settling basin that helps to remove damaging **sediment** from the water before it falls to a turbine via a pipeline called a penstock. The small turbine drives a generator that provides the electricity to the local community.

By not requiring an expensive dam and reservoir for water storage, run-of-the-river systems are a low-cost way to produce power. They also avoid the damaging environmental and social effects that larger HEP schemes cause, including a risk of flooding.

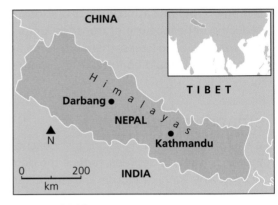

▲ Figure 25.19 Map of Nepal

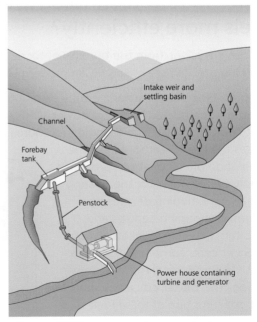

▲ Figure 25.20 Diagram of a micro-hydro scheme

Local communities are gradually becoming aware of the benefits that micro-hydro plants bring. Take the example of the rather inaccessible settlement of Darbang, located between Kathmandu and the Tibetan border (Figure 25.19). This now boasts a number of new industries. These include a metal and several furniture workshops, a cement block maker, a noodle factory, poultry farms and dairy farms. All these activities have sprung up since the Ruma Khola micro-hydro power plant came into operation in 2009. The 51-kilowatt micro-hydro supplies electricity to 700 households in five villages, including Darbang.

Over 1,000 micro-hydro plants have been built so far in 52 districts. The Nepalese micro-hydro project is meeting the energy needs of rural communities. In doing so, it is encouraging economic development. It is also making the point that even poor, rural communities can enjoy clean, renewable energy. The project promises a cleaner and more prosperous future.

Are there other sustainable sources of energy?

The harnessing of solar energy either by solar panels (for heating) or by voltaic cells (for electricity) offers other sustainable sources, but there are some possible limitations:

- the number of sunshine hours in this mountainous country
- the costs of the panels and cells
- the technology required to maintain the panels and cells.

Three other ways shown in Figure 25.21 are linked to low-tech ways of generating energy that are well-suited to an LIC such as Nepal, and are sustainable.

> ### → Activities
>
> 1 Why is it difficult for Nepal to import oil and natural gas?
> 2 Why is Nepal well-suited to development of micro-hydro plants?
> 3 How does a micro-hydro plant benefit agriculture?
> 4 Why might the use of solar energy be a less attractive option for Nepal than micro-hydro power?
> 5 Study Figure 25.21. Which one of the three biogas energy possibilities shown is best? Give reasons for your choice.
> 6 Make a case for and against building micro-hydro plants in the UK.

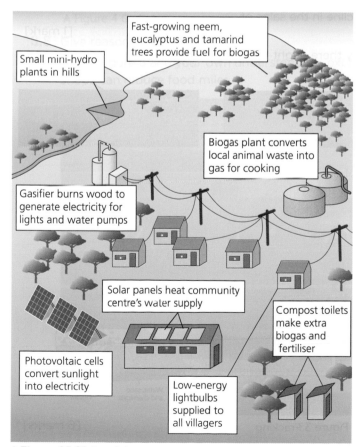

Small mini-hydro plants in hills

Fast-growing neem, eucalyptus and tamarind trees provide fuel for biogas

Biogas plant converts local animal waste into gas for cooking

Gasifier burns wood to generate electricity for lights and water pumps

Solar panels heat community centre's water supply

Compost toilets make extra biogas and fertiliser

Photovoltaic cells convert sunlight into electricity

Low-energy lightbulbs supplied to all villagers

▲ Figure 25.21 Possible energy sources in Nepal

The issue evaluation

The issue evaluation on Paper 3 is where you can really show how good a geographer you are. The paper is called 'Geographical applications' and it is a chance for you to apply all the geography you have learnt over the past two years.

You have been investigating a geographical issue – **Should a major new runway be built at London's Heathrow airport?** – using the resources in Sections 26.2 and 26.3. By now, you should have,

- made notes about each of the resources
- identified the main stakeholders with their views
- thought about which resources would help you to answer the question and why they would be helpful.

The resources you have been using are very similar to the ones you will find in the pre-release resources booklet for Paper 3 in your GCSE geography exam. Remember, that you will be given the booklet 12 weeks before the exam, giving you plenty of time to study the resources and make your decision about the issue. Here is a reminder of all the steps you need to go through (Figure 26.10).

You won't be allowed to take the pre-release resources booklet into the exam, or any of the notes that you made. Hopefully, you will re-member your notes. It is worth spending some time revising them, even though they may be your own ideas!

In the exam, you will be given a clean copy of the pre-release resources booklet, which you will be able to use to answer the questions. The resources will all be very familiar, so you shouldn't need to interpret them again. Many of the questions are also likely to be ones you have already thought about, including the main question you have to make your decision about.

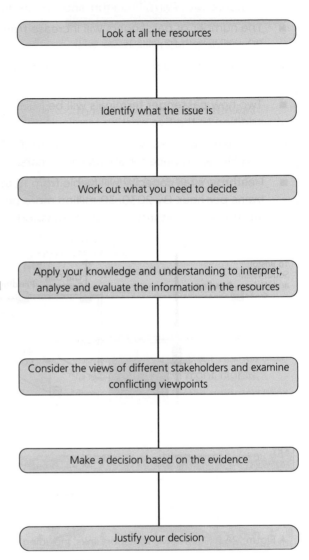

▲ Figure 26.10 Steps in the issue evaluation

Question practice: Paper 3

1.1 Study Figure 26.3 on page 386.

Please note that Paper 3 also includes a section on Fieldwork. In this section you will have questions on unfamiliar field-work as well as your own fieldwork.

In correct order, which are the three largest airports in the UK? Choose the correct answer from:

A Heathrow, Manchester, Birmingham

B Heathrow, Gatwick, Stansted

C Heathrow, Gatwick, Manchester [1 mark]

D Gatwick, Manchester, Heathrow

1.2 'London is by far the most important aviation hub in the UK.' To what extent do you agree with this statement? Give evidence to support your answer. [6 marks]

2.1 Give two reasons why the government might support the continued growth of the aviation industry. [2 marks]

2.2 Using Figures 26.4 and 26.5 on page 387, explain why the expansion of Heathrow airport is so controversial. [6 marks]

3.1 Study Figures 26.6 and 26.7 on page 388.

It is a good idea to explain the different viewpoints of more than one group of people in this question.

What is the approximate area of Heathrow airport now, as shown in Figure 26.7 on the OS map? Choose the correct answer from:

[1 mark]

A 5 km²   **B** 8 km²   **C** 10 km²   **D** 15 km²

3.2 'Heathrow airport is located on a suitable site for expansion'. Discuss this statement using evidence to support your answer. [6 marks]

'Discuss' means to put both sides of the case, the points which agree with the statement and those which disagree.

4.1 Study Figure 26.6 on page 388.

How will the M25 be affected by the building of a third runway at Heathrow? Choose the correct answer from:

A The M25 will stay on its original route.

B The M25 will be re-routed through a tunnel under the runway.

C The M25 will be re-routed on a bridge over the runway. [1 mark]

D The M25 will be re-routed away from Heathrow.

4.2 Do you think that the planned expansion of Heathrow airport should go ahead?

Yes ☐ No ☐ Tick **one** box to show your choice.

Use evidence from all information in the resources booklet and your own understanding to explain your choice. [9 marks] [+3 SPaG marks]

Sustainable development – development that meets the needs of the present without limiting the ability of future generations to meet their own needs.

Sustainable energy supply – energy that can potentially be used well into the future without harming future generations. Sustainable energy is the combination of energy savings, energy efficiency measures and technologies, as well as the use of renewable energy sources.

Sustainable food supply – food that is produced in ways that avoid damaging natural resources, provide social benefits such as good quality food and safe and healthy products, and contribute to local economies.

Sustainable urban living – a sustainable city is one in which there is minimal damage to the environment, the economic base is sound with resources allocated fairly and jobs secure, and there is a strong sense of community, with local people involved in decisions made. Sustainable urban living includes several aims including the use of renewable resources, energy efficiency, use of public transport, accessible resources and services.

Sustainable water supply – meeting the present-day need for safe, reliable and affordable water, which minimises adverse effects on the environment, while enabling future generations to meet their requirements.

Tectonic hazard – a natural hazard caused by movement of tectonic plates (including volcanoes and earthquakes).

Tectonic plate – a rigid segment of the Earth's crust which can 'float' across the heavier, semi-molten rock below. Continental plates are less dense, but thicker than oceanic plates.

Temperature range – the difference between the highest and lowest temperatures over a period of time.

Thermal expansion – when the sea expands and becomes larger as a result of increased temperature.

Thermal growing season – the period of time when temperatures are above 6°C and plants can grow.

Thermokarst – an uneven landscape of mounds and hollows, some of which may be water filled.

Tidal power – generating electricity using energy from the changing levels of the tide.

Tidal range – the difference in water level between high tide and low tide.

Till – an unsorted mixture of sand, clay and boulders carried by a glacier and deposited as ground moraine over a large area.

Traction – the rolling of boulders and pebbles along the river bed.

Trade – the buying and selling of goods and services between countries.

Traffic congestion – occurs when there is too great a volume of traffic for roads to cope with, so traffic jams form and traffic slows to a crawl.

Transnational corporation (TNC) – a company that has operations (factories, offices, research and development, shops) in more than one country. Many TNCs are large and have well-known brands.

Transpiration – the process by which plants lose water vapour through their leaves. Strong winds increase transpiration.

Transportation – the movement of eroded material.

Tropical rainforest – dense forest or jungle growing in a hot, wet climate near the Equator.

Tropical storm (hurricane, cyclone, typhoon) – an area of low pressure with winds moving in a spiral around the calm central point called the 'eye' of the storm. Winds are powerful and rainfall is heavy.

Truncated spur – a former river valley spur that has been sliced off by a valley glacier, forming cliff-like edges.

Tundra – the flat, treeless Arctic regions of Europe, Asia and North America, where the ground is permanently frozen. Lichen, moss, grasses and dwarf shrubs can grow here.

Typhoon – see Tropical storm.

Undernourishment – when people do not eat enough nutrients to cover their needs for energy and growth, or to maintain a healthy immune system.

Undernutrition – this occurs when people do not eat enough nutrients to cover their needs for energy and growth, or to maintain a healthy immune system.

Urban farming – the growing of fruits, herbs, and vegetables and raising animals in towns and cities, a process that is accompanied by many other activities such as processing and distributing food, collecting and reusing food waste.

Urban greening – the process of increasing and preserving open space such as public parks and gardens in urban areas.

Urbanisation – the process by which an increasing percentage of a country's population comes to live in towns and cities. Rapid urbanisation is a feature of many LICs and NEEs.

Urban regeneration – the revival of old parts of the built-up area by either installing modern facilities in old buildings (known as renewal) or opting for redevelopment (i.e. demolishing existing buildings and starting afresh).

Urban sprawl – the unplanned growth of urban areas into the surrounding countryside.

Vertical erosion – downward erosion of a river bed.

Volcano – an opening in the Earth's crust from which lava, ash and gases erupt.

Waste recycling – the process of extracting and reusing useful substances found in waste.

Waterborne diseases – diseases caused by microorganisms that are transmitted in contaminated water. Infection commonly results during bathing, washing, drinking, in the preparation of food, or the consumption of infected food, e.g. cholera, typhoid, botulism.

Water conflict – disputes between different regions or countries about the distribution and use of freshwater. Conflicts arise from the gap between growing demands and diminishing supplies.

Water conservation – the preservation, control and development of water resources, both surface and groundwater, and prevention of pollution.

Water deficit – this exists where water demand is greater than supply.

Waterfall – sudden descent of a river or stream over a vertical or very steep slope in its bed. It often forms where the river meets a band of softer rock after flowing over an area of more resistant material.

Water footprint – a water footprint is the amount of water you use in and around your home, school or office throughout the day. It includes the water you use directly (e.g. from a tap to drink or to shower). It also includes the water it took to produce the food you eat, the products you buy, the energy you consume and even the water you save when you recycle.

Water insecurity – when water availability is not enough to ensure the population of an area enjoys good health, livelihood and earnings. This can be caused by water insufficiency or poor water quality.

Water quality – quality can be measured in terms of the chemical, physical, and biological content of water. The most common standards used to assess water quality relate to health of ecosystems, safety of human contact and drinking water.

Water security – the reliable availability of an acceptable quantity and quality of water for health, livelihoods and production.

Water stress – water stress occurs when the demand for water exceeds the available amount during a certain period or when poor quality restricts its use.

Water surplus – this exists where water supply is greater than demand.

Water table – the level below which rock is saturated with water.

Water transfer – water transfer schemes attempt to make up for water shortages by constructing elaborate systems of canals, pipes, and dredging over long distances to transport water from one river basin to another.

Wave cut platform – a rocky, level shelf at or around sea level representing the base of old, retreated cliffs.

Wave power – generating electricity using energy from waves in the sea.

Waves – ripples in the sea caused by the transfer of energy from the wind blowing over the surface of the sea. The largest waves are formed when winds are very strong, blow for lengthy periods and cross large expanses of water.

Weathering – when rock is broken down in one place.

Wilderness area – a natural environment that has not been significantly modified by human activity. Wilderness areas are the most intact, undisturbed areas left on Earth –places that humans do not control and have not developed.

Wind energy – electrical energy obtained from harnessing the wind with windmills or wind turbines.

Index

abrasion (corrasion) 126, 152, 154, 182
aeroponics 340-1, 343
agribusinesses 323, 345
agriculture 48, 49, 50, 51, 54, 258, 356, 370: *see also* farming
agroforestry 82, 83
airports 304-5, 310, 386-9
Alaska 104-9
Antarctica 110-11
arches 134, 135
arêtes 186, 190
atmospheric hazards 2
attrition 126, 152, 154, 185
avalanches 2

bars 141
bays 130-1
beaches 136-7
beach nourishment 142, 144, 146, 147
beach recharge 146
beach recycling 146, 147
beach reprofiling 146, 147
berms 137
biodiversity 51, 54, 62, 179, 342
    cold climate regions 102, 103
    hot deserts 86, 87, 94
    tropical rainforests 69, 76, 78, 79
biological hazards 2
biomass (biofuels) 52, 251, 317, 328, 329, 331, 367, 372
biomass (living organisms) 61, 63, 68, 69
biotechnology 342
blow holes (gloups) 134
Brazil 72, 73, 74-7, 82
Bristol, UK 252-3
business parks 295

Cambridge, UK 296-7
carbon capture and storage (CCS) 52-3, 330, 377
carbon footprints 322, 376-7
car industry 298-9
carrying capacity 90, 106, 195, 197
caves 134
channel straightening 175
chemical weathering 124
China 10, 276, 285, 291, 292

economic development 259, 261, 264, 268, 269, 277, 335
    energy supplies 319, 368, 370, 371, 374
    megacities 206, 207
    trade links 284, 285, 310
    water supplies 354, 358, 360-1
cliffs 132-4
climate change 18, 40, 44-55, 76, 79, 336, 353
    adaptation to 54-5
    effects of 50-1, 64
    greenhouse effect 48-9
    mitigation 52-3
    and rainfall 42, 95
    and tropical storms 28-9
climate zones 22
coal mining industry 292
coastal erosion 126-7
coastal landscapes 122-51
    coastal realignment 148-51
    Dorset coast 128-33, 134
    hard engineering 127, 142-5
    marine processes 126-7
    mass movement 124-5
    soft engineering 146-7
    weathering 124-5
cold climate regions 100-15
    adaptations to 102-3
    Alaska 104-9
    Antarctica 110-11
    management of 114-15
    technology and 112-13
    see also polar environments; tundra
combined heat and power (CHP) systems 251, 377
commercial farming 74-5
Commonwealth 270, 308
composite volcanoes 8
concordant coastlines 128
conflict: and resources 337, 338, 357, 371
coniferous forests 66
conservative plate margins 10-11
constructive plate margins 4, 5, 6-7
constructive waves 123, 127, 141
continental crust 4

convection (tectonic plate movement) 4
convection cells 22-3
Coriolis effect 25, 26, 27
corries 186, 190
counter-urbanisation 300
Crossrail 234-5
Cumbria, UK 38-41
cycling 252-3
cyclones, *see* tropical storms

dams 174, 358-9
debt crisis 272, 287
debt reduction 80
debt relief 272-3, 287
deciduous forests 66
deforestation 3, 48, 49, 64, 72-7
de-industrialisation 292-3
demographic transition model (DTM) 262
deposition 122, 127, 154
depressions (low pressure systems) 23, 24, 36, 38, 42, 43, 168
desalination 55, 359
desert fringe areas 84, 94, 95, 96, 97, 98-9
desertification 87, 96-9
deserts 66: *see also* hot deserts; tundra
destructive plate margins 4, 5, 8-9
destructive waves 123, 137, 141
development 258-67
    demographic transition model (DTM) 262
    factors influencing 264-5
    and food resources 335
    industrial development 268-9
    intermediate technology and 271
    international aid and 270
    and international migration 267
    investment and 269
    social 260-1, 270
    uneven development 266-7
    world map of 259
differential erosion 159, 160
discordant coastlines 128, 130, 141
drought 3, 37, 41, 50, 336
drumlins 189, 191
dune succession 139

earthquakes 2, 8, 10–18, 20, 265
    Amatrice, Italy 12–13, 16
    Gorkha, Nepal 14–15, 16
    location 4–5, 6
East Village, London 250–1
ecological footprints 248
economic development 258–61, 264, 270, 354, 368
    Nigeria 276–89
economic migration 289
economic well-being 316
ecosystems 60–7
ecotourism 82, 83
embankments 176
energy conservation 376–7
energy gap 367
energy insecurity 366, 367, 370–1
energy security 108, 366, 369
energy supplies 74, 89, 219, 317, 319, 366–79
    economic/environmental issues 330–3
    energy mix 328–9
    factors affecting 368–9
    strategies to increase 372–3
    sustainability 376–9
    technology and 368, 369, 377
    UK 328–33
enhanced greenhouse effect 48
Epping Forest 62–3
erosion 77, 122, 152–3, 154, 159, 160, 182, 186–7, 339
erratics 189, 191
estuaries 166–7
European Union (EU) 309
eutrophication 326

Facebook 113, 311
Fairtrade 271
farming 18, 73, 74–5, 78, 88, 171, 192, 321, 346, 347: see also agriculture
fieldwork 393–5
    data collection 394–5, 398–9
    data presentation 396–7, 400–1
    evaluation 397, 401
    risk assessment 393
fishing industry 51, 104–5
flash flooding 36
flashy response hydrographs 173
flooding 2, 3, 28, 30, 50, 51, 336
    human causes of 170–1
    physical causes of 168–9
    in UK 36, 38–9, 40, 41
flood plains 165
flood plain zoning 178, 179
flood relief channels 177, 180–1

flood warnings 178, 179
food chains 61
food insecurity 334, 336–9
food loss 350–1
food miles 322, 346
food security 334
food supplies 316, 318, 321, 322, 334–51
    Almería, Spain 344–5
    conflict and 337, 338
    factors affecting 335, 336–9
    fish 347, 348–9
    flooding and 336
    Jamalpur, Bangladesh 348–9
    meat 347
    poverty and 337
    seasonality 346
    sustainability 346–9
    UK 320–3
    urban farming initiatives 346, 347
food waste 350–1
food webs 62, 64, 87, 103
foreign direct investment (FDI) 269
forest degradation 73
forestry 170, 174, 192, 194
fossil fuels 48, 49, 51, 79, 89, 328, 329, 330, 366, 367, 372, 377
fracking 330, 331, 375
freeze–thaw weathering 124, 125, 158, 183, 184, 186

gabions 143, 144, 176
geographical enquiries 392–401
geology 355, 368
geothermal energy 18, 373
Gini coefficient 266
glacial deposition 185, 188
glacial landscapes 182–99
    ice erosion 182, 186–7
    land use 192–5
    tourism 193, 194, 196–7
glacial movement 184
glacial outwash 185
glacial till 185, 189
glacial troughs 187
global atmospheric circulation 22–3
global warming 76, 79, 95, 115
GM crops 342–3
gorges 160–1
greenbelt 300
greenhouse effect 3, 48–9, 76
grey water 363
gross national income (GNI) 258, 259
groundwater management 362
growth corridors 294–5
groynes 142, 144

Hadley cells 23
hanging valleys 187
headlands 130–1
health issues 50, 336, 338, 356
Heathrow airport 386, 387–9
heatwaves 37, 50
high-income countries (HICs) 16, 30, 80, 224, 258, 259, 262
high-speed rail 302–3
Hjulstrom Curve 153
hot deserts 84–99
    adaptations to 86–7
    biodiversity 86
    characteristics 84–5
    desert fringe areas 94
    Western Desert, USA 88–93
housing 240–1
Human Development Index (HDI) 261, 288
hurricanes, see tropical storms
hydraulic power 126, 152
hydroelectric power (HEP) 74, 89, 174, 357
hydrographs 172–3
hydroponics 341, 344

ice cores 45
ice deposition 191
ice erosion 182, 186–7
industrial development 268–9
inequality 239, 318–19
interdependency 87, 102, 103
international aid 270, 286, 337
international migration 267
International Monetary Fund (IMF) 272
interlocking spurs 158–9
intertropical convergence zone (ITCZ) 23, 25
investment 269, 291
iron and steel industry 292
irrigation 340, 343

jet streams 23

Lagos, Nigeria 208–23
    challenges 213, 216–17
    energy supplies 219
    growth of 207, 210–11
    informal economy 215
    Makoko 216–17, 222–3
    opportunities 212, 214, 215
    pollution 219, 220
    rural–urban migration 211
    sanitation system 219
    squatter settlements 213, 216–17, 222
    traffic system 220–1

urban planning 222–3
water supplies 218
Lake District 190–1, 193
landslides 2, 30, 36, 124
land use conflicts 194–5
levées 164
life expectancy 238, 239, 260
logging industry 74, 82
London 224, 225, 226–47
    culture 230–1
    East Village 250–1
    employment 232–3
    green spaces 236–7
    housing 240–1
    inequality 239
    life expectancy 238, 239
    pollution 242–3
    poverty 238
    recycling 243, 251
    transport system 234–5
    urban regeneration 244–7
longshore (littoral) drift 127, 141, 146
low-income countries (LICs) 16, 80, 224,
    258, 259, 262, 264, 266

managed retreat 148–9
Mariana Trench 8, 9
marine processes 126–7
mass movement 124–5, 133
meanders 152, 154, 162
mechanical weathering 124
megacities 204, 207
microfinance loans 273
micro-hydro schemes 378–9
migration 228, 289
mining industry 19, 74, 83, 89, 105
mitigation 20, 34–5
moraines 184, 188, 189, 191
mud slides 124–5
multiplier effect 269, 330

natural gas 372, 374–5: see also fracking
natural hazards 2–3, 24
Nepal 14–15, 16, 378–9
net migration 228
newly emerging economies (NEEs) 16,
    80, 205, 258, 259, 262, 264, 266
Nigeria 285
    contexts 278–9
    debt 287
    economic development 276–89
    energy supplies 219
    international aid 286
    oil industry 281–3, 285
    population growth 276
    quality of life 288

TNCs in 282
trade relationships 284–5
non-governmental organisations (NGOs)
    30, 81
non-renewable energy 366, 367
nuclear energy 328, 330, 373
nutrient cycling 61, 63, 69

ocean currents 66
oceanic crust 4
Official Development Assistance
    (ODA) 286
oil industry 108–9, 281–3, 285, 369
Olympic Park, London 244–7, 250–1
organic farming 321, 346
over-abstraction 355
overgrazing 96, 97, 339
overpopulation 263
oxbow lakes 163

permaculture 346
permafrost 101, 103, 106–7, 110, 115
photosynthesis 49, 60, 61, 63, 70
plucking 182, 184, 186, 188
polar environments 100–1
pollution 298
    Lagos, Nigeria 219, 220
    London 242–3
    water pollution 77, 326–7, 355, 362
population growth 75, 96, 228–9, 263,
    276, 301, 335, 354, 368
ports 304
poverty 14, 238, 267, 271, 273
    desertification and 96, 98
    and food/water supplies 337, 340,
    355
    natural hazards and 15, 32
    Nigeria 280, 286, 287, 289
precipitation 23, 168: see also rainfall
pyramidal peaks 186

quality of life 258, 288, 316
quarrying 193
Quaternary period 44
quaternary sector (knowledge
    economy) 294
questionnaires 398–9

rainfall 23, 28, 30, 38–9, 42, 95
rapids 158–9
reclaimed water 363
recycling 243, 251, 363
reforestation 53
renewable energy 52, 74, 219, 328, 329,
    330, 331, 367, 372–3
reservoirs 174, 194, 355, 358–9
resource management 316–19

rewilding 64–5, 178, 179
ribbon lakes 121, 187, 190
rice–fish culture 348–9
ridge push 4
rift valleys 7
river landscapes 152–81, 187
    cross-profile changes 156–7
    discharge 155, 172–3
    erosion 152–3
    long profiles 154–5
    River Severn 154–5, 156, 158, 159,
    160–7
    velocity 155
river management
    hard engineering 174–7, 180–1
    soft engineering 178–9
river pollution 77
river restoration 178, 179
rock armour (rip rap) 142, 144
rock falls 125
rock slides 124
rural–urban migration 206, 211

Saffir-Simpson hurricane wind
    scale 29, 31
Sahel 84, 94, 95, 97, 98–9
saltation 137, 138, 153
salt weathering 124
San Andreas Fault 10, 11
sand dune regeneration 146, 147
sand dunes 84, 85. 98, 136, 137, 138–9, 144
sea levels: rises in 55
sea walls 142, 144
sediment
    coastal sediment 127, 130, 136, 139, 141,
    142, 144, 146, 166, 167
    glacial outwash 185
    ocean sediments 44, 45
    river sediment 92, 153, 154, 162, 164,
    166, 167, 174, 175
settlements 75
shield volcanoes 6
shipbuilding industry 292
slab pull 4
slash and burn 49, 73
slow response hydrographs 173
slumping 125
social media 40–1, 311
social well-being 316
soil erosion 77, 339
soil fertility 76, 77
solar energy 46, 52, 89, 99, 329, 379
solution (corrosion) 152, 153, 154
South–North Water Transfer Project
    (SNWTP), China 360–1